Modern Birkhäuser Classi

Many of the original research and survey monographs in pure and applied mathematics published by Birkhäuser in recent decades have been groundbreaking and have come to be regarded as foundational to the subject. Through the MBC Series, a select number of these modern classics, entirely uncorrected, are being re-released in paperback (and as eBooks) to ensure that these treasures remain accessible to new generations of students, scholars, and researchers.

BERNHARD RIEMANN
(1826–1866)

Bernhard Riemann
1826–1866

Turning Points in the Conception of Mathematics

Detlef Laugwitz

Translated by
Abe Shenitzer

With the Editorial Assistance of the Author,
Hardy Grant, and Sarah Shenitzer

Reprint of the 1999 Edition

Birkhäuser
Boston • Basel • Berlin

Detlef Laugwitz (Deceased)
Department of Mathematics
Technische Hochschule
Darmstadt D-64289
Gernmany

Abe Shenitzer (translator)
Department of Mathematics
and Statistics
York University
Toronto, Ontario M3J 1P3
Canada

Originally published as a monograph

ISBN-13: 978-0-8176-4776-6 e-ISBN-13: 978-0-8176-4777-3
DOI: 10.1007/978-0-8176-4777-3

Library of Congress Control Number: 2007940671

Mathematics Subject Classification (2000): 01Axx, 00A30, 03A05, 51-03, 14C40

Cover design by Alex Gerasev.

Printed on acid-free paper.

9 8 7 6 5 4 3 2 1

www.birkhauser.com

Detlef Laugwitz

Bernhard Riemann 1826–1866

Turning Points in the Conception of Mathematics

Translated by
Abe Shenitzer
With the Editorial Assistance of the Author,
Hardy Grant, and Sarah Shenitzer

The German-language edition, edited by
Emil A. Fellman, appears in the
Birkhäuser Vita Mathematica series

Birkhäuser
Boston • Basel • Berlin

Detlef Laugwitz
Department of Mathematics
Technische Hochschule
Darmstadt D-64289
Germany

Abe Shenitzer (Translator)
Department of Mathematics & Statistics
York University
Toronto, Ontario
Canada M3J 1P3

Library of Congress Cataloging-in-Publication Data

Laugwitz, Detlef.
[Bernhard Riemann. 1826–1866. English]
Bernhard Riemann, 1826–1866 : turning points in the conception of mathematics / Detlef Laugwitz : translated by Abe Shenitzer with the editorial assistance of the author, Hardy Grant, and Sarah Shenitzer.
p. cm.
Includes bibliographical references and index.
ISBN 0-8176-4040-1 (alk. paper). — ISBN 3-7643-4040-1 (alk. paper)
1. Bernhard Riemann, 1826–1866. 2. Mathematicians—Germany—Biography. 3. Mathematics—Germany—History—19th century.
I. Title.
QA29.R425L3813 1999
510'.92
[B]—DC21 98-17834
CIP

AMS Subject Classifications: 01Axx, 00A30, 03A05, 51-03, 14C40

Printed on acid-free paper.

Birkhäuser 

ISBN 0-8176-4040-1
ISBN 3-7643-4040-1

Reformatted by TEXniques, Inc., Cambridge, MA.
Printed and bound by Braun-Brumfield, Inc., Ann Arbor, MI.
Printed in the United States of America.

9 8 7 6 5 4 3 2 1

Contents

Preface

It is precisely in the case of Riemann, who has no equals in the intellectual penetration of mathematical problems, that it pays to trace the underlying integrated conception.

(Gerade bei Riemann, der an gedanklicher Durchdringung mathematischer Probleme nicht Seinesgleichen hat, lohnt es sich, der zugrunde liegenden einheitlichen Konzeption nachzuspüren.)

—H. Weyl (1925)

The degeneration of mathematics began with the ideas of Riemann, Dedekind, and Cantor, which progressively repressed the reliable genius of Euler, Lagrange, and Gauss.

(Die Entartung der Mathematik begann mit den Ideen von Riemann, Dedekind, und Cantor, durch die der solide Geist von Euler, Lagrange, und Gauss mehr und mehr zurückgedrängt wurde.)

—C.L. Siegel (1959)

One of the most profound and imaginative mathematicians of all time, he had a strong inclination to philosophy, indeed, was a great philosopher.

—H. Freudenthal (1975)

No one person is capable of a full analysis of Riemann's work, its history, its development and its influence on current mathematics.

—R. Narasimhan (1990)

The idea for a book on Riemann, to appear in the series *Vita Mathematica*, came from E. Fellmann exactly ten years ago on the occasion of the Leibniz Colloquium in Noordwijkerhout. One evening I discussed with Hans Freudenthal problems of the historiography of mathematics, and he told me with great enthusiasm about the intense joy he experienced when writing biographies, especially the ones of Cauchy and Riemann, which he had contributed to the *Dictionary of Scientific Biography*. The atmosphere of this encounter, marked in equal measure by themes of philosophy and of mathematics and its history, made me think that it was not entirely hopeless to try to write an essay that would be an approximation to Riemann's consistent conceptions, an essay that would bring together the available materials and prepare the ground for further work. This was the best I was able to strive for. Encouraging assistance came from the accounts of Weyl and Freudenthal as well as from more recent research, especially that of U. Bottazzini, J. Gray, and E. Scholz, and from R. Narasimhan's preface to N.

My objective, as well as my limitations, implied certain restrictions. I left out details of Riemann's mathematics that seemed to me unnecessary for the understanding of his global conception. This means that a mathematician may not find in my account certain parts of Riemann's work that strike him as significant for his own research. The need for selectivity applied as well to biography and to contemporary history. Here the choices were made easier by the fact that Dedekind's biography of Riemann, published in 1876, remains the most important source to this day. It is readily available in N., and so are additional materials bearing primarily on Riemann's schooling. Recent publications have changed next to nothing in the overall picture. Letters by members of the Dedekind family published by W. Scharlau were very helpful, and so too were individual references in the papers of E. Neuenschwander.

Riemann's qualifying papers — his doctoral dissertation of 1851 on complex analysis, his habilitation paper of 1854 on real analysis, and his habilitation lecture devoted to geometry and physics and rich in philosophical allusions — suggested a natural ordering of the contents of the book. I thought it necessary to supplement the brief introductory account of Riemann in his time with a sketch of analysis in the decades prior to his creative period. It is remarkable that the views of mathematicians in the early part of the nineteenth century were so different from our present conceptions that without some familiarity with these views it is hardly possible to understand the turning points in Riemann's conception that I tried to present in the last chapter.

This book will have achieved its purpose if it encourages others to produce further commentaries on Riemann and his work, as well as on the intellectual history of mathematics.

I could not have written the book without manifold assistance. I recall my Göttingen teachers Theodor Kaluza (senior) and Gerhard Lyra, who taught me as a young student to appreciate Riemann's world of thought in the spirit of Hermann Weyl and Richard Courant, and I remember the lectures of Carl Ludwig Siegel, in which he expressed his admiration for Riemann's analytical techniques. During the various stages of writing the book I profited from discussions with, and advice from, my colleagues U. Bottazzini, H. Harborth, W. Luh, E. Neuenschwander, and, in Darmstadt, P. Dintelmann, Dr. L. Schönefuss, and Dr. Th. Walter. The latter also rendered expert assistance during the final stages of editing. In dealing with archival materials I was helped by Dr.

E. Neuenschwander, by Dr. Rohlfing in the manuscript department, and by Dr. U. Hunger in the university archive in Göttingen. My daughter Annette Laugwitz, M.A., in Hamburg, prepared materials on Riemann's home country which contribute to an understanding of his scientific development. I am grateful to the editor, Dr. E. Fellmann, for his constant encouragement and professional advice. He and the publisher were able to surmount unexpected difficulties. Mrs. R. Jaschik patiently retyped the frequently modified manuscript.

I wish to take this opportunity to express my thanks to all persons mentioned in the preface. I will thank those not mentioned here in other ways.

Baltrum/Nordsee, August 1994 DETLEF LAUGWITZ

* * *

I wish to thank Birkhäuser for their willingess to publish an English translation of my book. My warm thanks to Abe Shenitzer and his "team" for a careful and fluent translation. As a result of our fruitful collaboration, a few passages in the translation are clearer than the corresponding passages in the original.

Darmstadt, July 1997 DETLEF LAUGWITZ

Note to the Reader

Some of the most important sources have been collected in the work published by R. Narasimhan in 1990. It is referred to in the sequel as N. followed by a page number (see the bibliography). When reading the present book the reader should, if possible, have ready access to Narasimhan's work. This work includes a reprint of the second edition of Riemann's collected works, published in 1892, with unaltered pagination. The second edition is referred to in the sequel as W. followed by a page number. The first edition of Riemann's collected works, published in 1876, is difficult to come by and is not referred to.

The bibliography includes only works that are frequently referred to. Infrequently mentioned source materials and items that go beyond our coverage of particular topics are quoted in the main text. N. 869–910 contains extensive bibliographies by W. Purkert and E. Neuenschwander. I am grateful to E. Neuenschwander for giving me the opportunity to consult another bibliography he is working on.

As a rule, when quoting texts I have retained the spelling of the original.

As usual, the *Journal für die reine und angewandte Mathematik* is referred to as *Crelle* or *Crelle's Journal.*

DETLEF LAUGWITZ

List of Illustrations

Sources of the illustrations

1–9: Original photographs by Annette Laugwitz

10 and 15–17: Hans-Heinrich Himme, *Stich-haltige Beiträge zur Geschichte der Georgia Augusta in Göttingen*, Göttingen: Vandenhoeck & Ruprecht 1987

12, 14, 29: Prof. Dr. H. Harborth

13: Universitätsbibliotheck Göttingen

20: Euler-Archiv Basel

23, 35: Niedersächsische Staats- und Universitätsbibliotheck Göttingen, department of manuscripts

25: Universitäts-Archiv, Göttingen

All others: Archiv Birkhäuser Verlag

Translator's Remarks

On this book

I don't always agree with the author, but I find him stimulating and enlightening. The book is an intellectual panorama of mathematics from Leibniz to Bourbaki.

On usage

(a) I hope that readers are not baffled by the following sentence:

Already in his dissertation, Riemann introduced a topological invariant.

Try the same thing without using "already" in this way. I think that one needs quite a few words to do the job. (By the way, this is *not* a Yiddishism. "So go already" is.)

(b) I say "what follows is stories" rather than "what follows are stories."

(c) Eric Partridge, author of *Usage and Abusage* (last revised in 1957) prefers "besides other things" and "in addition to other things" to "among other things." I follow Partridge.

German spelling

Some of the German texts go back to the 18th century. The reader who keeps this in mind won't jump to the conclusion that *bey* should be *bei*, *direct* should be *direkt*, and so on.

Thanks

In addition to the people whose help is acknowledged on the title page, I wish to thank my friend Abe Achtman for calling my attention to a number of linguistic rough spots.

ABE SHENITZER

0. Introduction

0.1 Bernhard Riemann in his time

0.1.1 His life and the development of his personality

The external circumstances of Riemann's short life can be quickly set down. He was born on 17 September 1826 into the family of the pastor of Breselenz near Dannenberg in the kingdom of Hanover. He attended the gymnasiums in Hanover (1840–1842) and in Lüneburg (1842–1846) and studied at Göttingen (1846–1847, 1849–1851) and at Berlin (1847–1849). He obtained his doctorate in 1851, and habilitated, i.e., took the qualifying examination for lecturing at a university, in 1854, both at Göttingen under Gauss. He became an associate professor (at Göttingen) in 1857 and a full professor in 1859. In June of 1862 he married Elise Koch, a friend of his sisters. Always in poor health, he spent a large part of the years that followed in Italy. His only daughter, Ida, was born there. In 1866, in the first days of the war with Prussia, he again decided to go south. He died on 20 July 1866 in Selasca on Lago Maggiore.

We will try to supplement this summary of the main events of his life so as to form an image of his personality.

To understand his development one must keep in mind that until the age of 14 Riemann lived in the circle of his family in the isolated Hanover Wendland, a thinly settled and undeveloped area on the Elbe river. His father, Friedrich Bernhard Riemann, came from Boizenburg, on the Mecklenburg shore of the Elbe. A few years after Bernhard's birth he took over the parish in Quickborn, not far from Breselenz. Riemann's mother was a daughter of a Hofrat in Hanover. In Section 1.4 of the Introduction we will comment on photographs of his native place.

All his life Riemann found it difficult to associate with people. Both as a student and as a docent he was always attracted by the security and solitude of Quickborn. He lost this anchorage when his father died in 1855.

He seems not to have sought contacts with people. In fact, he seems to have stubbornly discouraged attempts at greater closeness on the part of those — such as Eisenstein in Berlin — who were close to him in spirit. Dedekind, who occasionally managed to get him out of the loneliness of his room, has nothing to report about the concrete contents of scientific contacts. Riemann's scientific thought reflects his introvert makeup. He becomes absorbed in self-contemplation and wants to arrive at the "explanation of existence and of historical development" (die "Erklärung des Daseins und der geschichtlichen Entwicklung") by relying on the "inner perception" of the "laws of spiritual events" (aus "der inneren Wahrnehmung" der "Gesetze geistiger Vorgänge") (W. 511). He wants to find the universe reflected in his soul as in a Leibniz monad. He seeks refuge in speculative contemplation even when the need to complete mathematical papers, or to engage in robust activity in connection with a physical experiment, requires his total involvement. He always calls himself to duty and is under double compulsion to vindicate himself. The "daily self-test before the face of God" (Die "tägliche Selbstprüfung vor dem Angesicht Gottes"), in his own words the "main thing in religion" (die "Hauptsache in der Religion") (W. 557/558), is accompanied by the earthly call of duty. It is characteristic that following one of his rare escapes from his lonely life as a doctoral student, a trip in the company of a larger group of scientists, he almost apologetically reports to his father that he "was therefore all the more diligently at work in the morning" (er sei "dafür des Morgens um so fleissiger bei der Arbeit gewesen") and progressed as much as if he had been sitting in front of his books all day long (W. 546).

His mode of life and thought had physical parallels. The world saw him as a hypochondriac; his widow would later implore Dedekind to counteract this image in his biography (Dugac 1976). In fact, his constitution was always frail. For a long time he suffered from chronic constipation (Neuenschwander 1981a, 238). In the fall of 1862 he contracted pleurisy which resulted in permanent lung damage. There followed long stays in Italy. These did not heal him physically but brought him spiritual liberation. Of this Dedekind writes (W. 555/556) that the years of stay in Italy "were a true luminous point in his life; it was not only that looking at the glory of this enchanting land, of nature and art, made him endlessly happy. There he felt that he was a free man vis-à-vis other people, free of the inhibiting concerns he thought he had to have regard for in Göttingen at every step; all this, and the beneficial influence of the splendid climate on his health, made him often gay and happy and let him spend there many happy days" (bildete "einen wahren Lichtpunkt in seinem Leben; nicht allein, dass ihn das Schauen aller Herrlichkeit dieses entzückenden Landes, von Natur und Kunst, unendlich beglückte, er fühlte sich dort auch als freier Mensch dem Menschen gegenüber, ohne alle die hemmenden Rücksichten, die er in Göttingen auf Schritt und Tritt nehmen zu müssen meinte; dies alles und der wohlthätige Einfluss des herrlichen Klima auf seine Gesundheit stimmte ihn oft recht froh und heiter und liess ihn dort viele glückliche Tage verleben").

What do we know about Riemann's interests outside of mathematics, physics, and philosophy? There are hardly any clues. Dedekind writes rather impersonally that the newly married convalescent took "great interest" in the "art treasures and antiquities" of Italy when he travelled early in 1863 from Sicily to Göttingen. In Italy, contacts with mathematicians soon came to the fore. This was similar to Riemann's earlier visits to Berlin (1859) and to Paris (1860), which also served exclusively scientific interests. Dedekind writes (W. 554) that during Riemann's stay in Paris, which lasted a full month, the raw weather "often [made] the viewing of the local attractions virtually impossible" (habe... die "Besichtigung von Merkwürdigkeiten oft geradezu unmöglich" gemacht).

Outside his family, Riemann had close ties only to his immediate colleagues. Thus it is not surprising that he could find a wife only in the circles of their acquaintances. Neuenschwander reports (Riemann 1991, 48) that for a number of years Riemann hoped to marry a niece of the physicist Wilhelm Weber, but in the end he married a friend of his sisters. The marriage seems to have been happy.

Dedekind's biography of Riemann appeared ten years after the latter's death. It was influenced by consideration for Riemann's wife and sister. Dedekind's letters to his own family, written when he and Riemann were in Göttingen, contain valuable information which we will discuss in Section 3 of this Introduction.

The few comments by his students and by people who knew him more closely attest without exception to the respect and helpfulness shown to him.

0.1.2 The political and economic situation

In the 19th century agriculture played the most important role in the North-German plain, but industrialization and the development of communications came in Riemann's lifetime. Customs boundaries gradually lost their importance. Like Gauss, Riemann seems to have taken little interest in political change. The union of Hanover with England expired in 1837, but the subsequent suppression of liberal trends had less effect on rural parishes and gymnasiums than on Göttingen University, seven of whose eminent professors had to leave.

The 1848 revolution occurred when Riemann was a student at Berlin. Dedekind reports (W. 544) that "The great political events of the year 1848 moved [Riemann] deeply; he witnessed the March revolution, and as a member of the corps formed by the students was on guard duty at the royal palace from 9 in the morning on March 24 until 1 in the afternoon of the next day" ("Die grossen politischen Ereignisse des Jahres 1848 ergriffen auch ihn mächtig; er war Augenzeuge der März-Revolution und hatte als Mitglied des von den Studenten gebildeten Corps die Wache im königlichen Schlosse vom 24. März

Morgens bis zum folgenden Tage Mittags 1 Uhr"). But we do not find out what his thoughts were. He died before the annexation of Hanover by Prussia on 3 October 1866.

At that time the landless had things other than politics to worry about. A family had to provide for its old and uneducated members, such as unmarried daughters or sisters. There were no old-age or surviving dependents' pensions. Pastors, teachers, and university professors stayed in their jobs for life. A commendable exception, essentially due to Gauss, was the fund for widows of Göttingen instructors. And Riemann's later decision to turn down an offer of a position at Pisa may well have been influenced by the thought of losing the Göttingen support. Academics often had to rely on secondary income derived from tutoring, from taking in schoolboys and students to board, and from licenses to brew and serve beer — a rather common practice even in the 19th century. Higher education for sons was expensive, and it was obviously expected that the family would benefit before long. (Bernhard Riemann depended on his father for an unusually long time.) Then there were the inadequate provisions for health care. Someone who, like Riemann, always suffered from ill health, could hardly shoulder additional obligations, such as tutoring, on top of his own scientific work; on the other hand, Dedekind reports that as a docent he himself relied on such sources of income.

Bernhard had a brother, Wilhelm, who worked for the post office, and four sisters. The vicarage could accommodate the large family, and living in the country made it easier to provide for the modest daily needs. But school education, and all the more so university education in a distant place, meant huge sacrifices on the part of the family; later, there might have been the prospect of theological study with the support of the church authorities. While the grandmother in Hanover was alive, she provided accommodation near the school. But the stay in Lüneburg entailed costs. The Hebrew teacher G.H. Seffer, in whose house Riemann stayed from Easter 1844, reports that, following the request of the school director Schmalfuss, he took him in to board, along with others, "at a reduced rate"(N. 849).

At that time, even very gifted students and instructors had to lead very modest lives. This was accepted as self-evident. We know that Richard Dedekind, who, like Riemann, habilitated at Göttingen in 1854, submitted a humble petition to the state authorities in Braunschweig for the relatively modest sum of 25 thaler a term for food (Freitisch) (communicated by H. Harborth). At the time, Riemann's income from tuition fees was extremely limited, and when his brother died in Bremen in the fall of 1857 and he had to provide for his three surviving sisters, he was elated to be appointed an associate professor with a yearly salary of 300 thaler. We mention for comparison's sake that upon his appointment at Zürich in April 1858 Dedekind was offered a yearly salary equivalent to 850 thaler, and that when he became professor at Braunschweig in 1864 his yearly salary was 1480 thaler (according to Knus in Harborth 1982, 52, 57).

0.1.3 Upbringing and education

It is striking that in the 19th century North Germany produced a number of outstanding mathematicians who significantly influenced their science not only by their results but also by their new ideas and fundamental conceptions. This cannot be explained by claiming that North Germans have a special inclination towards mathematics, since no comparable development occurred at other times. People have tried to explain this phenomenon as the result of the combination of Prussianism, post-Kantian philosophy, and Protestantism, and this may be correct in an indirect sense. One must not exaggerate the importance of the aftereffect of the flourishing of the Berlin Academy between 1741 and 1786, with luminaries such as Euler, Lambert, and Lagrange — none of them a North German. Another factor in a general sense was the spirit of the Enlightenment.

A source of direct influence was the reorganization of higher education that originated in Prussia. This was the only German state which, after the defeats of 1806/1807, retained a certain independence of Napoleon in its provinces east of the Elbe and concentrated on long overdue internal reforms. In 1809/1810 the head of the section for education and culture in the ministry for internal affairs was the philosopher, linguist, and diplomat Wilhelm von Humboldt. In the space of just 14 months he effected a reorganization of the Prussian educational system which was to serve as a model for the rest of North Germany. The key element was the neohumanistic ideal of all-round education. Already in earlier philosophical writings on the state Humboldt expressed the idea that the state should be rendered unnecessary and conditions should be created that would promote the free development of man's personality.

Berlin University, founded in 1810, was thought of as a center for research and learning, whereas the old universities were just centers for learning. The general preliminary education, the quadrivium, was removed from the university curriculum and assigned to the new high schools which replaced the old Latin schools. The teachers in the Latin schools were members of the clergy. The future gymnasium teachers were to be educated at the new philosophical faculties of the universities, which were thus assigned a well-defined task.

(Our primary sources for what follows are F. Paulsen, *Das deutsche Bildungswesen in seiner geschichtlichen Entwicklung* (*The German educational system in its historical evolution*), 3. Aufl. Teubner, Leipzig-Berlin 1912, and G. Schubring 1990.)

This development entailed that mathematical research and the study of mathematics became firmly established at the universities, and mathematics was assigned an important role as one of the four main subjects at the ten-year gymnasiums. According to the first Prussian draft, the number of hours per week assigned to various subjects, summed over ten years, was to be: Latin 76, Greek 50, German 44, mathematics 60, history and geography 30, natural science 20, religion 20. Initially there was no provision for the study of foreign languages. The main theme was "permeation with the spirit of antiquity through exposure to its greatest writers, especially Greek..." ("die Durchdringung mit dem Geist des Altertums durch Verkehr mit seinen hervorragendsten Schriftstellern, besonders den griechischen...") (Paulsen, p. 126). Another figure: until 1860, the total number of mathematics majors who graduated in a year from all Prussian universities was about 20 (Schubring, footnote on p. 266).

The turn to an idealized version of antiquity signified a withdrawal from the problems of the present. In the case of mathematics it translated into a renunciation of topical applicability. Freedom from state interference was purchased by withdrawal to an ivory tower. This fitted perfectly into Humboldt's view of science. On 16 July 1816 he wrote to Goethe: "I have busied myself here with science a great deal, although I have not worked [at it] a lot myself, but I have deeply felt the power antiquity has always wielded over me. The new disgusts me ... " ("Ich habe mich hier viel mit der Wissenschaft beschäftigt, obgleich nich viel selbst gearbeitet, aber so recht wieder die Gewalt gespürt, die das Altertum an mir immer ausgeübt hat. Alles Neue ekelt mich an ...").

In addition to neohumanism with its idealized image of antiquity, it was German idealism and romantic natural philosophy that decisively determined the spirit of the new university.

It took time for the reforms to be generally accepted in practice. In the first decades the professors of mathematics were not what we would now call first-rate research mathematicians. The important Berlin mathematicians were at the Academy. They could teach at the university but were not allowed to examine students. Initially teachers in the new gymnasiums were mostly members of the clergy from the old Latin schools, and gymnasium education was formal education. This was true of the teaching of Latin and Greek as well as of the teaching of mathematics.

When Wilhelm von Humboldt's brother Alexander (1769–1859) returned to Germany after a 20-year sojourn abroad, spent mostly in Paris, he tried to carry over to the German universities some of the French scientific-mathematical orientation and to establish institutions on the model of the Paris Ecole Polytechnique. But it took decades for all the German princes to embellish their residence cities with polytechnics.

C.G.J. Jacobi, himself one of the first graduates of Berlin University, described the situation very aptly in a letter to Legendre dated 2 July 1830: "Fourier reproaches us, me and Abel, for not having chosen to study heat conduction. It is true that Fourier was of the opinion that the principal object of mathematics is public use and the explanation of natural phenomena; but a philosopher like him ought to have known that the sole object of the science is the honor of the human spirit, and that on this view a problem in [the theory of] numbers is worth as much as a problem of the system of the world."

Riemann was exposed to the German neohumanistic gymnasium for only the last six of the required ten years. Before that, his father was practically his only teacher. The gymnasium may have promoted his philosophical inclinations. His view of mathematics was also influenced by its connection with physics. In this respect he resembled Gauss, who belonged to a different era, and Dirichlet, who studied in Paris for five years. Riemann's special new conception of mathematics may have been due to a combination of physical and mathematical influences. Incidentally, Richard Dedekind, who was born in 1831, was a gymnasium student for eight of the ten years and spent the last two years at the Braunschweig Collegium Carolinum (later an Institute of Technology (Technische Hochschule)), which was strongly inclined towards mathematics and the natural sciences. When he came to Göttingen in 1850 he was so well prepared that he could get his doctorate, under Gauss, at age 21.

All we know with certainty about the education Riemann received from his father is its success: before he turned 14, he became a fourth-year student at a Hanover gymnasium. Dedekind reports (W. 541) that the village teacher Schulz gave the boy a good introduction to arithmetic and geometry. We know next to nothing about what his father taught him and we know just as little about what reading matter was available to the boy at home.

Two letters, by Schmalfuss and Seffer respectively, shed a great deal of light on Riemann as a gymnasium student in Lüneburg (1842–1846). But they must be interpreted with caution, for they were meant to be used in an obituary for a former student who later acquired great fame (N. 849–853). Riemann was deeply unhappy in Hanover. He was morbidly shy and homesick. Dedekind writes (W. 543) that while Lüneburg was just slightly more than 60 km from Quickborn, "getting there and back, largely on foot, involved an effort his body was not always up to" ("die Hin- und Herreise, die zum grössten Theil zu Fuss gemacht wurde, [war] mit Anstrengungen verbunden, denen sein Körper nicht immer gewachsen war"). The school director Schmalfuss, who soon became aware of Riemann's mathematical talent, supplied him with relevant reading matter (N. 852). But when it came to written assignments, Riemann's perfectionism made things difficult for himself and for his teachers. In 1844 Seffer was asked to take him in to board and to see to the "prompt delivery of his compositions" ("prompte Ablieferung seiner Aufsätze"), because "the staff meeting, faced by the school rules, was in despair on his account" ("die Lehrer-Conferenz den Schulgesetzen gegenüber seinetwegen in Verzweiflung war") (N. 849). One does not get the impression that Riemann enjoyed school very much! If one considers the fact that the biographies of great mathematicians of the past almost invariably mention, in addition to furtherance by the home, a sympathetic teacher who spotted special talent in good time, then it is a safe guess that the "school rules" of that time may well have thwarted the unfolding of many a talent.

What must also have oppressed young Riemann was the thought that he was a source of worry for his father — by that time his mother was dead — and his well-meaning teachers. The daily self-test before the face of God must have often been an ordeal for him.

As a child, Riemann was used to freedom, and the coercion involved in fitting into the strict rules of behavior must have been as hard on him as the spiritual coercion students had to put up with until the age of 20. The letters of his two teachers indicate that the main reason he managed to endure the anguish of school was that he did not want to disappoint his father and his well-meaning teachers. What did school give him that had the quality of permanence?

The long exposure to ancient languages was not very productive. In connection with Riemann's doctoral application Dean Ewald commented on 15 November 1851: "The Latin in the application and in the vita is clumsy and barely tolerable" ("Das Latein in dem Gesuche und der Vita ist ungelenk und kaum erträglich"), and went on to say that one would have expected something better of a candidate who wanted to pursue a university career.

Riemann retained in memory only the things that interested him. We will see over and over again that he rediscovered results and proofs without going back to what he had read earlier. The drill in writing Latin and German compositions that frequently made Seffer work with him late into the night was of little profit. Free and unaffected command of German prose always eluded him. The German in his later papers is precise but difficult to translate into other languages. He always wrote with great effort. His notes on introductory theorems in the lectures show a constant struggle for a perfect mode of expression. We will have occasion to see examples of this. Speaking of Riemann as a schoolboy, Schmalfuss observed (N. 852) that he found it difficult "to develop his thoughts in a free-flowing lecture" ("in fließendem Vortrage seine Gedanken zu entwickeln").

Schmalfuss writes with great enthusiasm (N. 851/853) about Riemann's mathematical aptitude, which he tried to promote in a pedagogically reasoned manner, in particular by lending him books. He mentions Euclid editions and commentaries, works of Archimedes and Apollonius, and Newton's *Arithmetica universalis*, and emphasizes at the same time that he tried at first to keep Riemann away from "the transcendental," so as to prevent him from "mistaking form for substance, although such caution may not have been necessary in his case" (damit Riemann nicht "die Form für das Wesentliche nähme, obwohl bei ihm solche Vorsicht wohl nicht erforderlich gewesen wäre"). This is undoubtedly a reference to the then rampant treatment of transcendental functions by the formal methods of algebraic analysis. Characteristic of Riemann's later mathematics is his steadfast avoidance of formulas and their rule-bound transformations, and their replacement by thinking in concepts. Speaking of Descartes' analytic geometry, Schmalfuss observes that the "mechanical method" did not appeal to Riemann. In connection with Legendre's *Number Theory* he notes that Riemann returned it in less than a week with the comment:

> "This is a wonderful book; I know it by heart,"
>
> ("Das ist ja ein wundervolles Buch; ich weiß es auswendig.")

Two years later, during the oral part of the *Abitur*, he demonstrated his familiarity with the book in spite of the fact that he had not consulted it in the interim. Number theory held a special attraction for him. Schmalfuss does not mention Gauss' *Disquisitiones*. Incidentally, in 1859, when he was writing his paper on prime numbers, Riemann seems to have forgotten Legendre, because he does not mention Legendre's prime number formula but does mention the almost equivalent formula of Gauss.

"Already at that time he was a mathematician next to whose wealth the teacher felt poor" ("Schon damals war er ein Mathematiker, neben dessen Vermögen der Lehrer sich arm fühlte"), recalled Schmalfuss in a thoroughly credible manner. It must have pained him to refrain from advancing this youngster mathematically so as not to keep him from completing tasks he did not care for but had to fulfill.

0.1.4 Riemann's homeland

The country where Riemann was born and where he spent his childhood has been known since the beginning of the 18th century as Wendland. For a long time little attention was paid to this region in the northeastern part of the kingdom of Hanover (today's Lower Saxony), bounded by the river Elbe in the north and east, by the Lüneburg Heath in the west, and by the then Prussian province of Saxony in the south. More attention began to be paid to it and to the Lüneburg Wends (Lüneburgische Wenden) when Leibniz, then in Hanover, prompted research into their folklore and Slavic language as part of his manifold scientific interests. The Wends had settled in this area beginning around 600, and immigrants from (lower) Saxony arrived around 1200.

Because of the history of its origins, the Wendland landscape presents certain singular features, including varied and striking surface formations. The land alternates between roughly ordered, dome-shaped hills in northeastern Drawehn and the wide, slightly wavy, high plateau of the Göhrde in the west. During the penultimate ice age this strip of land was covered by ice for about 10000 years longer than western Lower Saxony, so that erosion due to water, wind, and frost set in rather late. Water from the melting ice of the last Vistula ice age, flowing south and west, formed the Elbe-North Sea river, and its bed became a huge primeval river valley which determines the direction of all of the Wendland's bodies of running water. Schering (N. 829) speaks mistakenly of the "Elbmarsh " ("Elbmarsch").

Breselenz, Riemann's birthplace, lies at the edge of a range of hills of a moraine ridge. The land drops gently to the east and goes over into the Elbe-Jeetzel valley, a part of the primeval river land. Riemann's family moved to this area a few years after his birth. The Elbe, the great trade route, flows nearby. It was at first the only visible connection between his home country and the rest of the world. The dykes, whose construction began in the 12th century and ended with the one built after the disastrous flood of 1962, emphasize the extent to which the water flows determine the landscape. The latter is also molded by agriculture and forestry, the main sources of employment for the population. Even today there are hardly any large industrial and manufacturing establishments in the area. With 41 inhabitants per square kilometer, this is the most sparsely inhabited area of Lower Saxony. This, and the location in the immediate vicinity of the border with the former GDR, were the main factors that hampered change in the region.

Most of the settlements are villages. Only a few, such as Lüchow, Dannenberg, and Hitzacker, have become towns. The layout of many of the villages is very original. They conform to the Wendish pattern of a radial village. The peasant homes are built around a circular green and the house gables are directed towards its center. Most of them are timber-framed houses with bookrack construction. The frames are filled in with red bricks that often form decorative patterns. The church is frequently located at the village's edge. Many

1: House in Breselenz where Riemann was born; view from the front

2: House in Breselenz where Riemann was born; view from the back

3: Church in Breselenz

of the churches have massive towers which have earned them the sobriquet "fortified churches" ("Wehrkirchen"). Since most of them were built during the Romanesque period, it is more likely that what inspired their solid architecture was the then current view of the role of the church as "God's stronghold" ("Gottesburg").

4: Memorial tablet in Breselenz

Such a Romanesque church is found in Quickborn, where Riemann's father was vicar until his death in 1855. The church is a squat brick structure on a fieldstone base. Baroque modifications have considerably altered its interior.

Riemann was born in Breselenz, in Lower Drawehn. The name of the place derives from the Slavic word for a birch tree; this hints at the nature of the landscape. Around 1870 the church in which Riemann's father once served as vicar was radically modified and the Riemann family home was torn down.

5: Church in Quickborn

6: Vicarage in Quickborn

7: Gymnasium Johanneum in Lüneburg, now a school of a different type

8: The Seffer house in Lüneburg

Old photographs of the house are in the possession of Dr. A. Kelletat. At his suggestion, and with the support of professor Georg J. Rieger of Hanover, a memorial stone was set up in the churchyard on the occasion of the celebration

9: Memorial tablet affixed to the Seffer house in 1886

of Riemann's 150th birthday. A local street is named after him and a school in Dannenberg reminds us of him.

To come back to Quickborn. Today it is still a small place with partly unpaved streets, big roadside oaks, and the original tillage. It contains the only windmill in the region (the sails are now just mockups). The other mills, once water-driven, are also idle. The Quickborn church is an exception in that it stands on a little knoll at the center of the village rather than at its edge. Nearby is the vicarage, rebuilt in 1866. A timber-frame house takes the place of the one in which Riemann grew up and which he later visited many times.

Beginning in 1842, Riemann attended the Gymnasium Johanneum in Lüneburg, 60-odd kilometers west of Quickborn. The school, still used as such today, was built in 1829. It is close to the St. Johannis church, which in turn is close to two other churches, St. Michael and St. Nicolai. These three key Lüneburg churches, whose tall towers exemplify the so-called "North German Brick-Gothic" ("Norddeutsche Backsteingotik"), help to determine the town's character. The houses of middle-class citizens and the city hall, rebuilt in the 18th century, also attest to the wealth of this Hanseatic town in the 14th century. Commerce flourished. The crane at the river port reminds one of that time. The main commercial product was salt. Salt production and the salt trade, as well as staple rights and compulsory transshipment rights, secured for the city a kind of economic independence. Lüneburg was granted city rights by Henry the Lion (Heinrich der Löwe) (1129–1195). The city's economic decline began in the 17th century; few new structures were built and few old ones were rebuilt. Most streets are narrow and are still paved with cobblestones. In one of them, named Auf dem Kauf, is the house of Riemann's teacher Seffer, where Riemann lived from 1844 to 1846. (The memorial tablet near the entrance gives the incorrect dates 1842–1846.)

0.1.5 Göttingen and Berlin as centers of study

By the summer term of 1846 Riemann was finally free. At the university in Göttingen he soon turned to mathematics rather than take up the study of theology, suggested earlier by his father.

The university of Göttingen was founded in 1734 and was opened in a solemn ceremony in 1737. It is named Georgia Augusta after George II August, prince elector of Hanover and king of England and Ireland from 1727 to 1760. In Hanover, government affairs were run by Baron G.A. von Münchhausen, who influenced the new university in the spirit of the Enlightenment. Next to Halle (in Prussia), where rationalism was championed by Chr. Wolff, a universalist scholar and popularizer of Leibniz' philosophy brought back (from Marburg) to Halle by Frederic II after his ascent to the throne in 1740, Göttingen became a center of rationalism and enlightenment in Germany. The leading role of theology came to an end, the teachers of mathematics and of the natural sciences were strongly urged to do research, and their lectures were not limited to trivium and quadrivium topics. And while the glory of Euler outshone all others, one must not underestimate the significance of Göttingen scholars. G.Chr. Lichtenberg (1742–1799; professor from 1769) was a physicist and brilliant experimenter. A.G. Kästner (1719–1800; professor at Göttingen from 1756) was more influential by virtue of his textbooks of above-average quality and, later, of his *Geschichte der Mathematik* (*History of Mathematics*), than by virtue of his lectures, which were smiled at condescendingly — not by Gauss alone. Both are remembered today primarily for their literary contributions. J.A. Segner (1704–1777; professor of mathematics, physics, and chemistry at Göttingen from 1735 and at Halle from 1755) and the astronomer Tobias Mayer (1723–1762; professor at Göttingen from 1751) were held in the highest esteem by Euler. G.S. Klügel (1739–1812), a student of Kästner's, became a professor at Helmstedt in 1767 and at Halle in 1787. He wrote advanced textbooks and, beginning in 1803, published his famous *Mathematisches Wörterbuch* (*Dictionary of Mathematics*). J.Fr. Pfaff also studied with Kästner and Lichtenberg and became a professor at Helmstedt in 1788. While there, Pfaff was Gauss' "Doktorvater" (doctoral advisor). After the dissolution of the university at Helmstedt in 1809 Pfaff went to Halle.

B.F. Thibaut, G.K.J. Ulrich, and M.A. Stern were professors of mathematics at Göttingen. Their respective tenure dates are 1802–1832, 1817–1879, and 1848–1883. One would look in vain for their names in the *Lexikon bedeutender Mathematiker* (*Encyclopedia of Outstanding Mathematicians*), but in the case of the first two one can hardly speak of historical injustice. Gauss became a professor at Göttingen in 1807. Of course, he was a full-time professor of astronomy.

Riemann's choice of Göttingen as a place of study had little to do with the presence there of Gauss. After all, he was initially supposed to study theology and Göttingen was the only university within the sphere of the Hanover church. Riemann immediately took courses in mathematics and physics. In the summer of 1846 he took Stern's course on the numerical solution of equations and in the winter his course on definite integrals. The two added up to an introduction to analysis. The presentation was almost certainly permeated by the algorithmic spirit, which Stern rated highly. In the same winter Riemann also took Gauss' course on the method of least squares, which included an introduction to linear algebra.

We will encounter Carl Friedrich Gauss' (1777–1855) name in this book many times and so, at this point, we will say just a few words about him. There are many biographies of Gauss and they complement one another. His collected

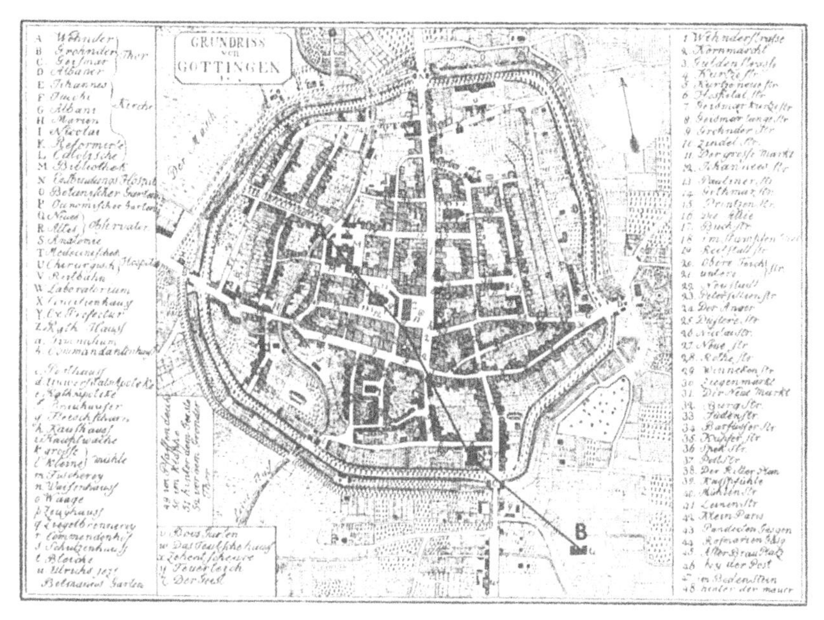

10: Plan of Göttingen showing the telegraph line of 1833

works, including his *Nachlass* and correspondence, are readily available in libraries.

Gauss came from a humble background. His family lived in Braunschweig. Attentive teachers quickly recognized his great talent and the duke provided generous stipends even for "study abroad" at Göttingen. Beginning in 1795, Gauss got free board and a yearly stipend of 158 thaler. We recall that when Riemann was appointed an associate professor (in 1857) his initial yearly salary was 300 thaler. Two youthful achievements provided the basis for Gauss' fame. His *Disquisitiones arithmeticae*, written between the ages of 19 and 21, appeared in 1801. This was the first systematic account of number theory, and it included solutions of many open problems. In the same year Gauss developed for the determination of the orbits of planets a method that enabled astronomers to find the "lost" planetoid Ceres. In 1807 he was appointed professor of astronomy and director of the observatory. He significantly advanced mathematics, astronomy, geodesy, and physics. He regarded university teaching as an annoying waste of time. This once eminently sociable young man became in old age unapproachable and withdrawn. When the shy Riemann wanted to learn from Gauss he had only his writings to fall back upon. Incidentally, Moritz Abraham Stern was Gauss' first doctoral student (in 1822).

Already in the gymnasium Riemann was used to self-study and learned a great deal in this way. He continued this habit as a university student by reading the advanced literature. Gauss taught the method of least squares and nothing else, so that for someone like Riemann, lectures at Göttingen were of little use. Berlin was the right choice for the next two years. Dirichlet, rather than Jacobi or the young Eisenstein, praised gushingly by Gauss, must have seemed to him from the beginning his predestined teacher. For Riemann, Gauss and Dirichlet were always authorities above criticism.

Dirichlet must have been a fascinating personality. In view of his importance for Riemann we say more than a minimum about him. He was born on 13 February 1805 in Düren, near Aachen. His family was of French origin. What shaped him as a scientist was his stay between 1822 and 1827 in Paris, where he made a living by tutoring. He was put in touch with the circle around Fourier (1768–1830), which provided him with lasting stimulation in mathematics and physics. He also studied on his own, paying special attention to Gauss' *Disquisitiones arithmeticae*. As a foreigner, he had no access to Cauchy's courses at the Ecole Polytechnique. Visible fruits of his Paris years of study were two papers which appeared in *Crelle's Journal* in 1829 and in 1830 respectively. The second paper is of a rather physical nature and is connected with Fourier's theory of heat. The first contains the famous theorem on the representability of a function by its Fourier series and solves satisfactorily a problem that Poisson, Cauchy and others, including Gauss, contended with. In this paper Dirichlet surpasses Cauchy, the inventor of epsilontics, in the precise use of this technique.

In Paris, Alexander von Humboldt found out about the young man and was so impressed by him that he set him up as a docent, first in Breslau (1827) and then in Berlin (1829). Dirichlet ignored many an antiquated custom of the German university, and did so in a confident and self-assured manner. In particular, he persistently balked at demonstrating sufficient knowledge of Latin. From a scientific viewpoint, the Berlin Academy was anyway far more important than the newly (1810) founded university, whose neohumanistic spirit and Hegelian philosophizing threatened to make it sleep through developments such as Parisian mathematics and physics. While it is true that the Berlin full professor Dirksen had read Cauchy's books, it is with an amused smile that one is likely to read his paper in vol. 4 of *Crelle's Journal*. It follows Dirichlet's paper, but the only thing the two have in common is the topic. Of Dirksen's paper Riemann said (W. 238) politely that it contains a number of inaccuracies.

Dirichlet was the founder of analytic number theory, which Riemann developed in essential ways (see Section 1.4). We will encounter numerous instances of Dirchlet's manifold influence on Riemann.

11: Peter Gustav Lejeune-Dirichlet

Dirichlet's lectures must have been very impressive. If the publisher's (Arendt 1904) claim that he reproduced his Berlin lectures on definite integrals without a change is literally true, then these lectures exemplify a masterful way of dealing with language and with mathematics that Riemann must have strived to emulate.

Riemann took Dirichlet's courses in number theory, the theory of definite integrals, and partial differential equations; Jacobi's in analytical mechanics and higher algebra; and Eisenstein's in the theory of elliptic functions. This is reported by Dedekind (W. 544), who adds that little is known about Riemann's life in Berlin from his letters to his family. F.G.M. Eisenstein (1823–1852) was just a few years older than Riemann. He obtained his doctorate in 1847. In principle, there could have been a lively exchange of ideas between him and Riemann. In fact, they initially discussed, besides other things, complex numbers and functions. But, as Riemann later reported (according to Dedekind, W. 544), "Eisenstein stopped at formal computation" ("Eisenstein sei bei der formellen Rechnung stehengeblieben"), whereas he, Riemann, realized that what counts is the partial differential equation. Moreover, the idea of closer personal relations with Eisenstein may not have appealed to Riemann. Biermann's (1988, 62ff.) descriptions are instructive. Like Riemann, Eisenstein was a hypochondriac. Jacobi valued him highly but was forced to call him, both in private and in public, a literary thief who took his main ideas from oral communications of other people or from unpublished course notes. These facts may have persuaded Riemann to be reserved in personal contacts with Eisenstein. Gauss thought of Eisenstein as highly as he did of Dirichlet and regarded him as a prodigy. On 24 April 1852, following a motion by Dirichlet, Eisenstein took Jacobi's place at the Academy. But he was seriously ill, and "On 11 October 1852 Eisenstein succumbed to his illness; a life, joyless in spite of all the recognition, ended in solitary desolation" ("Am 11 October 1852 erlag Eisenstein seiner Krankheit; ein trotz aller Anerkennung freudenarmes Leben endete in einsamer Verlassenheit"), writes Biermann (1988, 67). In letters to Stern in Göttingen, Eisenstein evinced his continued interest in Riemann and expressed regret that the latter had avoided him.

When he returned to Göttingen in 1849, Riemann initially took up the study of physics and philosophy. He may have done this not only because of an interest in these subjects but also for practical reasons: he had to have this kind of background if he was to become a gymnasium teacher. These activities extended beyond his doctorate in 1851 and paid off in scientific results of a very high order (see Ch.3).

0.1.6 Riemann as full professor: 1859–1866

The decade from 1850 to 1859 was the golden period of Riemann's career as a mathematician. We will discuss this period in the next section. Here we will talk about the last years of his life.

In his lifetime Riemann was recognized as an important mathematician primarily because of his "Theorie der Abel'schen Functionen" (W. 88–142), which appeared in 1857 in Vol. 54 of *Crelle's Journal*. While it is true that the method he used in this paper had been sketched (in an admittedly cryptic form), and its results hinted at, in his dissertation of 1851, it was in this paper that he solved problems that had been handled with only partial success from the time of Abel and Jacobi. Consequently, he could compete on an equal footing with Weierstrass, who used different methods for dealing with the same group of problems. This is yet another confirmation of the fact that, when it comes to recognition, the mathematical world honors palpable results bearing on "hot" or on unsolved problems that have been around for a long time, and not deliberations, however profound, dealing with concepts and methods.

The papers to be discussed in Chapters 2 and 3 were unknown in Riemann's lifetime or received scant attention.

By the time he was appointed full professor, in 1859, Riemann had presented his principal work and had acquired a superb and self-assured command of the techniques of analysis. This is clear from notes in the *Nachlass*, such as the computations on the zeta function published by Siegel in 1932. The short but important paper on theta functions (published in 1865; W. 212–224) completed his principal work. Further applications of his methods could have been carried out by gifted students. In this respect, regrettably, Göttingen was hardly the promised land. It was sometimes possible to gain recognition, and an appropriate income, by publishing works less profound than Riemann's. Dedekind's early papers are a case in point.

Riemann had few personal contacts with nonlocal mathematicians. In the fall of 1858 the Italian mathematicians Brioschi, Betti, and Casorati came for a few days to Göttingen as one of the important places that they visited to gather information about mathematical life. In September 1859, buoyed by his appointment as corresponding member of the Berlin Academy, Riemann stayed with Dedekind in Berlin. Dedekind reports that Riemann was very cordially received by Kummer, Kronecker, and Weierstrass, as well as by Borchardt, who, after Crelle's death, became the editor of the latter's important journal. In the spring of the following year Riemann spent four weeks in Paris and got to know personally such important mathematicians as Serret, Bertrand, Hermite, Puiseux, Briot, and Bouquet. The fact that Dedekind reports these contacts (W. 554/555) and gives a detailed list of names may perhaps be taken as a confirmation of the impression that such personal contacts were rare in Riemann's life. More precisely, they were rare until 1862, after which he spent a great deal of time in Italy and had lively social contacts with Italian mathematicians.

It is well to remember that at Göttingen, with the single exception of the yearly meeting of the society of natural scientists and medical doctors, there were then, unlike today, no professional meetings, and that colloquium lectures by visiting scientists were virtually unknown. Göttingen hardly reminded one of Paris with its lively scientific discussions. Nonlocal members of Academies had publication rights. Riemann was a corresponding member of the Bavarian Academy from 28 November 1859 and a regular member from 28 November 1863. The Paris Academy appointed him a foreign member on 19 March 1866, and the Royal Society on 14 June 1866 (W. 554). The news failed to reach him.

From 1862 on Riemann must have felt that his health was setting ever narrower bounds to his work and that he had to decide on priorities. Creative individuals often stop working in an area that is the source of their recognition in the world because they regard other pursuits as more important. Goethe's theory of colors is the best-known example of this kind. In the last years of his life Riemann devoted what energy he had to the study of natural philosophy, leaving unfinished things that we may regard as more important.

A telling example is Riemannian geometry, of such great importance for posterity. Riemann published nothing on this subject. There was an opportunity to write about it in connection with the prize paper he submitted to the Paris Academy. The problem had to do with the propagation of heat, but Riemann concentrated quite generally on the treatment of partial differential equations in curvilinear coordinates and also developed, in a by-the-way and sketchy manner, the principal aspects of his geometry. Application to the posed problem is just hinted at near the end of the paper. Today we realize that this short paper (W. 391–404) contains *in nuce* tensor analysis and Riemannian geometry, and the rest is almost all drudgery. But Riemann's sketchy arguments failed to satisfy his contemporaries. Had the paper not been his, it would not have been taken seriously. (Notwithstanding the anonymity of the submissions, the judges were very well aware of the identity of the author.) No one was awarded the prize and Riemann's paper languished.

He was still fully capable of doing scientific work. There is no doubt that he thought it important to make his theoretical-scientific and methodological considerations intelligible in a concrete case of research in natural science. In the last months of his life, in a dispute over Helmholtz's new *Lehre von den Tonempfindungen* (*The Science of Tone Sensations*), he tried his hand at the mechanics of the ear (W. 338–350).

From the middle of 1862 his health declined precipitously. After November 1862 he spent more time in Italy than in Göttingen. His lung condition could be palliated but not cured. He established mathematical and personal contacts with Enrico Betti, who had visited Göttingen in 1858 and was to pursue his topological ideas. The progress of the unification of Italy promoted the sciences and benefited the universities. Here Riemann's ideas fell on rich soil. The place where he stayed before his death is now part of the town of Verbania on Lago Maggiore. In 1991 the town commemorated the 125th anniversary of his death, and his influence on Italian science was thereby acknowledged (Riemann 1991).

0.2 The golden 1850s in Göttingen: From Gauss and Dirichlet to Riemann and Dedekind

0.2.1 Riemann and Dedekind: Personal circumstances

In retrospect it is clear that the 1850s at Göttingen was an extraordinary time, a time that prepared to a considerable extent the mathematics of the 20th century. True, Gauss was no longer very active, and Dirichlet, who after Gauss' death in 1855 was brought from Berlin to Göttingen, lived for just four more years,

12: Richard Dedekind (ca. 1854)

but the most important works of Riemann and the first papers of Dedekind came into being as a result of their influence. Dedekind was destined to be closely connected with the life and work of Riemann.

Richard Dedekind was born in Braunschweig on 6 October 1831. There are references to a "multiplicity of scientific and aesthetic influences" ("Vielfalt von wissenschaftlichen und schöngeistigen Einflüssen") in the family (Scharlau 1981, 2). He attended the gymnasium in his home town (until 1848) and then the Collegium Carolinum, which later became an Institute of Technology. From 1850 he studied at Göttingen. He obtained his doctorate under Gauss in 1852, just one year after Riemann, who was his senior by five years, and habilitated just a few weeks after him. Judged by the richness or depth of their ideas, Dedekind's two papers cannot possibly be put in the same category as Riemann's. His dissertation *Über die Theorie der Eulerschen Integrale* (*On the theory of Eulerian integrals*) contained nothing new for Gauss, and his short habilitation paper *Über die Transformationsformeln für rechtwinklige Koordinaten* (*On the transformation formulas for rectangular coordinates*), while original, could have been mistaken for the final-examination paper of a prospective gymnasium teacher (Staatsexamensarbeit). But in his habilitation lecture Dedekind self-confidently announced thoughts on the development of the number systems which partly anticipated future developments.

13: The Göttingen university library (around 1854)

Contacts between Dedekind and Riemann were established rather quickly. They certainly occurred in Wilhelm Weber's experimental physics course, where Riemann was an assistant. Dedekind's restrained and sympathetic biography of Riemann (W. 541–558) has for a long time been the most important source for the understanding of Riemann's development. But the Dedekind family letters that have recently become accessible paint a somewhat different picture, which we will have to take into consideration.

At the time there were hardly any mathematics students at Göttingen, and the two young instructors could often expect their courses to be cancelled for want of listeners. Dedekind attended the lectures of Dirichlet and Riemann and taught courses in probability and geometry (winter term 1854/55), higher algebra, Galois theory, cyclotomy and group theory (winter term 1856/57, winter term 1857/58). Riemann began his teaching activity with a course on partial differential equations and their applications to physics (winter term 1854/55) and an introductory analysis course on definite integrals (summer term 1855). Beginning in 1855/56, he lectured regularly in the winter term, sometimes into the summer term, on his new complex function theory. In this connection topics in mathematical physics came up. The function-theoretic courses gave rise to his two substantial and significant papers of 1857. Unlike his virtually overlooked doctoral dissertation, these papers were destined to call the attention of the mathematical world to him. In these papers Riemann pursued the most topical problems of mathematics, tackled by Gauss, Abel, and Jacobi. Not only did he handle them in a way that was novel and very different from the way followed at the same time by Weierstrass, but also partly solved them in definitive form. Now a few more students came to Göttingen. The situation suggests a comparison with another golden decade of mathematics, the decade of the 1820s in Paris, when the reshaping of analysis was begun by Cauchy, and mathematical physics flourished with Laplace, Fourier, Poisson, and Legendre. But unlike Göttingen, Paris had no shortage of students and gifted young researchers.

It was only in connection with his lectures that Riemann again concerned himself thoroughly with the questions introduced in his dissertation. One reason for the gap may have been the lack of encouragement on Gauss' part. Be that as it may, in the interim he regarded as his "principal work" ("Hauptarbeit") his investigations into "a new conception of the well-known laws of nature" ("eine neue Auffassung der bekannten Naturgesetze") (W. 507). The habilitation paper on trigonometric series and the habilitation lecture on the foundations of geometry both got the short end of the stick. This is also indicated by the family letters consulted by Dedekind. Riemann undertook the habilitation paper because of a certain amount of pressure on Dirichlet's part. There are no notes bearing on either of these two sets of issues and Riemann published neither of these manuscripts. As a result, he was classified in his lifetime as a function theorist who also made a few contributions to mathematical physics.

Riemann's father died in October 1855 and the sisters moved from Quickborn to Bremen, where their brother Wilhelm was a post office clerk. The usual long stays in the home country that must have been for Bernhard a source of strength came to a sudden end. In 1857, after he had completed the two function-theoretic papers that engaged him completely, periods of depression set in. Dedekind speaks of spiritual fatigue and an utterly dark mood (W. 552). His letters from that time show a touching concern for his colleague and friend (Scharlau 1981, 44–45; I wish to thank Herr W. Scharlau and Vieweg-Verlag for their kind permission to quote from Dedekind's letters. (D.L.))

On 13 August 1857 Dedekind wrote to his sister in Harzburg, where his family had a summer home. He mentions that he informed his cousin Dr. Mägde about Riemann; that he did not want Riemann to travel to Harzburg by himself, and so prevailed on his next-room neighbor Ritter to accompany him. (At that time, August Ritter (1826–1909) began to teach at the Polytechnic in Hanover. From 1870 he was a professor of mechanics at the newly founded Institute of Technology in Aachen and later became one of the prominent pioneers of science at the Institutes of Technology.)

"In my stead, I commend to you doctors Riemann and Ritter; the first, about whom I told you so much, is now very unhappy; all summer long, until recently, he stayed in Bremen in order to complete for publication certain extremely

14: The Dedekind house in Harzburg

important papers; but his solitary life and, in addition, his physical suffering have made him extremely hypochondriac and mistrustful of other people and of himself, although he seems outwardly to be quite friendly. Incidentally, he has completed these papers, and it is to be hoped that a peaceful stay in Harzburg and innocuous dealings with people will have a very good effect on him. I have written about this to Mägde for I had no idea that you too were still in Harzburg. The other gentleman, whom I know less well, has decided, for old friendship's sake, to keep Riemann company in Harzburg for some time; he is a very interesting, though gruff and even repelling, person, who has seen much of the world. I may return to Braunschweig sometime in the middle of next week by way of Harzburg and spend there at least one day with Riemann; or I may go directly to Braunschweig and then, one or two days later, make an excursion to Harzburg. One must do all one can to save so excellent and scientifically most important a person as Riemann from his now utterly unhappy state; but he must not notice this intention too clearly; it has always been difficult to do him a favor. And the only way one could get him to accept a favor was to persuade him that one did it as much for one's own sake as for his; he hates to trouble other people. He has done the strangest things here only because he believes that nobody can bear him, and so on. And so I hardly need to ask you to concern yourselves deeply about him."

("Statt meiner empfehle ich Euch für die wenigen Tage die beiden Herren Doctoren Riemann und Ritter; der erstere, von dem ich Euch so viel erzählt habe, ist jetzt sehr unglücklich; er ist den ganzen Sommer bis vor Kurzem in Bremen geblieben, um dort gewisse höchst bedeutende Arbeiten für den Druck zu vollenden; aber sein einsames Leben und dazu körperliche Leiden haben ihn im höchsten Grade hypochondrisch und misstrauisch gegen die Menschen und gegen sich selbst gemacht, wenn er auch äusserlich ganz freundlich erscheint. Er hat übrigens diese Arbeiten vollendet, und es steht zu hoffen, dass ein ruhiger Aufenthalt in Harzburg und ein harmloser Umgang mit Menschen sehr günstig auf ihn einwirken wird. Ich habe darüber an Mägde geschrieben, da ich gar nicht wusste, dass Ihr auch noch in Harzburg wäret. Der andere Herr, den ich weniger genau kenne, hat sich aus alter Freundschaft entschlossen, Riemann eine Zeit lang in Harzburg Gesellschaft zu leisten; er ist ein sehr interessanter, wenn auch schroffer, ja zurückstossender Mensch, der schon viel von der Welt gesehen hat. Möglich ist es, dass ich meinen Rückweg nach Braunschweig, etwa in der Mitte der nächsten Woche, über Harzburg nehme, und dort wenigstens einen Tag mit Riemann verleben werde; vielleicht fahre ich aber auch direct nach Braunschweig, und mache dann gleich an einem der ersten Tage einen Ausflug nach Harzburg. Man muss Alles aufbieten, um einen so vortrefflichen und wissenschaftlich höchst bedeutenden Menschen wie Riemann aus seinem jetzt höchst unglücklichen Zustande herauszureissen; aber er darf die Absicht nicht gar zu deutlich merken; es war von jeher schwer, ihm einen Gefallen zu thun, und nur dann gelang es, ihn zur Annahme irgend einer Gefälligkeit zu bringen, wenn man ihn uberzeugen konnte, dass man ebensowohl aus Rücksicht auf sich selbst als auf ihn handelte; er hasst es andern Menschen Mühe zu machen. Er hat hier die wunderlichsten Dinge gemacht, blos aus solchen Gründen, weil er glaubt, Niemand mag ihn leiden u.s.w. Also brauche ich wohl kaum noch darum zu bitten, dass Ihr Euch recht um ihn bekümmert.")

This detailed description of Riemann's hypochondria is not included in Dedekind's biographical account that is part of the edition of Riemann's collected works. The omission was probably motivated by consideration for Riemann's widow. Dugac (1976) published Elise Riemann's letter to Dedekind in which she writes that it pains her to think that Riemann would always be talked about as a hypochondriac. We conclude from this that Riemann's condition was permanent, rather than a one-time passing crisis of the kind suggested by Dedekind's remarks in the biography. One may also guess that the moving description of Riemann's gentle death (W. 557), so unlike Dedekind's matter-of-fact style, was the result of the widow's insistence.

After a few weeks Dedekind decided to follow Riemann to Harzburg. Ritter had stayed there for just a few days. Dedekind speaks of numerous walks as well as of longer excursions to the Harz mountains (W. 553). Riemann seems to have stayed longer in Harzburg, for Dedekind continues: "On 9 November 1857, shortly after his return to Göttingen, he was appointed an associate professor in the philosophical faculty, and his pay was increased from 200 to 300 thaler. But almost at the same time he was profoundly shocked by the death of his deeply beloved brother Wilhelm. He now shouldered completely the burden of care for his three surviving sisters and insisted earnestly that they should join him in Göttingen even before the end of the winter; the move took place at the beginning of March 1858, but only after death deprived them of the youngest sister, Marie. After so many blows of fate, living together with his sisters has contributed substantially to an improvement in his depressed mood, and the fact that now, although slowly, his works are gaining recognition in wider circles has gradually lifted his sunken self-confidence and has enabled him to find fresh courage for new work." ("Bald nach seiner Rückkehr nach Göttingen wurde er am 9 November 1857 zum ausserordentlichen Professor in der philosophischen Facultät ernannt, und seine Remuneration von 200 Thaler auf 300 Thaler erhöht. Aber fast gleichzeitig erschütterte ihn auf das tiefste der Tod seines innig geliebten Bruders Wilhelm; er übernimmt nun ganz die Sorge für seine drei noch lebenden Schwestern und dringt inständig darauf, dass sie noch im Laufe des Winters zu ihm nach Göttingen übersiedeln; dies geschah auch im Anfang März 1858, aber erst nachdem ihnen die jüngste Schwester, Marie, noch durch den Tod entrissen war. Nach so vielen Schicksalsschlägen trug das Zusammenleben mit den Schwestern wesentlich zur Besserung seiner tief niedergedrückten Gemüthsstimmung bei, und die Anerkennung, welche von nun an, wenn auch langsam, seinen Werken auch in weiteren Kreisen zu Theil wurde, hob allmählich sein gesunkenes Selbstvertrauen und liess ihn frischen Muth zu neuen Arbeiten finden.")

But let us go back to the fall of 1857. On 18 October Dedekind reported from Göttingen (Scharlau 1981, 46): "I also had an audience with Frau Dirichlet; there was much talk about Riemann, who actually remained here but often behaved in a strange way." ("Dann hatte ich auch Audienz bei Frau Dirichlet; es wurde viel über Riemann gesprochen, der nun wirklich hier geblieben ist, sich aber schon häufig sonderbar benommen hat"). One senses here disappointment that the care provided in Harzburg had no permanent effect.

It is almost certain that Dedekind made a sacrifice for Riemann by staying so long in Harzburg. He repeatedly reports that, while in Göttingen, he had almost daily contact with Dirichlet, and his own scientific interests were almost exclusively devoted to number theory. If significant mathematical conversations with Riemann had taken place, we would have certainly been told about them. All Dedekind mentions is that Riemann's interests were again focussed on natural philosophy (Newton).

We will see in the sequel how Dedekind later established a connection with Riemann's work. But it seems that there was not much chance for direct collaboration between the sociable younger man and his introvert friend. In due course we will discuss their different views of mathematics, but first we must sketch the course of events insofar as the Göttingen mathematicians are concerned.

Gauss died on 23 February 1855 and Dirichlet was very quickly appointed his mathematical successor. His wife Rebecka, sister of Felix Mendelssohn-Bartholdy, with whom, incidentally, Dirichlet was on very good terms, tried to bring to Göttingen some of the atmosphere of the Berlin salons. Dedekind thought initially that she was not nearly as well liked as Dirichlet (letter of 26 January 1856). The musical Dedekind seems to have been a very popular figure in the social life of Göttingen and sometimes complained of an excessive number of invitations. On 14 February 1856 he mentions a "gigantic party of 60–70 persons" ("Riesengesellschaft von 60–70 Personen") at the Dirichlets', at which he tirelessly provided music for the dancers. Needless to say, Riemann's lifestyle was very different! Of course, as the son of a respected and well known Braunschweig family Dedekind could easily establish connections in Göttingen. Not so Riemann. But after Dirichlet's move to Göttingen, Riemann, whom he regarded highly, would have likewise found all doors open had his makeup been different.

The letters written by Dedekind to his sisters in the summer of 1857 show at times a measure of dissatisfaction with Göttingen (quoted after Scharlau 1981, 28 ff.). He is occasionally depressed by the lack of listeners: "the second stage of my upbringing has been completed here, in Göttingen; but I am looking forward to the third stage, to being somewhat effective, and the prospects here are not very good" ("hier in Göttingen ist der zweite Teil meiner Erziehung vollendet; aber ich habe Lust zum Anfang des dritten Theils, auch Etwas zu wirken, und das scheint hier nicht gut möglich") (loc. cit. 44).

The change came very soon. Already in the summer term of 1858 Dedekind became a professor at the Zürich Polytechnic. Riemann was one of the many applicants for this position. Like Dedekind, he described himself as a student of Dirichlet. The appropriate Swiss official came to Göttingen and listened to lectures by the two applicants. The decision in favor of Dedekind was reached very quickly. His lecturing style won the day. In this case Riemann's greater scientific importance was not very relevant.

Speaking of lecturing, the usually discreet Dedekind states openly (W. 550) that Riemann had great difficulty in this respect. The *Nachlass* contains pages with detailed versions of the introductory sentences in his lectures. We will later give a few examples. Contrary to the impression conveyed by Dedekind, Riemann's lecturing style failed to improve significantly over the years, as evidenced by, besides other things, a letter (Neuenschwander, 1987, 9) written on 20 May 1861 by the student Ernst Abbe: "one could see his tiredness and lassitude, his thoughts frequently failed him and he was unable to explain the simplest things" ("man merkte ihm die Ermüdung und Abspannung sehr an, es gingen ihm oft die Gedanken ganz aus, und er konnte die einfachsten Dinge nicht herausbringen"). Abbe was a devoted student of Riemann. He came to Göttingen for his lectures and carefully wrote them down. Whether or not Riemann's difficulties were entirely due to his poor health must remain an open question. To judge by Dedekind's vivid description of the Zürich institution, it is safe to say that Riemann would have been utterly unhappy there. He would have had to deal with a heavy teaching load of elementary subject matter and with poorly motivated "students" whose behavior was "disrespectful."

Dedekind did not stay in Zürich very long. In 1862 he was appointed full professor at the polytechnic of his hometown, and he remained there rather than take advantage of offers from more prestigious institutions. He died in 1916, and so survived his friend Riemann by almost 50 years.

After Dedekind's departure, Riemann's stimulating cooperation with Dirichlet was also destined to be shortlived. In the fall of 1858 Dirichlet suffered a heart attack from which he recovered very slowly. Rebecka Dirichlet died quite unexpectedly on 1 December 1858, and he died just a few months later, on 5 May 1859.

Dirichlet's succession was decided very quickly: on 30 July Riemann was appointed full professor. (Incidentally, so was his old teacher Moritz Stern.) Further honors were bestowed on him: on 11 August 1859 he became a corresponding member of the Berlin Academy and in December 1859 a member of the Gesellschaft der Wissenschaften (Science Society) at Göttingen. The Berlin honor was the occasion for a trip to Berlin in Dedekind's company. A worthy conclusion of the golden 1850s was the publication by the Berlin Academy of Riemann's paper on the zeta function and the distribution of primes. At the age of 34 he had gained recognition and an assured livelihood.

15: The Göttingen observatory

0.2.2 Towards change in mathematics

So far we have described the outer framework of mathematics at Göttingen during the decade we wish to designate as the golden one. It is not initially clear that it deserves this name. Nevertheless, profound mathematical changes have their roots here and one can speak of a scientific revolution in Kuhn's sense. Riemann was aware of this, Dedekind may have been, but to others such awareness came very much later. Gentle revolutions can have more lasting effects than sudden cataclysms. They build rather than destroy. We quote Dedekind's fellow countryman and contemporary Wilhelm Raabe (1831–1910), who opened his collection of lives *Alte Nester* (Old nests), published in Braunschweig in 1880, with these words: "When a flower opens it makes no noise ... Unnoticed comes all that will endure in this changeable, noisy world of false heroism, false happiness, and artificial beauty" ("Eine Blume die sich erschließt, macht keinen Lärm dabei ... Unbemerkt kommt alles, was Dauer haben wird in dieser wechselnden, lärmvollen Welt voll falschen Heldentums, falschen Glücks und unechter Schönheit").

In this book we will try to follow the initiation of modern mathematics. Today we are familiar with this version of mathematics, but some of Riemann's contemporaries found it baffling or difficult and others simply missed the point. All we can do in this introduction is indicate what we will have to pay attention to.

In a lecture attended by Dedekind, Riemann spoke of *a turning point in the conception of the infinite* (*von einem Wendepunkt in der Auffassung vom Unendlichen*). In this well-thought-out formulation, supported by examples, he summarized one aspect of this newness which he viewed as essential.

Other novel aspects appear clearly in the work of Riemann and Dedekind, and occasionally we find them earlier, especially in the work of Gauss and Dirichlet. Their traditions were carried forward by Riemann and Dedekind: Riemann advanced the analytic-geometric aspect and Dedekind the number-theoretic-algebraic aspect. The view of mathematics as a form of *thinking in concepts* (*Denken in Begriffen*) is second nature to us but was not to 19th-century mathematicians. Freudenthal (1975, 448) emphasizes as a singular feature of Riemann's style that it was conceptual rather than algorithmic, and more so than the style of any of his predecessors. This style implies a new mode of thought: mathematical objects are no longer given primarily as formulas but rather as carriers of conceptual properties. Proving is no longer a matter of transforming terms in accordance with rules but a process of logical deduction from concepts. Leibniz had both possibilities in mind, but the algorithmic approach triumphed beginning with Euler. Then came the day of the conceptual approach. Bolzano and Cauchy (1817/1821) casually announce the notion of continuity of functions, and Dirichlet (1829) deduces the representability by Fourier series for a class of functions defined by concepts. Riemann defines a complex function by its property of differentiability rather than by an analytic expression, which he regards as a secondary matter. In algebra, Dedekind moves from expressions made up of terms to the concepts of ring, field, and ideal.

Leibniz was aware of, and used, concepts intensionally as well as extensionally. Viewing a mathematical object as a carrier of conceptual properties emphasizes *intensio*, the conceptual content. The mathematician does not want to think of objects abstractly, as bundles of concepts. He prefers to view them somewhat less abstractly as elements of a manifold, of a system (Dedekind), of a *set* whose elements share a definite conceptual property. The concept is represented by its extension, or scope. Single objects are no longer represented by terms. But then the road to proof by transformation of terms is blocked; proofs must be conceptual.

The Gaussian number plane replaces the expressions $a + b\sqrt{-1}$ by a "set with a structure." In the *Disquisitiones arithmeticae* of 1801, Gauss introduced residue classes as objects of elementary number theory. They are represented by sets of integers that are "the same" with respect to an equivalence relation. (We take the liberty of using here the term "set" although it was first used explicitly only later.) Riemann employs n-dimensional spaces with an additional structure, the distance between two points. The idea of a Riemann surface of a holomorphic function makes possible a kind of investigation that is largely independent of specific expressions and provides a basis for new (topological) concept formations and methods of proof. Dedekind represents ideal numbers, used symbolically by Kummer, "concretely" as sets (ideals). Together with Heinrich Weber, he is able to construct an arithmetical, i.e., nontopological, version of parts of Riemann's function theory. But the new methods of proof were long regarded as imprecise, because people wanted to hold on to term transformations as the only admissible procedure.

The view of mathematics as the study of sets with structures, a view canonized by Bourbaki in the middle of the 20th century, can be regarded as a consistent continuation of the Dedekind–Emmy Noether–van der Waerden line in algebra and the Riemann–Hausdorff–Fréchet line in set-theoretic topology. This view is a consequence of the gentle revolution due to the Göttingen golden decade. (Cantor's transfinite arithmetic, which was championed less gently, is irrelevant for this development. Riemann's view of the infinite has the necessary load-carrying capacity for analysis; Cantor's alephs and omegas are not needed.)

We shall see that Riemann was initially unaware of the full scope of his innovations. In 1851 he still thought that the concept of an analytic expression of a function has the same scope as the concept of ("epsilontically" defined) continuity. When he realized his error he held firmly to the primacy of the conceptual as against the algorithmic.

In the 1850s Göttingen was a center of gravity not only in mathematics but also in physics. In this sense it resembled the Paris of the 1820s, except that the number of active individuals was far smaller. In physics too everything was linked with Gauss: in the realm of theory with his investigations of potential (which also influenced Riemann's dissertation), and in the realm of phenomena with his investigations of terrestrial magnetism, besides other things. Wilhelm Weber (1804–1891), who as one of the "Göttingen seven" was dismissed in 1837, regained his Göttingen professorship in 1849. As early as 1833 Gauss and Weber developed the first electromagnetic telegraph. Weber was behind such measuring instruments as a magnetometer (1837) and an electrodynamometer (from 1840). In 1856 Weber and Kohlrausch used electrical measurements to determine the velocity of light. Wilhelm's older brother was the Leipzig physiologist Ernst Heinrich Weber. His relevance to our story is that some of Riemann's interests extended in some measure to his professional area.

Incidentally, for a number of years Riemann hoped to marry his daughter Laura (Neuenschwander in Riemann 1991, 48).

Riemann's contributions to physics were definitely not limited to theory and are valued by experts. Their judgment may, to some extent, reflect the fact that these contributions are due to a famous mathematician. There is no doubt, however, that his theoretical paper on shock waves anticipated certain future developments. Some of Riemann's speculations on physics, to which he attached great importance, would today be classified as attempts to devise a unified field theory of all physical phenomena. Like his deliberations on the physiology of the senses, referred to earlier, these were the tangible basis for his formulation of a philosophy of nature. But it seems that he hardly ever discussed these issues with anyone at Göttingen.

0.2.3 Snapshots by an English observer

We want to share with the reader the few available descriptions of some of the people Riemann dealt with at Göttingen. They come from the diary of the Englishman Thomas Archer Hirst (1830–1892) and were published in part by J. Helen Gardner and Robin J. Wilson in their paper "Thomas Archer Hirst – Mathematician Xtravagant III. Göttingen and Berlin," *Amer. Math. Monthly* 100 (1993), 619–625. In 1852 Hirst obtained a doctorate at Marburg with a dissertation on the triaxial ellipsoid. He spent just two weeks in Göttingen, but his impressions may be reliable. He attended lectures by Wilhelm Weber, "a curious little fellow [who] speaks in a shrill, unpleasant and hesitating voice," and by Moritz Stern, whose lectures on integrals and on mechanics he found to be "beautifully clear."

On 6 August he writes: "To-day I called on Weber again. He received me kindly and explained to me a new method of his of determining the inclination of the Magnet. He speaks and stutters on unceasingly; one has nothing to do but to listen. Sometimes he laughs for no earthly reason, and one feels sorry at not being able to join him."

Subsequently, he visited Stern: "Stern is a stern fellow — not an atom of unnecessary ceremony about him, but the greatest plainness and character. We got a bit of brown bread and butter together, as he would have taken himself, with a glass of water to it, and then a cigar. At first we talked about a dark point in his lecture, then on a multitude of topics."

On 12 August came the highlight of his Göttingen trip, a visit to Gauss: "Personally he is a venerable, fine old fellow, with a contented manly expression. There is an extraordinary aspect of power about him and his every word: without effort he suggests to every one the presence of manly might. He is about 80 years of age, but not a trace of superannuation is to be seen about him. He can even read without spectacles ... No sooner was the first word spoken than I felt perfectly at ease." Hirst said he regretted that he had not had the opportunity of hearing a lecture by Gauss, for this was one of his main motives for coming to Göttingen, a kind of curiosity that is perhaps excusable for a lover of science. Gauss replied that in this term there were too few students, and that anyway he had other work to finish. After a short silence Hirst asked: "Have you ever been in England, Professor?" Gauss replied that he never got farther than Belgium. The rest of the conversation was a polite chat. "So we chatted quite comfortably for the three-quarters of an hour, and then I bid the old veteran good-bye and thanked him heartily. I left him copies of some of Tyndall's memoirs and of my own dissertation."

The conversation would perhaps have been more fruitful if Hirst had mentioned that before he embarked on his studies he had been for many years a land surveyor in connection with railroad construction in Yorkshire. The founder of higher geodesy and the "tough old veteran" with a lively interest in railroad construction would have loved it! But at the very least the report shows that Gauss could be very friendly and sociable as long as contacts did not threaten to involve longterm obligations. Nor did the young visitor notice any health problems in his host (who incidentally was only in fact 75 at the time of Hirst's visit). In this connection we mention that when sometime in 1853 Alexander von Humboldt inquired about Gauss' health, the latter replied (10 May): "In my younger years I suffered a great deal from stomach and abdominal troubles of which I am by now almost free ... But in the last 6 or 7 years their place has been taken by other troubles that I did not know before: congestion in the chest and throat, shortness of breath when taking a walk that exceeds my usual (small) scope, heart flutter and sleeplessness; the result of all of these being that the hours fit for working up scientific investigations are becoming ever rarer" ("Verschleimung in Brust und Schlund, Ausgehen des Atems bei Bewegung zu Fuß, die mein gewöhnliches (kleines) Maß überschreiten, Herzklopfen und Schlaflosigkeit, alles zusammen dahin wirkend, daß die zur Verarbeitung wissenschaftlicher Untersuchungen geeigneten Stunden immer seltener werden"). One year later, on 21 May 1854, he again writes to Humboldt and mentions the worsening of all these afflictions and the onset of others, such as swelling of the legs and a vehement heart flutter provoked by the least physical strain and by stretches of uninterrupted speech. At the time of the Riemann and Dedekind habilitations Gauss was indeed a very sick man (Biermann 1990, 201).

During the winter term of 1853/54 Hirst studied in Berlin. His remarks about Dirichlet are especially interesting. On 13 October Dirichlet received him very cordially. Hirst wrote, "He is a rather tall, lanky-looking man, with moustache

16: Wilhelm Weber

17: The physics laboratory

and beard about to turn grey..., with a somewhat harsh voice and rather deaf: it was early, he was unwashed, and unshaved (what of him required shaving), with his 'schlafrock', slippers, cup of coffee and cigar .... " They each sat at an end of the sofa, smoked their cigars, and talked in an animated way, and Hirst thought: "If all be well, we will smoke our friendly cigar together many a time yet, good-natured Lejeune Dirichlet."

18: Carl Friedrich Gauss

He seems to have visited Dirichlet many times. He also thought highly of him as a teacher. Already on 31 October Hirst wrote: "Dirichlet cannot be surpassed for richness of material and clear insight into it: as a speaker he has no advantages — there is nothing like fluency about him, and yet a clear eye and understanding make it dispensable: without an effort you would not notice his hesitating speech. What is peculiar in him, he never sees his audience — when he does not use the blackboard at which time his back is turned to us, he sits at the high desk facing us, puts his spectacles up on his forehead, leans his head on both hands, and keeps his eyes, when not covered with his hands, mostly shut. He uses no notes, inside his hands he sees an imaginary calculation, and reads it out to us — that we understand it as well as if we too saw it. I like that kind of lecturing."

The reader is not likely to have heard of Hirst and so we wish to point out that he was not devoid of importance. In addition to Germany, he travelled in France and Italy, was a fellow of the Royal Society at 31, succeeded De Morgan as professor at University College, London, and held leading positions in the English scientific establishment. It is to be hoped that his diaries — there are about 5000 pages in typescript — will be put to further use.

0.3 Influences in the final years: Riemann between Germany and Italy

After the departure of Dedekind and the death of Dirichlet, Riemann became isolated as a mathematician at Göttingen. True, he was used to being alone with his thoughts, and, as we saw, he had earlier sought such aloneness, but now a different situation set in. As a full professor of high scientific reputation he could expect to attract students. Also, now that he had essentially prepared the foundations of his function theory, it must have been important to him that the expected students should further develop this theory by working on relevant concrete questions.

But he was to be doubly disappointed. The students were few and their quality low. In many ways he had to rely on K. Hattendorff, a well-meaning and faithful student who could do computations correctly but was not gifted enough to develop Riemann's mathematics independently. His books of elaborations of Riemann's lectures, and his reworked version of Riemann's manuscript on minimal surfaces (W. 301–333 and Section 1.3.5 of our book), show the immense difference, both in the treatment of the foundations and in style, between student and teacher.

It would probably have been better for Riemann to have had the opportunity to leave Göttingen for good in the late 1850s. Neither the human nor the meteorological aspects of the climate in that small town agreed with him. As a student and as an instructor he took advantage of every opportunity to escape to Quickborn, and not only during official vacations. When, after his father's death, the family was forced to leave Quickborn and ended up in Bremen, he was anxious to go there. From early 1858 he had his two surviving sisters stay with him, and there were no longer family reasons for escaping from Göttingen.

No evidence on Riemann's health before 1862 is available. During those years he was especially active in study and research, and he lived with, and was cared for, by his sisters, whose friend he married on 3 June 1862. "Already in July of that year he became ill with pleurisy. He seemed to recover quickly, but it left the seed of a lung disease which was to cause his early death" ("Schon im Juli desselben Jahres befiel ihn eine Brustfellentzündung, von welcher er scheinbar sich rasch erholte, welche aber doch den Keim zu einer Lungenkrankheit zurückliess, die sein frühes Ende herbeiführen sollte"), wrote Dedekind (W. 555). The doctors — Dedekind uses the plural — advised a long stay in the south, and this recommendation was taken so seriously that Weber and Sartorius prevailed on the government to provide for Riemann, a man of the greatest scientific merit, a leave of absence and money for a stay in Italy during the winter term. With the help of his friend Jäger, the consul at Messina, Sartorius arranged for Riemann to stay in warm Sicily, together with his young wife, from November until 19 March 1863. Dedekind describes Riemann's visits to Italy as "points of light in his life" ("Lichtpunkte in seinem Leben") (W. 556).

19: Bernhard Riemann in 1863

The return journey was delayed until June 1863, which gave Riemann the opportunity to admire the old and new treasures of Italy. He made contacts with scientists in many towns and was soon to enter into close and friendly collaboration with the slightly older Enrico Betti in Pisa. After just two summer months in Göttingen he was given a leave of absence to go once more to Italy. Together with his family, he spent the unusually cold winter of 1863/64 in Pisa, where he would later stay repeatedly for longer periods. His conscientious attempt to resume work in Göttingen was soon commuted to another leave of absence to be spent in Italy.

At the time Italian mathematicians pursued Riemann's mathematical research areas more intensely than did others. Casorati worked in function theory, Betti in topology, and Beltrami, Bianchi, Ricci, and Levi-Cività in differential geometry. It is reasonable to guess that Riemann's personal influence was at work here. We shall find it necessary to investigate this matter and we shall get some surprises in the process. We begin by sketching the external circumstances.

After the Congress of Vienna in 1815 the many states of Germany were largely independent. Their pride and competition benefited education. Berlin University, founded at the time of resistance to Napoleon, and the Berlin Academy were models to be followed. In Prussia, Humboldt strongly promoted mathematics, albeit primarily in the service of the neohumanistic ideal of a well-rounded education. The progress of industrialization favored the establishment of institutions for advanced technical education, and no king or grand duke would miss the chance of adorning his residence city with such an institution. These schools were the forerunners of the Technical Universities (Technische Hochschulen), modeled after the Paris Ecole Polytechnique. In addition to other subjects, they became centers for the cultivation of mathematics, of function theory next to geometry and mechanics. This went so far that late in the century the engineers rebelled against the takeover by "theta functions." The establishment of the Reich in 1871 had very little effect on this cultural federalism, so propitious for the sciences.

The situation was very different in Italy. After the Congress of Vienna, that country remained divided into territories determined by foreign power. The rulers supported the venerable universities only to a minor extent. The development of internationally significant mathematics was out of the question. In particular, this was also true of the Papal States, the last region to join the Kingdom of Italy. This took place in 1870, following the sudden termination of France's support for the pope due to the Franco-Prussian war.

The trip made in 1858 by Betti, Brioschi, and Casorati, which took them not only to Paris and Berlin but also to Göttingen, is regarded by the Italians as a watershed in the European view of their mathematics: from now on, Italian mathematics was granted European recognition. It was also the beginning of Riemann's contacts with Italy. His Italian trips took place during the years that followed the temporary conclusion of the unification process of 1859, when the former great dukedom of Tuscany, including Florence and Pisa, formed the core of the new kingdom prior to the annexation of the Papal States. After his first stay at Pisa, Riemann was offered the chair of geodesy. He declined, and explained in a letter to his friend Betti that his decision was dictated by the deterioration of his health. He added:

> "Non potrei parlare ad alta voce che con molta fatica."
>
> ("I cannot speak loudly without a great deal of fatigue.")

He felt that he could not impose on his Italian friends by accepting a position without being able to lecture.

During his long visits to Pisa, even after turning down the job offer, he had (health permitting) almost daily mathematical conversations with Betti. Bottazzini (1991, 35) mentions as topics the (topological) connectivity of space (we recall the Betti numbers), the foundations of real and complex analysis, and questions of mathematical physics, such as terrestrial magnetism, the air flow around a body moving in the atmosphere, and a method for the determination of the force of attraction of a homogeneous elliptical cylinder.

In February 1864, Eugenio Beltrami (1835–1900), a student of Brioschi from Pavia, took over the geodesy professorship turned down by Riemann. He was to become one of the foremost mathematicians of his country. He was primarily a geometer, but he also made significant contributions to analysis as well as to the theory of heat and to mechanics. Moreover, he was the author of the first model of the non-Euclidean plane, which he published in 1868. Given the connection between his research interests and those of Riemann, it has been conjectured that Riemann must have influenced him. But recently Bottazzini (1991, 36/37) arrived at the conclusion that there could hardly have been any contacts between the two. The letters of Italian mathematicians do not mention Riemann's habilitation lecture before its publication in 1868. Moreover, in his letters to Betti, Riemann mentions the names of friends and colleagues he was sending regards to, but Beltrami's name is not among them. Beltrami later translated Riemann's lecture and circulated the translation among his friends. Of course, he marked it *per uso interno*, i.e., for private use.

Riemann was revered in Italy as an authority, but, as Bottazzini puts it, his influence during his stay at Pisa was of a rather general nature: it promoted the notion of the close connection between mathematics and knowledge of the physical world.

To be sure, the Italian mathematicians were particularly receptive to Riemann's new ideas in analysis and in geometry. But the main reason for this may have been that, given the historical situation of a new beginning described earlier, novel ideas had a better chance to be accepted by the new Italian group than by the established systems in Germany, France, and England, which were encumbered with scientific traditions.

0.4 Competing conceptions of analysis before Riemann

0.4.1 Riemann in the historical evolution of analysis: An overview

Modern analysis, both real and complex, has been influenced by Riemann to such an extent that one may find it difficult to form an opinion of his contribution to its foundations. To do so, one must consider the state of mathematics that confronted Riemann and the reasons for its historical evolution. It is only thus that one will be able to appreciate the difficulties he had to overcome and the motives behind his fundamental upheaval in the conception of analysis.

His work belongs completely to analysis in our present understanding of the term. Algebraic geometry appears in it as part of complex analysis; he treats number theory with methods of complex function theory; he subsumes physical applications, as was usual at the time, into partial differential equations; he replaces the usual axiomatic conception of geometry, inherited from Euclid, by his novel (Riemannian) geometry, which is part of real analysis of several variables; and he develops the topology of manifolds as a new discipline derived from analysis.

Riemann definitely accepts the epsilontic justification of limit analysis, as shown by his well-known introduction of the concept of the integral in his habilitation paper of 1853 (W. 239), and, more fully, by this paper as a whole. It also became clear to him in connection with his lectures that the only way to achieve complete clarity in analysis was to reduce concepts such as continuity, infinity, etc. to equalities and inequalities (N. 709).

As is well known, epsilontics became generally accepted as the only valid method of proof in analysis following Weierstrass' Berlin lectures (1860 ff.). As such it is usually labeled "Weierstrassian rigor" ("Weierstrasssche Strenge"). Before Riemann we encounter it only infrequently; it appears, for example, in the lectures of Riemann's teacher Dirichlet. Beginning in 1821 Cauchy occasionally used epsilontic arguments in his lectures. Riemann spells out what every student experiences in the very first lectures:

> "This makes the proofs of the fundamental theorems of infinitesimal calculus ... somewhat more involved."
>
> ("Es werden die Beweise der Fundamentalsätze der Infinitesimal-rechnung dadurch ... etwas umständlicher") (N. 709).

Of course, when it comes to obtaining new insights, epsilontics is largely irrelevant; it is an algorithm for securing previously found insights, canonically prescribed by habitual consensus. Other innovations, pursued purposively by Riemann, are far more important for the progress of mathematics. Here we must mention, above all, the conception of a function as a mapping, which he used consistently in his dissertation of 1851. It was only much later that this conception, which we regard today as fundamental, was generally accepted. As late as 1928 Hermann Weyl wrote in his *Philosophie der Mathematik und Naturwissenschaft* (*Philosophy of Mathematics and of Natural Science*): "Nobody can explain what a function is. But 'A function f is given if to every real number a there is associated, in some definite, legitimate way a number *b* ...' " ("Niemand kann erklären was eine Funktion ist. Aber 'Eine Funktion ist gegeben, wenn auf irgendeine bestimmte gesetzmäßige Weise jeder reellen Zahl *a* eine Zahl *b* zugeordnet ist ... ' ") (p. 8).

Riemann showed no interest in the "justification" ("Begründung") of the real numbers. In this he differed from his friend Dedekind and from Weierstrass. Here Dedekind succeeded and Weierstrass failed, in spite of spending many weeks of lectures devising unprofitable setups.

Every commitment to precisely delimited methods and concepts implies the renunciation of other, previously available, possibilities. Riemann's work shows, and this will have to be made clear in the present book, how fruitful *his* restrictions turned out to be. We will examine the picture of analysis in his time and the historical origins, but are not in a position to attempt a complete history.

In the middle of the 19th century there were many competing conceptions of analysis, and individual researchers and textbook writers sometimes used more than one of them. We mention the algorithmic conception of so-called algebraic analysis; the infinitesimal conception; the geometric foundations; the limit conception; and the formulations of epsilontics. All of these conceptions can be traced back to beginnings in the work of Leibniz and Newton, when two *geometric* problems provided essential stimuli for development. The insight that the "inverse tangent problem" is equivalent to the computation of area was the most important source of the connection between differentiation and integration spelled out in the so-called fundamental theorem of the calculus. (This remains true today, when we first embark on the study of analysis.) But soon came the separation from geometry, in history as well as in today's teaching.

This confusing multiplicity of approaches was in part due to the 18th century's pragmatic way of dealing with the calculus. Here the prominent personality was Leonhard Euler (1707–1783). Both in analysis and in its applications, Euler used and developed on any given occasion the method that struck him as most appropriate. And he was invariably successful. The great mathematicians and scientists of the 18th century did not teach students, and thus had no need to justify the foundations they adopted for methods and concepts. But they too wished for reliable conceptual justifications. Lagrange, who became prominent at the Berlin Academy after Euler's departure in 1766, launched in 1785 a prize competition for the provision of such a solid foundation. He was not satisfied with any of the two dozen submissions; but even so his authority made the problem a subject for discussion.

20: Joseph Louis Lagrange

0.4.2 Algebraic analysis

The *algorithmic conception* of the so-called *algebraic analysis* aimed at the elimination from analysis of limit and infinitesimal considerations. Today this objective strikes us as unbelievable, but until about 1850 it played an important role. As a young student Riemann too wrote a paper on this topic (W. 353–366), and we must pay a measure of attention to this phase of his development. Lagrange saw in this conception his own answer to the prize question and embodied it in his influential *Théorie des fonctions analytiques* of 1797. The core of the matter is the observation that power series can well be meaningful even if they do not converge. This is obvious to a modern mathematician if he views a formal series $\sum_{k=0}^{\infty} a_k z^k$ as a symbol for the sequence $(a_0, a_1, ...)$ and adds, subtracts, and multiplies such formal series like ordinary series. In this way one obtains a commutative ring with unit element and without divisors of zero. This ring has a field of quotients. It is natural to ask if this construct is of any use in analysis. Here is the view of the 20-year-old Riemann:

> "The assertion has been put forward that one cannot base reliable conclusions on series in general but only under the condition that we assign to the quantities appearing in such a series numerical values that render it convergent ... " But the partial sums of these series are precisely determined, at least when there is a definite law of

formation of the coefficients. Riemann protests emphatically "against the sentence of condemnation pronounced on divergent series."

("Man hat wohl die Behauptung aufgestellt, man könne auf die Reihen im Allgemeinen gar keine sicheren Schlüsse gründen, sondern nur unter der Bedingung, dass man den darin vorkommenden Grössen solche Zahlenwerthe beilege dass die Reihe convergire ... " Es seien aber, jedenfalls bei Vorliegen eines bestimmten Bildungsgesetzes für die Koeffizienten, die Teilsummen der Reihe genau bestimmt. Riemann protestiert nachdrücklich "gegen das Verdammungsurtheil, welches man den divergenten Reihen gesprochen hat") (W. 355, 358).

A similar view is expressed by A. Speiser in his introduction (1945) to Euler's *Opera omnia* (I) p.X. Actually, Euler's use of divergent series was extremely fruitful. We will return to his results on the zeta function in 1.4. His *Introductio in analysin infinitorum* of 1748 was very influential, and a German translation was widespread. In Ch. 16 Euler justified additive number theory, and to this end introduced power series as a methodological tool without reference to their convergence. He also made successful use of not necessarily convergent trigonometric functions (see 2.2). Contrary to the judgment that turns up repeatedly in the literature, Euler had a very clear idea of the concept and of the significance of convergence. As early as 1734 he knew and applied the convergence criterion later named after Cauchy. Unfortunately, such considerations are absent from books available to the wider public, so that nothing prevents people from making uncritical use of formal power series and from referring to Euler for its justification.

Great researchers test new methods as tools for obtaining results or ponder their significance for the foundations of mathematics. Euler comes under the first of these two headings, and so does Laplace with his use of generating functions in probability. Lagrange's attempt to provide an algebraic justification of analysis comes under the second heading. This attempt has been largely a failure. In the hands of other people methods become independent. Deprived of their original sense and purpose they become arid drill-algorithms. At different times the school curricula have been plagued by instances of this phenomenon, such as degenerate versions of Euclidean triangle constructions, spherical trigonometry, descriptive geometry, and (of late) set theory. The same thing happened with the algebraic analysis of symbolic expressions, which for a long time determined the mathematical material taught in gymnasiums. Following Humboldt's education reform in Prussia, algebraic analysis gradually became an essential component of the curriculum. It became the pre-eminent elementary model of pure mathematics. Not only in Prussia, but also in Hanoverian

21: Isaac Newton

Göttingen and elsewhere, the training of prospective gymnasium teachers was largely in the hands of professors who were not researchers but votaries of algebraic analysis. This scenario also applies to Riemann's school time.

Today algebraic analysis strikes us as uninteresting. The subject and its role have long been neglected by historians, but this has recently begun to change. We refer the interested reader to H.N. Jahnke, "Algebraic analysis in Germany 1780–1840," *Historia Mathematica* 20 (1993) 265–284, and the bibliography in that paper.

At Göttingen, M.A. Stern (1807–1894), who taught Riemann introductory analysis, was one of the last champions of this subject. His textbook on algebraic analysis, published in 1860, gives us an idea of how he focussed his lectures. He emphasized the significance of the so-called polynomial theorem, where by a polynomial he generally meant a formal power series. "We call ... the expansion of the law of formation of the terms of the product corresponding to a power of the polynomial, the polynomial theorem, and the outcome (Produkt) itself, the polynomial series" ("Man nennt ... die Entwickelung des Gesetzes, nach welchem die Glieder der des einer Potenz des Polynoms entsprechenden Produkts gebaut sind, den polynomischen Lehrsatz und dieses Produkt [Ergebnis (D.L.)] selbst die Polynomialreihe") (loc. cit. p. 40).

C.F. Hindenburg (1739–1808) was the founder of the so-called combinatorial school. In 1796 he published in Leipzig a collection of articles whose title begins with the words: "*Der polynomische Lehrsatz, das wichtigste Theorem der ganzen Analysis...*"(*The polynomial theorem, the most important theorem in all analysis ...*). Like Newton, who regarded his binomial formula as valid for nonintegral exponents, the adherents of algebraic analysis took the same general view of the polynomial theorem. In 1755, using differential equations and recursion formulas, Euler had announced this theorem in his differential calculus (Part II, §202) not only for positive integral exponents but for negative and fractional exponents as well. But the combinatorial school wanted to introduce a "purely algebraic" conception, i.e., one independent of analysis.

One of the "professorial" followers of Hindenburg in Göttingen was B.F. Thibaut, author of *Grundriss der allgemeinen Arithmetik oder Analysis zum Gebrauch bey academischen Vorlesungen* (*Outline of general arithmetic or analysis for use in university lectures*). His followers at Berlin included E.H. Dirksen (1792–1850) and, above all, Martin Ohm (1792–1872), a brother of the physicist. Riemann does not seem to have had close contacts with the latter two. By the time he got to Berlin he had outgrown the algorithmic phase of his development.

And yet Riemann's youthful paper shows some interesting things about him. He handles the gamma function in a secure and self-confident way and deals with differential equations and recursions in Euler's manner. He beats the champions of algebraic analysis at their own game in that he considers power series whose exponents vary from $-\infty$ to $+\infty$, carefully noting

> "that there are series of this form that are equal to 0 or to a constant for every value of x"
>
> ("dass es Reihen von dieser Form giebt, die der Null oder einer Constanten, für jeden Werth von x, gleich sind") (W. 358).

As an auxiliary result he obtains

$$\frac{x^{\mu}}{\Pi(\mu)} = \sum_{\alpha=-\infty}^{\alpha=\infty} \frac{b^{\mu-\alpha}}{\Pi(\mu-\alpha)} \frac{(x-b)^{\alpha}}{\Pi(\alpha)}.$$

Here, more explicitly, for a fixed α_0, $0 < \alpha_0 < 1$, $\alpha = \alpha_0 + k$, k an integer, $k = -\infty$ to $+\infty$, Π denotes the Gauss version of the gamma function, $\Pi(\alpha) = \Gamma(\alpha - 1)$, so that $\Pi(n) = n!$ for natural n (W. 358). For $x = 1+h$, $b = 1$, $\alpha_0 = 0$ we obtain the binomial series.

The paper ends with a perceptive announcement (W. 366):

> "In conclusion, we observe that the constructed theory can be just as reliably extended to the case where one assigns to the relevant quantities imaginary values."

("Schliesslich bemerken wir noch, dass sich die aufgestellte Theorie mit derselben Sicherheit auch auf den Fall ausdehnen lässt, wo man den in Rede stehenden Grössen imaginäre Werthe beilegt.")

This paper must have pleased Stern. He may have had it in mind when he said much later to Felix Klein that "already at that time Riemann sang like a canary" ("Riemann sang damals schon wie ein Kanarienvogel") (Klein 1926, 249). In the main part of the paper Riemann goes beyond the actual, narrower algebraic analysis. The paper is titled "Versuch einer allgemeinen Auffassung der Integration und Differentiation" ("Essay toward a general conception of integration and differentiation"). For a real r he wants to define the r-th derivative $\partial_x^r z$ of the real function $z(x)$. At first he adopts completely the mode of thought of algebraic analysis. He refers to Lagrange when introducing the operation ∂_x^r and works with formal series of powers $(x-b)^\alpha$, $\alpha = \alpha_0 + k$, k integral. Already for negative integral r the operation of r-tuple integration yields not a unique function but an r-dimensional manifold with r integration constants. For nonintegral r Riemann obtains infinitely many "arbitrary constants" ("willkürliche Constanten"). For every $r < 0$ a special solution of the problem is given by

$$\partial_x^r z = \frac{1}{\Pi(|r|-1)} \int_k^x (x-t)^{|r|-1} z(t)\, dt \,,$$

and from this one obtains solutions for positive r by repeated ordinary differentiation (W. 363). In a footnote Riemann also gives the power series expansion (again for a special solution) valid for an arbitrary real r:

$$\partial_x^r z = \sum_{n=0}^{\infty} z^{(n)}(k) \frac{(x-k)^{n-r}}{\Pi(n-r)} \,.$$

He is aware that the vector space of solutions of each of the homogeneous problems which can be added to this special solution is spanned by the functions

$$\frac{x^{-r-n}}{\Pi(-r-n)}\,, \qquad n = 1, 2, \ldots \,.$$

The special solutions, whether in the form of an integral or in that of a series, are obvious generalizations of well-known formulas for integral r, and all that is required here is knowledge of the gamma — or Pi — functions as generalizations of the factorial. With these Riemann became acquainted in Stern's course on definite integrals or earlier. These formulas have been repeatedly rediscovered or re-proved by mathematicians from Liouville to Mikusiński, and a treatment undoubtedly acceptable to Riemann is provided by the Laplace transformation. But this is a side issue. What counts here is that Riemann treated the homogeneous problem with persistence and analytic skill, and that as a young student he showed a great deal of liking for the modes of thought and methods of algebraic analysis.

The preference for representation by integrals rather than by series signals his later development. The advantage of the integral is its "greater pliability" ("grössere Geschmeidigkeit"). We encounter this judgment later in his complex function theory and in his treatment of partial differential equations.

In his 1917 paper "Bemerkungen zum Begriff des Differentialquotienten gebrochener Ordnung " ("Remarks on the differential quotient of fractional order") (*Ges. Abh.* I, 663–669) H. Weyl went on from Riemann's youthful paper and gave applications to Fourier series. He writes: "One must say that, if stated in tenable form, which is possible, Riemann's concept extension is completely natural; in fact, it is the only possible natural one, and it should certainly be credited with more than just formal importance." ("[Es] ist zu sagen, dass die Riemann'sche Begriffserweiterung, wenn man sie nur in haltbarer Weise formuliert — und dies ist möglich —, völlig naturgemäss, ja die einzig naturgemässe ist und ihr keineswegs eine bloss formale Bedeutung zukommt").

In Riemann's time, algebraic analysis was still running riot in the schools and in university textbooks, but it hardly played any role in the progress of science. This was due to the insight, acquired around 1820, that it breaks down in the case of trigonometric series. At the beginning of his treatment of the series named after him, and before he discovered his integral formulas for the coefficients as the "correct" first step, Fourier made extensive use of divergent power series and obtained in this way many convergent Fourier series. It pays to read the faithful account of his progress as set down in his *Théorie analytique de la chaleur*, published in 1822. In order to illustrate the breakdown of the method we mention here the historically important example known as the Poisson paradox, discussed with passion for more than a decade beginning in 1811.

Already Euler gave the equation

$$(2\cos x)^m = \sum_{k=0}^{\infty} \binom{m}{k} \cos{(m-2k)x}\,,$$

obtained from $2\cos x = e^{ix} + e^{-ix}$ and the binomial series. For natural m this is a correct (terminating) series. But for $m = \frac{1}{3}$ and $x = \pi$ Poisson obtained an incorrect "equality," namely, $(-2)^{1/3}$ on the left and the arithmetical mean of the two complex values of $(-2)^{1/3}$ on the right. Those familiar with regions of convergence and Riemann surfaces realize what is at issue. Initially such examples had a shocking effect. Votaries of algebraic analysis such as Martin Ohm tried hard to find an explanation. One insight, still in effect in Stern's book, was that one must distinguish formal equality from value equality. (For details see H.N. Jahnke, *Arch. Hist. Ex. Sci.* 37 (1987), 101–182.) But this made the method literally valueless!

The dominant discipline in France was mathematical physics, which required numerical and functional values that could be compared with results of measurements. But in England, as well as in Germany, the algebraic-formal thinking could hold its own. The primary objective of the Analytical Society, formed at Cambridge in 1813 and headed by George Peacock (1791–1858), was to close the gap between British and continental mathematics, a gap due to the British refusal to adopt the Leibniz differential notation and to the rigid adherence to Newton's dot notation ("d-ism versus dot-age"). But the calculus of Leibniz had its own seductive formal-algebraic charms, and so British mathematicians, in their enthusiasm for the Leibniz notation, narrowly avoided missing out on the French preference for the derivative function as against the differential quotient. Cambridge continued to be a source of the tendency that championed rule-bound term transformations in mathematics. (Boole's algebra of logic is a product of this tendency.)

We have given a short account of these circumstances because it is necessary to be aware of the mathematical climate in which analysis developed in Riemann's youth. It is not at all obvious that a young mathematician of that time would turn away from the algorithmic and the formal-algebraic and find his way to thinking in concepts.

One must not overlook the fact that there are solid intrinsic reasons why the wish to reduce analysis to algebra and, as we say today, to eliminate the topological element, is reborn in every generation since Leibniz. Algebra, in the sense meant here and looked at from the viewpoint of the predicate calculus, transpires on a lower level, is simpler in principle. We will see that Riemann's friend Dedekind also tried immediately to formulate his (i.e., Riemann's) function theory as much as possible in algebraic terms. And the modern function theorist R. Remmert honored his teacher Karl Stein with a lecture titled "Die Algebraisierung der Funktionentheorie" ("The algebraization of function theory") (*DMV-Mitteilungen* 4-1993, 13–18). What is involved here is the function theory of several variables, and the algebra which must be employed has left term transformations and formal series far behind. But Lagrange's old dream is still alive.

0.4.3 Infinitesimal analysis

Cauchy's *Cours d'analyse* of 1821 is subtitled "Analyse algébrique," and its content (as well as, incidentally, the content of Stern's book of 1860) is modelled on Euler's *Introductio*.

One feature the *Cours* shares with Euler's book, and with the later book of Stern, is that it presents the elementary functions and infinite series without referring to either differentiation or integration. But Cauchy does much more, and the part of this "much more" that interests us is his approach to a theory of infinitely small quantities. Following Leibniz, one proceeded, roughly speaking, as follows: given a functional expression $f(x)$, one transformed the difference $f(x+dx) - f(x)$ so as to factor out dx, $f(x+dx) - f(x) = g(x, dx)\,dx$. The value $g(x, 0)$ was the differential quotient; the first to speak of "derivatives" was Lagrange.

In his first publication Leibniz gave the root rule. Using the procedure just mentioned we obtain it as follows:

$$\sqrt{y+dy} - \sqrt{y} = \frac{dy}{\sqrt{y+dy} + \sqrt{y}} = dx \cdot \frac{dy}{dx} \cdot \frac{1}{\sqrt{y+dy} + \sqrt{y}}\,.$$

Since $dy = \frac{dy}{dx}\,dx$ vanishes together with $dx = 0$, it follows that

$$\frac{d\sqrt{y}}{dx} = \frac{dy/dx}{2\sqrt{y}}\,.$$

The procedure worked not only for algebraic functions but also for infinite expressions such as power series, but there was always an air of missing justification about it. And when Cauchy freed himself from the notion of an expression as the defining property of a function and interpreted $y = f(x)$ as a correspondence between numerical values, he saw himself compelled to construct a rigorous theory of infinitesimals. He defined them as variables with limit 0. Since he divided variables into dependent and independent ones, we can identify them with the functions, provided that we do what was then usually done and classify the independent variable as the identity function. Cauchy states that, having chosen a basis i, we obtain the system of infinitesimals as the totality of $f(i)$, provided that f is a function (of small, positive arguments t) such that $\lim f(t) = 0$ for $\lim t = +0$. Cauchy conceives his "system" in a completely abstract way and thinks of its elements, after the choice of the basis i, as represented by the $f(i)$, a viewpoint readily understood by 20th-century mathematicians. The field of real numbers is extended by the addition of infinitesimals. Cauchy did not develop his system consistently in his introductory textbook (for engineers!) or in later publications. Small wonder that it was hardly accessible to his contemporaries. When trying to interpret his work we must contend with the difficulty that, like Euler, he is pragmatic in the sense that he uses different methodological approaches in parallel to one another. Thus the critic is free to interpret what he reads according to his own preferences.

On the basis of his infinitesimal conception Cauchy introduced the following three innovations, adopted independently of the particular approach they grew out of:

1. A function $y = f(x)$ is said to be continuous on an interval if in this interval, for every infinitely small α, $f(x+\alpha) - f(x)$ is infinitely small (Cauchy 1821).
2. A function $f'(x)$ is called the derivative of $f(x)$ on an interval if in this interval, for every infinitely small α
$$\frac{f(x+\alpha) - f(x)}{\alpha} - f'(x)$$
is infinitely small (Cauchy 1823).
3. The definite integral is introduced, based, as in Leibniz's case, on infinite sums of infinitely small summands, and the integrability of continuous functions and the fundamental theorem of the calculus are proved (Cauchy 1823).

Thus the fundamental role of infinitely small numbers is shifted away from the differential quotient, which is consistently replaced by the derivative function, and towards continuity. Infinitely small numbers continue to be important as a tool for proofs, for example in connection with the concept of the integral. Cauchy's definition of continuity is to be thought of as extending over intervals and therefore includes what later came to be known as uniform continuity. Incidentally, his definition of the derivative implies the continuity of $f'(x)$, though he does not say this explicitly. Another of his valid theorems is the continuity of the limit of a sequence of continuous functions that converges throughout an interval, for here too, as we explain in the next paragraph, uniform continuity follows implicitly from the infinitesimal conception.

The result just mentioned was bound to be questioned. Indeed, Fourier series supplied a host of examples of series of continuous functions with discontinuous sums. Of course, Cauchy was aware of this and did not respond to objections such as Abel's. *Cauchy's theorem is true if we require convergence for all $x + \alpha$, x real and α infinitesimal, and false if all we are given is convergence for all real x in the interval!* Since Cauchy did not bother to issue a clarification and the use of infinitely small quantities seemed to be questionable, they were gradually eliminated. But they continued to play a role in education. Specifically, following the Cauchy tradition, they continued to be used at the Paris Ecole Polytechnique (in Duhamel's textbook), and, following the Euler tradition, in Russia.

Dirichlet, Riemann, and Weierstrass speak of a sequence, or a function, becoming infinitely small. But this is just another expression for convergence to zero, or for the limit zero. It took more than a century for nonstandard analysis to provide a recognized justification of the infinitesimal methods and modes of thought of the older mathematicians. For the foundations of Cauchy's analysis the reader can consult Laugwitz 1989a, especially sections 1 and 6, and for Cauchy's numbers, Laugwitz 1991. Another relevant reference is the two-part article by Laugwitz in the column "The evolution of ... " in the May 1997 and the August–September 1997 issues of *The American Mathematical Monthly*.

0.4.4 Geometric deliberations: Fourier

Undoubtedly, even after completion of the founding stage of analysis, we always encounter the geometric interpretation of the integral as an area, and of the differential quotient, and later of the derivative, as the slope of the tangent, but these are regarded as applications of analysis to geometry rather than as a basis for a significant justification of the calculus. The fundamental theorem of the calculus, which describes integration as the inverse of differentiation, was viewed as a definition of the indefinite integral, and this worked as long as one stayed in the realm of familiar functions; here "familiar" refers to the algebraic conception of Euler and his contemporaries. The lack of a justification of integration caused problems already in connection with elliptic integrals, but it became glaring when Fourier found it necessary, beginning in 1807, to integrate "arbitrary" functions. Actually, Riemann's definition of the integral took off directly from this situation (see Chapter 2).

Fourier required the following theorem, usually associated with Riemann's name: if $f(x)$ is piecewise continuous and bounded on the interval $a \leq x \leq b$, then

$$\lim_{n\to\infty} \int_a^b f(x) \sin nx \, dx = 0 .$$

Obviously, we need only consider the intervals of continuity. Fourier applied a mixed geometric and infinitesimal method. For n infinitely large consider three successive zeros, $x_1 < x_2 < x_3$, of $\sin nx$. In view of the continuity of $f(x)$ we can regard it as constant on the infinitely small interval from x_1 to x_3. Moreover, in view of the changes of sign of the sine function, the area from x_1 to x_2 between the curve $f(x) \sin nx$ and the x-axis is oppositely equal to the corresponding area from x_2 to x_3. The theorem follows.

This geometric argument motivated Poisson and Cauchy to try to formulate a nongeometric definition of a definite integral. As mentioned in the previous section, Cauchy managed to do this in 1823. There is no doubt that geometric arguments provided at least heuristic help.

Grattan-Guinness (1990, 55–57) maintains that three modes of thought, the algebraic, the geometric, and the analytic, coexisted in analysis in the early part of the 19th century, and that this can be seen by focussing on the concept of the integral. Next to Fourier, it was Monge who successfully used a geometric approach to develop his theory of partial differential equations.

Riemann viewed (complex) functions, consistently and at all times, as (conformal) mappings, and characterized them by means of differential equations. This was a new turn in the geometric mode of thought in analysis.

Throughout the 19th century almost all renowned analysts were of the opinion that the geometric mode of thought is inadmissible in questions of foundations. Even clarifying figures were frowned upon. This was no different in the case of Euler and Lagrange. Cauchy's textbooks of 1821 and 1823 contain just one geometric proof. Namely, to establish the intermediate value theorem for a continuous function Cauchy argues (Cauchy 1821, 44) as follows: if $f(x_0) < b < f(x_1)$, then the straight line $y = b$ must intersect the curve $y = f(x)$ between x_0 and x_1. But he immediately sets this proof aside and adds that, incidentally, this theorem can be proved in a direct and purely analytic way as is done in Note III (at the end of the book). The proof in question uses the well-known nested interval method.

0.4.5 The limit conception: Newton

We encounter it today in its epsilontic form, but this is historically a later development. While the different versions of the foundations of analysis discussed so far can be attributed in one way or another to Leibniz, limits are determinative for Newton's conception of the calculus.

Isaac Newton (1643-1727) did far more in mathematics than invent his version of the calculus. He himself regarded the general binomial series as his most important mathematical contribution. In his early deliberations on the tangent and area problem he argued using infinitesimals, but later he dropped this approach.

In his greatest work, *Mathematical Principles of Natural Philosophy* (1687), Newton describes the limit conception as follows (Book I, Section I): "Quantities, and the ratios of quantities, which in any finite time converge continually to equality, and before the end of that time approach nearer to each other than by any given difference, become ultimately equal." – "those ultimate ratios with which quantities vanish are not truly the ratios of ultimate quantities, but limits towards which the ratios of quantities decreasing without limit do always converge; and to which they approach nearer than by any given difference, but never go beyond, nor in effect attain to, till the quantities are diminished *ad infinitum*."

Our trained eye reads "approach nearer to each other than by any given difference" as "for every ϵ there is ... ," but one will hardly find such a numerical translation of Newton's verbal definition in the 18th century. Newton immediately dilutes this clear-sounding formulation by the addition: "but never go beyond, nor in effect attain to."

Many mathematicians and philosophers thought it unreasonable that, basically, Newton let all variables depend on a single auxiliary variable, namely time. After all, physics has no business in mathematics! Newton did not use the differential notation; rather, one should interpret the transition to the limit in $\frac{y-y_0}{x-x_0}$ for x tending to x_0 as

$$\lim \frac{y(t)-y(t_0)}{x(t)-x(t_0)} = \frac{\dot{y}}{\dot{x}}$$

when $\lim t = t_0$. If, when passing to the limit, we say carelessly that a quantity tends to another quantity, then, without being aware of it, we use the notion of time. Cauchy's notion of infinitely small quantities as variables that converge to zero also contains a "hidden parameter" which these variables are thought to depend on, for he admits as variables only the dependent and independent variables, i.e., the functions, including the identity function.

Newton's dot notation certainly has advantages and remains standard in physics. But the Newtonian foundation, the metaphysics of the matter, has not met with approval. In his prize competition Lagrange found it as wanting as the infinitesimal conception.

0.4.6 Towards epsilontics: Cauchy and Dirichlet

Epsilontics must now seem to us like a cleansing broom in the jumbled mess of algebraic-formal, infinitesimal, geometric, and (vague) limit conceptions. As Riemann said, it reduces everything to equalities and inequalities. As Leibniz would occasionally say, it marks a return to the ancients, to Archimedes and his exhaustion procedure, which anticipated, in the special case he considered, the general method: if we are given two sequences of numbers a_n and b_n such that $a_n < a_{n+1} < b_{n+1} < b_n$, and if for every positive ϵ there is an N such that $b_N - a_N < \epsilon$, then there is at most one number r such that $a_n < r < b_n$ for all n. That there is such a number, that the real numbers are complete in this sense, is an additional property which the reader can easily establish if he starts out from the decimal representation of the reals; for mathematicians before Cantor and Dedekind this completeness property must have been obvious, if only because of the ease with which it could be proved. (It was not obvious to Eudoxus and Euclid, for the segments constructible by Euclidean means do not give "all" real numbers.) Of course, the uniqueness can be proved indirectly: if there were two such numbers, denote them by r and s, with $a_n < r < s < b_n$ for all n, then we could obtain a contradiction by taking $\epsilon = s - r$.

In a letter dated 2 February 1702 Leibniz made a clear statement along these lines. Such a method, the future epsilontics, would provide a rigorous justification for his calculus. For if an opponent were to claim that a result is false, then one could challenge him to specify the error (in the form of a number), and then one could always show that the error is less than the specified number. Varignon, the Paris champion of Leibniz' calculus, immediately published the letter. It was later reprinted many times and was generally available. But Leibniz' argument could hardly prevail, in part because it was inserted between popular and metaphysical arguments neither of which could satisfy a mathematician. Lagrange and the respondents to his prize question seem not to have paid attention to it.

The great cleanup began with Cauchy's *Cours* of 1821. The infinitesimals were spared, but Cauchy intended to define them rigorously. If the values assigned successively to a variable come ever closer to a fixed value so that they ultimately differ from it by arbitrarily little, then this value is said to be their limit (*la limite*). This is the verbal version of the definition of convergence of a sequence: a variable a takes on successively values $a_1, a_2, a_3, \ldots$ and a fixed value A is their limit if, given an arbitrary $\epsilon > 0$, we have ultimately, i.e., for all $n \geq$ some N, $|a_n - A| < \epsilon$. Cauchy (1821, 4) gave just the verbal definition, and this is also how he proceeds (1821, 123 ff.) when defining the sum s of a series as the limit of the sequence of its partial sums s_n. Here he also gives the following equivalent definition: for infinitely large values of the number n the s_n must differ from the limit s by infinitely small quantities. Similarly, continuity of functions is defined by means of infinitesimals rather than epsilontically.

Still, there are good reasons for regarding Cauchy as the pioneer of epsilontics, which he uses whenever his infinitesimals fail him. An example that even includes the letter ϵ is found on p. 48, in the proof of the following theorem: if, for increasing values of x, the difference $f(x+1) - f(x)$ converges to a definite limit k, then the fraction $f(x)/x$ tends to the same limit. Indeed, choose a value of h such that for x greater or equal to h the difference lies between $k - \epsilon$ and $k + \epsilon$. It follows that the same is true for

$$\frac{f(x) - f(h)}{x - h},$$

which implies that

$$\frac{f(x)}{x} = \frac{f(h)}{x} + \left(1 - \frac{h}{x}\right)(k + \alpha)$$

with α between $-\epsilon$ and $+\epsilon$. In the last equality keep h fixed and let x tend to ∞. Then we see that the limit stays between $k - \epsilon$ and $k + \epsilon$ regardless of the smallness of ϵ. The conclusion follows.

What is more important for us than his occasional epsilontic proofs is that Cauchy was probably the first author of a textbook to provide a detailed discussion of the handling of inequalities. This is contained in the 22-page Note II which is an appendix to his *Cours*.

Both Cauchy and Abel must have been aware that epsilontics is a general tool for making rigorous proofs in analysis, but neither of them, to say nothing of their contemporaries, used it consistently. The first to do so was Dirichlet, and the relevant paper is his 1829 account of Fourier series, to which we will return below. For more on Cauchy we refer the reader to Bottazzini 1986 and 1992.

When dealing with the history of epsilontics, the problem we run into is that epsilontics played a less important role in research papers than in teaching, and we know little about the lectures of important mathematicians. For the most part, textbook writers were not, at that time, mathematicians in the forefront of research. It is a piece of great good luck that in 1904 Arendt published Dirichlet's lectures given at Berlin in 1854, i.e., 50 years earlier. (Incidentally, Riemann, for whom Dirichlet was a model as a researcher and as a teacher, did not take Dirichlet's introductory course.) The Arendt notes give us an idea of Dirichlet's style. What is somewhat surprising is that, notwithstanding his commitment to epsilontics, Dirichlet occasionally relapses into earlier conceptions.

Dirichlet defines continuity pointwise: for a fixed x and steadily decreasing h the difference

$$f(x + h) - f(x)$$

must converge to zero (p. 3). Then he immediately gives a precise account of the fundamental property that later came to be known as uniform continuity on closed intervals: "for an arbitrarily chosen absolute quantity ρ it is always possible to find another small quantity σ proportional to it such that the function y varies by at most ρ on every subinterval that is $\leq \sigma$" (Es "besteht immer die Möglichkeit, zu einer beliebig klein gewählten absoluten Größe ρ eine zweite ihr proportional kleine Größe σ von solcher Beschaffenheit zu finden, dass in jedem Teilintervall, welches $\leq \sigma$ ist, die Funktion y sich um nicht mehr als höchstens ρ ändert"). For proof he chooses in the interval $a \leq x \leq b$ numbers $c_0 = a$, c_k and c_{k+1}, $c_{k+1} > c_k$, such that

$$|f(c_{k+1}) - f(c_k)| = \rho\,, \quad \text{but} \quad |f(x) - f(c_k)| < \rho$$

for all x, $c_k < x < c_{k+1}$. Then he proves indirectly that one can choose finitely many c_k such that the corresponding finitely many subintervals cover the interval, and specifies the length of the shortest subinterval corresponding to σ. Next he proves the existence of the definite integral. On p. 4 there is the curious statement: "For a function to be continuous we must also require that its values be finite throughout" ("Zur Stetigkeit einer Funktion wird auch noch erfordert, dass alle ihre Werte durchaus endlich seien"). The proof of the fundamental property implies boundedness on the closed interval. This property is later used but not explicitly noted.

On p. 145 Dirichlet states the following true theorem:

"If in the definite integral

$$\int_a^b f(x,k)\,dx \tag{1}$$

... $f(x,k)$ is itself a continuous and everywhere finite function of k, then the integral (1) is also a continuous function of k." ("Wenn in dem bestimmten Integral

$$\int_a^b f(x,k)\,dx \tag{1}$$

... $f(x,k)$ selbst eine stetige, überall endliche Funktion von k ist, dann ist das Integral (1) auch eine stetige Funktion von k").

Now comes a proof by infinitesimals:

"We must show that to an infinitesimal change ∂k of k there corresponds a likewise infinitesimal change of the integral (1). Let

$$f(x, k + \partial k) - f(x,k) = \partial f(x,k)$$

denote the change, infinitesimal by assumption, in the function $f(x,k)$ which corresponds to the increment ∂k. Then with the new functional value $f(x, k + \partial k)$ the integral (1) becomes

$$\int_a^b (f(x,k) + \partial f(x,k))\,dx = \int_a^b f(x,k)\,dx + \int_a^b \partial f(x,k)\,dx\,,$$

so that the change it sustains when the parameter is increased by ∂k is expressed by the integral

$$\int_a^b \partial f(x,k)\,dx\,.$$

But since the integrand $\partial f(x,k)$ is infinitesimal everywhere and the distance $b-a$ between the limits is finite, the latter is itself infinitesimal.

Geometric intuition makes this result even clearer ..."

("Es ist nachzuweisen, dass einer unendlich kleinen Änderung ∂k von k auch eine unendlich kleine Änderung des Integrals (1) entspricht. Setzen wir die dem Zuwachs ∂k zugehörige, nach Voraussetzung ebenfalls unendlich kleine Änderung der Funktion $f(x,k)$:

$$f(x,k+\partial k) - f(x,k) = \partial f(x,k),$$

so wird für den neuen Funktionalwert $f(x,k+\partial k)$ das Integral (1):

$$\int_a^b (f(x,k)+\partial f(x,k))\,dx = \int_a^b f(x,k)\,dx + \int_a^b \partial f(x,k)\,dx,$$

so dass die Änderung, welche es erleidet, wenn der Parameter um ∂k zunimmt, durch das Integral:

$$\int_a^b \partial f(x,k)\,dx$$

ausgedrückt ist. Dasselbe ist aber, da die Integralfunktion $\partial f(x,k)$ überall unendich klein und die Entfertung der Grenzen $b-a$ endlich ist, selbst unendlich klein.

Noch einfacher erhellt das Resultat aus der geometrischen Anschauung ...")

No epsilontic proof is given. In fact, Dirichlet resorts to infinitesimals and geometry. The argument fails in both cases. If one admits infinitesimal ∂k, then, for the sake of consistency, one must be able to use infinitesimal ∂x. But then the proof fails for, say, the function

$$f(x,k) = \frac{2kx}{k^2+x^2}, \qquad f(0,0)=0,$$

considered, for example, on $0 \le x,k \le 1$; this in spite of the fact that the assumptions of the theorem are satisfied. The function is continuous in each variable and bounded by 1. But for infinitesimal $\partial x = \alpha$, $f(\alpha,k)$ is not infinitesimal everywhere; for example, take $\partial k = \alpha$. The geometric argument is also invalid.

At first sight, the infinitesimal argument looks like one possibly due to Cauchy, but it is false in his conceptual system. As the first application of his definition of continuity Cauchy proves (1821, 37–39) the following — epsilontically incorrect — theorem: If $f(x,y)$ is continuous as a function of each variable separately, then it is continuous as a function of two variables.

Indeed, for infinitesimal α and β we have:

$$f(x+\alpha, y+\beta) - f(x,y) =$$
$$[f(x+\alpha, y+\beta) - f(x, y+\beta)] + [f(x, y+\beta) - f(x,y)].$$

Here the expression in the first bracket on the right is infinitesimal because of the continuity in x with y fixed. Similarly, the second bracket on the right is infinitesimal because of the continuity in y with x fixed. The reader will note that $\frac{2\alpha x}{\alpha^2+x^2}$ is not continuous in x for infinitesimal α, for an infinitesimal change from $x = 0$ to $x = \alpha$ results in a finite change in the functional value. Cauchy's theorem is so simple because the assumptions are very strong, whereas Dirichlet needs weaker assumptions for his applications.

We may wonder at the assumption that the function be "finite everywhere" ("überall endlich"). This turn of phrase is also a product of the infinitesimal conception and is best replaced by "bounded" ("beschränkt"). If we put $g(0,0) = 0$ and

$$g(x,k) = \frac{2kx}{k^2 + x^4}$$

otherwise, then g is always continuous for *real* x and k but not for infinitesimal x. This function is also continuous in each of the variables separately, but the integral over x from 0 to 1 does not depend continuously on the parameter k; in fact, its value is $\arctan \frac{1}{k}$ for $k \neq 0$ and 0 for $k = 0$.

It is hardly possible that Arendt deviated from Dirichlet's lecture. For one thing, he emphasizes that he followed Dirichlet faithfully. For another, had he deviated from the lecture, he would have published in 1904 a very different account. Incidentally, Arendt put stars in the text whenever he was not sure that he had followed Dirichlet accurately, and there are no stars in this part of the text. When it comes to other nontrivial theorems, such as differentiation under the integral sign, Dirichlet uses them without proof.

We were not out to cavil at Dirichlet's proofs but wanted to give an example of how the earlier conceptions failed in Riemann's time in arguments presented even by eminent mathematicians.

Recent findings show that Dirichlet frequently used infinitesimal arguments "in private." In particular, this is true of his lectures on probability (H. Fischer, "Dirichlet's contributions to mathematical probability theory," *Historia Mathematica* 21 (1994), 39–63). In particular, Fischer found Dirichlet's note from 1842 with the significant heading:

"Proof that for a modulus k infinitely close to 1 the elliptic integral [satisfies the equality]

$$\int_0^1 \frac{dx}{\sqrt{(1-x^2)(1-k^2x^2)}} = \log(\frac{4}{k'})(1+\rho),$$

where ρ is infinitesimal" ("Beweis daß für einen der Einheit unendlich nahen Modul k das elliptische Integral

$$\int_0^1 \frac{dx}{\sqrt{(1-x^2)(1-k^2x^2)}} = \log(\frac{4}{k'})(1+\rho)$$

wo ρ unendlich klein ist"). (Here $k' = \sqrt{1-k^2}$.)

It turns out that in his lectures on Fourier *integrals* Dirichlet was far less "epsilontically rigorous" than in his 1829 paper on Fourier *series*. What he did was go over from series with *infinite* period to the integrals. Incidentally, a similar remark applies to his paper of 1830.

It is certain, however, that Dirichlet exceeded his contemporaries in mathematical rigor. We quote (Biermann 1986, 46) from Jacobi's letter to A.v. Humboldt dated 21 December 1846: "Only Dirichlet, not I, not Cauchy, not Gauss, knows what a perfectly rigorous proof is, but we learn it only from him. When Gauss says he has proved something, I think it very likely; when Cauchy says it, it is a fifty-fifty bet; when Dirichlet says it, it is *certain*; I prefer not go into these delicate matters" ("Dirichlet allein, nicht ich, nicht Cauchy, nicht Gauß, weiß, was ein vollkommen strenger Beweis ist, sondern wir lernen es erst von ihm. Wenn Gauß sagt, er habe etwas bewiesen, so ist es mir sehr wahrscheinlich, wenn Cauchy es sagt, ist ebensoviel pro als contra zu wetten, wenn Dirichlet es sagt, ist es *gewiß*; ich lasse mich auf diese Delikatessen lieber gar nicht ein").

We recall that around that time Riemann studied with Dirichlet and Jacobi, and that he was more impressed by the former than by the latter.

1. Complex Analysis

1.1 The genesis of complex analysis up to Riemann's time

1.1.1 Preliminary remarks

In Section 4 of the Introduction it became clear that the state of analysis in the first half of the 19th century hardly justified referring to it as systematically ordered. In particular, this was true of complex analysis, whose evolution up to Riemann's time we are about to describe. If our account is to be reasonably systematic, then we cannot fully adhere to the chronological development: some things that came into being prior to the firm establishment of the complex numbers were tacked on to complex function theory much later, and impressive results involving complex integration were obtained before the need arose for a justification of complex differentiation. The historical evolution of complex analysis was different from the genetic construction we feel tempted to make up today.

An autonomous systematic account of the foundations of complex analysis is found for the first time in Riemann's lectures, beginning in 1855. One way of obtaining an essentially correct version of the developments that preceded him is to adopt his view of these matters, and this is what we will do. The advantage of this approach is that it gives us the opportunity of familiarizing ourselves with Riemann's mode of thought in the context of subjects that he could assume to be known, or that he needed to sketch only briefly. The disadvantage is that we must give up strict adherence to chronology. This is a price we are willing to pay.

Riemann wrote no textbook. Fragments of his lectures became known through publications by other people, and many of the notes taken by his listeners have been preserved (see Neuenschwander in N. 855–867). As pointed out by Neuenschwander (1987), there is little difference among the variants of the introductory passages written by Riemann between 1855 and 1861, when he last lectured on the foundations of complex analysis. Neuenschwander's paper of 1987 is a careful transcription of the relevant notes, mostly due to Ernst Abbe (1840–1905; professor of physics at Jena as of 1870; cofounder of the famous Jena works). It is surprising to find that the structure of Riemann's lectures has an almost canonical form, and that they could be used even today without essential modifications.

Neuenschwander has pointed out that notes written by Riemann and containing the beginnings of individual lectures, set down word for word, have been preserved at Göttingen. Facsimile copies of some of these pages have been published (see Neuenschwander 1987, and partly also Noether 1902 (in N. 709, besides others)). A few samples are reproduced below with the kind permission of the department of manuscripts at Göttingen. In our account we hold to the construction of Riemann's lectures, but use reliable texts taken from his original papers rather than from the — admittedly insignificantly different — notes of his listeners. For a complement to our meager presentation, we expressly refer the reader to Neuenschwander 1987.

The reader will note that much of the content of Section 1.1 goes back to Cauchy, who, in fact, never produced a systematic account. Such an account was written by his students Briot and Bouquet. Riemann's initial lectures were given before its publication, but in his later lectures he did refer to their "Etudes des fonctions d'une variable imaginaire," *J. Ecole Polytechnique*, Cah. 36 (1856).

1.1.2 The complex numbers

Riemann used the complex plane in a way familiar to modern students. But this representation of the complex numbers, which we take for granted, came at the end of a long historical path that we are about to sketch.

We sometimes encounter complex numbers when we use the rule for solving quadratic equations; but then there are no real roots. This state of affairs is of little interest. Things are different in the case of cubic equations. In Girolamo Cardano's (1501–1576) solution formulas, real solutions are sometimes represented by complex numbers. In his algebra of 1572 Bombelli dealt with the equation $x^3 = 15x + 4$. Using Cardano's formula he obtained

$$x = \sqrt[3]{2 + \sqrt{-121}} + \sqrt[3]{2 - \sqrt{-121}},$$

and since he knew that

$$(2 \pm \sqrt{-1})^3 = 2 \pm 11\sqrt{-1},$$

he found that

$$x = 2 + \sqrt{-1} + 2 - \sqrt{-1} = 4.$$

Leibniz recalled in 1702: "When I called the attention of the late Herr Huygens to the fact that $\sqrt{1 + \sqrt{-3}} + \sqrt{1 - \sqrt{-3}}$ is equal to $\sqrt{6}$, he found this so remarkable that he said that there was something in this that defied our understanding." ("Als ich den verstorbenen Herrn Huygens darauf aufmerksam machte das $\sqrt{1+\sqrt{-3}}+\sqrt{1-\sqrt{-3}}$ gleich $\sqrt{6}$ ist, so fand er dies so wunderbar, dass er mir erwiderte, es läge darin etwas für uns Unbegreifliches"). Computing experience slowly paved the way for the acceptance of the Fundamental Theorem of Algebra, i.e., for the insight that using the square root of -1 one can decompose higher-order polynomials into linear factors. Nevertheless, as late as 1702, Leibniz, who was interested in the integration of rational functions, and thus in their decomposition into partial fractions with denominators involving just linear and quadratic factors, was convinced that $\frac{1}{x^4+a^4}$ admitted no such decomposition. His conviction was based on the fact that one cannot obtain real quadratic factors by combining any two of the four linear factors $x \pm a\sqrt{\pm\sqrt{-1}}$. The seemingly obvious decomposition $x^4 + a^4 = (x^2 + a^2)^2 - (\sqrt{2}xa)^2$ escaped him. Thus it seemed plausible that one needed other "imaginary" or "impossible" numbers in addition to $\sqrt{-1}$.

Euler (1707–1783) was a master at computing with complex numbers and knew the polar form $z = r(\cos\varphi + i\sin\varphi)$ of z. This suggests that he may have thought about the geometric representation of complex numbers, but we have no proof that he did. Certainly the idea was not in the public domain. Even Cauchy hesitated for a long time to use equations involving complex numbers, and justified their use by regarding them as abbreviations of pairs of equations over the reals. It took him a long time to accept a geometric interpretation of the complex numbers, and this delayed his important discoveries in complex analysis.

As late as 1854, three years after publication of the dissertation of his friend Riemann, Richard Dedekind was to say in his habilitation lecture, in the presence of Gauss, that "It is well known that an unobjectionable theory of imaginary numbers is either not yet available, or, at least, has not yet been published" ("Bis jetzt ist bekanntlich eine vorwurfsfreie Theorie der imaginären ... Zahlen entweder nicht vorhanden, oder doch wenigstens nicht publiziert") (Dedekind, *Math. Werke* 3, 434).

It may occasion some surprise that the concept of the number plane, so obvious to us, did not become fruitful until the middle of the 19th century. To dispel the surprise we must keep in mind that complex numbers turned up first in connection with algebraic, and then also with analytical problems, and that prominent mathematicians, such as Euler, Lagrange, and Cauchy, adhered to the dogma that algebra and analysis must not derive their powers of proof (Beweiskraft) from geometry. This was carried to the point that intuitive figures were tabooed for fear that one might succumb to the temptation to draw algebraic or analytical conclusions from geometry. The idea that plane analytic geometry deals with number pairs (x, y) and yields an isomorphic model of the complex numbers for a suitable choice of computational rules was apparently foreign to these mathematicians. It was, however, obvious to Riemann.

But what of Gauss? Dedekind's reproach was not entirely unjustified, for in this matter, just as in other fundamental matters, Gauss shared his own clarity of thought with the mathematical public only reluctantly and partially. In his dissertation of 1799 he dealt with the Fundamental Theorem of Algebra, and his argument is easy to explain with the help of the number plane: if we substitute $z = r(\cos\varphi + \sin\varphi)$ in a polynomial $P(z)$, then it splits into a real and imaginary part, $P = R + iS$, and what we are looking for in the plane with polar coordinates r, φ is the points of intersection of the curves $R = 0$ and $S = 0$.

But already in connection with the lemma that $R = S = 0$ implies that $P(x)$ has the factor $x^2 - 2xr\cos\varphi + r^2$, Gauss said explicitly: "In most cases, the present theorem is proved by means of complex quantities, cf. Euler, Introductio ...; I thought it worth the effort to show how it can be just as easily deduced without using them" ("Der vorstehende Satz wird meistens mit Hilfe imaginärer Größen bewiesen, vgl. Euler, Introductio ...; ich hielt es der Mühe für wert zu zeigen, wie er auf gleich leichte Art ohne Hilfe derselben abgeleitet werden könne"). It was only 50 years later, in connection with his doctoral jubilee, that he resumed the proof and explained: "The whole domain of complex quantities is represented by an unbounded plane in which every point with coordinates t and u with respect to two perpendicular axes is viewed as corresponding to the complex quantity $x = t + iu$ ..." ("Das ganze Gebiet der komplexen Größen wird vertreten durch eine unbegrenzte Ebene, in welcher jeder Punkt, dessen Koordinaten in Beziehung auf zwei einander rechtwinklig schneidende Achsen t, u sind, als der komplexen Größe $x = t+iu$ entsprechend betrachtet wird ...). The paper appeared in 1850 and the students Riemann and Dedekind were probably familiar with it.

On numerous occasions, including his 1854 lecture on geometry (in connection with the concept of a manifold), Riemann referred to Gauss' paper of 1831 on biquadratic residues (Gauss, *Werke* 2, 93–148 and 169–178). Since Gauss used a number-theoretic paper as the vehicle for discussing the complex number plane, he had no need to publicly violate in it the taboo against the use of geometry in analysis. For him personally this taboo had ceased to exist much earlier. In his famous letter to Bessel of 18 December 1811 (*Werke* 8, 90–92, also Remmert 1991 I. 130), which admittedly became known only much later, he even anticipated Cauchy's integral theorem. This letter includes the following passage on the number plane: "just as we can think of the whole realm of real numbers in terms of an infinite straight line, so too we can convey to the senses the whole realm of all quantities, real and imaginary, by means of an infinite plane, in which every point determined by abscissa $= a$ and ordinate $= b$ represents, as it were, the quantity $a + bi$ " ("so wie man sich das ganze Reich aller reellen Grössen, durch eine unendliche gerade Linie denken kann, so kann man das ganze Reich aller Grössen, reeller und imaginarer Grössen sich durch eine unendliche Ebene sinnlich machen, worin jeder Punkt, durch Abscisse $= a$, Ordinate $= b$ bestimmt, die Grösse $a + bi$ gleichsam repräsentirt").

The letter could not have influenced Riemann. But he knew Gauss' position from other sources and shared it. The essential thing was the primacy of "geometry": *the* straight line, *the* plane, were assumed to be directly given mathematical objects whose points represent the appropriate numbers. Gauss regarded the whole matter as a completely philosophical — or, as people said at the time, metaphysical — issue (*Werke* 10.1, 404): "All this notwithstanding, as long as their foundation consisted in just a fiction, the complex quantities were tolerated in mathematics rather than regarded as legitimate inhabitants, and their status was hardly the same as that of real quantities. But now that the metaphysics of imaginary quantities is seen in the proper light, and it has been shown that just like the negative numbers they too have real objective significance, there is no longer any basis for such discrimination" ("Bei allem dem sind die imaginären Grössen, so lange ihre Grundlage immer nur in einer Fiction bestand, in der Mathematik nicht sowohl wie eingebürgert, als vielmehr nur als geduldet betrachtet, und weit davon entfernt geblieben, mit den reellen Grössen auf gleiche Linie gestellt zu werden. Zu einer solchen Zurücksetzung ist aber jetzt kein Grund mehr, nachdem die Metaphysik der imaginären Grössen in ihr wahres Licht gesetzt, und nachgewiesen ist, daß diese, ebenso gut wie die negativen, ihre reale gegenstandliche Bedeutung haben").

We recall that in his 1854 habilitation lecture Dedekind, Gauss' last student, disagreed completely with this metaphysics. Later he developed his famous theory of real numbers without the assumption of a straight line. The introduction of complex numbers as ordered pairs of reals with appropriate rules of computation goes back to 1835 and is due to Hamilton. To Gauss, Hamilton's approach may have seemed different from the usual introduction of complex numbers in the plane. To us the two approaches are rather similar. These developments had no significance for Riemann and we do not pursue them further.

The "Riemann number sphere" is not mentioned in any of Riemann's publications, but was introduced and used by him in his lectures (N. 679, lecture given in the winter term 1858/59):

"It is easy to find the point on the sphere corresponding to a point in the plane if we think of the sphere as touching the z-plane at the origin ... All we need do is join the other pole to z and find the point of intersection of Pz with the sphere ..."

("Man kann den Punkt der Kugelfläche, welcher einem Punkt der Ebene entspricht, leicht finden, wenn man sich die Kugel die z-Ebene im Nullpunkt berührend denkt ... Man hat nur den anderen Pol mit z zu verbinden und den Schnittpunkt von Pz mit der Kugel aufzusuchen ...")

Riemann notes that this mapping is conformal, i.e., angle-preserving.

1.1.3 Complex functions and their derivatives

If a real function is given by an expression as a rational function or a power series, then in general this expression is meaningful for complex values of the argument. Moreover, the assumption that the variable is real has no effect on the rules for differentiation, and the expressions for derivatives are formally the same in the complex realm as in the real. For example, if $f(z) = z^n$, then the quotient $[f(z+h) - f(z)]/h$ yields the derivative $f'(z) = nz^{n-1}$ no matter how the complex h tends to zero. In the case of algebraic expressions, difficulties arise not in connection with differentiation and integration but in connection with multiple values. This problem will be discussed in the sequel.

In the case of real functions, it turned out that one could not make do with the vague notion of an "expression"; this matter will be discussed in Chapter 2. Cauchy realized that for real analysis the concept of continuity makes it possible to deal with a class of functions without assuming the presence of "expressions." Similarly, Riemann recognized that differentiability is the corresponding concept for complex functions. In his dissertation (W. 6) Riemann wrote $z = x + yi$ and $w = f(z) = u + vi$ with real x, y, u, v. Then, for fixed x and y,

$$\frac{dw}{dz} = \frac{du + dvi}{dx + dyi} = \frac{(u_x + v_x i)dx + (v_y - u_y i)dyi}{dx + dyi}$$

is independent of dx and dy if and only if

$$u_x = v_y \quad \text{and} \quad v_x = -u_y\,. \tag{I}$$

(Here we are writing u_x for Riemann's $\frac{\partial u}{\partial x}$.) The relations (I) came to be known as the Cauchy-Riemann differential equations. Riemann wrote them later (W. 68, 88) in condensed form as

$$w_y = i w_x\,. \tag{II}$$

If (I) or (II) hold, then the derivative can be written as

$$\frac{dw}{dz} = \frac{\partial w}{\partial x} = -i\frac{\partial w}{\partial y}.$$

In his dissertation (W. 5) Riemann set down the following definition:

> "A variable complex quantity w is called a function of another variable complex quantity z if w varies with z so that the value of the differential quotient $\frac{dw}{dz}$ is independent of the value of the differential dz."
>
> ("Eine veränderliche complexe Grösse w heisst eine Function einer andern veränderlichen complexen Grösse z, wenn sie mit ihr sich so ändert, dass der Werth des Differentialquotienten $\frac{dw}{dz}$ unabhängig von dem Werthe des Differentials dz ist.")

He adhered to it later as well.

Riemann observed immediately that

$$u_{xx} + u_{yy} = v_{xx} + v_{yy} = 0. \tag{III}$$

We compare this account with that of Cauchy, given in the *Comptes Rendus* of the Paris Academy for February to April 1851, which Riemann referred to in his promotion theses. We change Cauchy's $u = v + iw$ and D_x to $w = u + iv$ and $\frac{\partial}{\partial x}$ respectively.

On 10 February 1851 Cauchy stated: "in general, the differential quotient of w with respect to z depends not only on the real variables x and y ... but also on the differential quotient of y with respect to x." But then he immediately obtains for $dz = dx$

$$\frac{dw}{dz} = \frac{\partial u}{\partial x} + i\frac{\partial v}{\partial x},$$

and for $dz = idy$

$$\frac{dw}{dz} = \frac{\partial v}{\partial y} - i\frac{\partial u}{\partial y},$$

and notes that if the differential equations (I) hold, then $\frac{dw}{dz}$ depends only on x and y. Then he also obtains (III). In fact, in the note of 7 April 1851 he refers to functions with a direction-independent differential quotient as a special case and calls them monogenic functions, but adds immediately that his old results, in particular the residue theorem, apply to these monogenic functions. Incidentally, there is rather close agreement between Section 4 of Riemann's dissertation (W. 6–7), in which he gives (II), and Cauchy's text of 10 February 1851.

At the very beginning of his dissertation (Section 3, W. 5–6) Riemann notes that the mapping of the z-plane A into the w-plane B effected by such a (holomorphic) function is "similar in the smallest parts" ("in den kleinsten Theilen ähnlich"). The notion of a mapping is foreign to Cauchy. Riemann considers $z + dz$ and $z + d\tilde{z}$, two values "infinitely close" ("in unendlicher Nähe") to z, and the images $w+dw$, $w+d\tilde{w}$, and w, and writes $dz = \epsilon e^{\varphi i}$, $d\tilde{z} = \tilde{\epsilon} e^{\tilde{\varphi} i}$, $dw = \eta e^{\psi i}$, $d\tilde{w} = \tilde{\eta} e^{\tilde{\psi} i}$. (We have changed the notations slightly.) He asserts that the equality

$$\frac{dw}{dz} = \frac{d\tilde{w}}{d\tilde{z}}$$

implies that $\frac{\tilde{\eta}}{\eta} = \frac{\tilde{\epsilon}}{\epsilon}$ and $\psi - \tilde{\psi} = \varphi - \tilde{\varphi}$, i.e., that the angle at z in the z-plane and the angle at w in the w-plane are

> "equal, and the sides that enclose them are proportional. Thus two corresponding infinitely small triangles, and therefore also the smallest parts in the A-plane and their images in the B-plane, are similar."
>
> ("gleich und die sie einschliessenden Seiten einander proportional. Es findet also zwischen zwei einander entsprechenden unendlich kleinen Dreiecken und folglich zwischen den kleinsten Theilen der Ebene A und ihres Bildes auf der Ebene B Ähnlichkeit statt.")

In this connection Riemann cites Gauss' paper "Allgemeine Auflösung der Aufgabe: Die Theile einer Fläche so abzubilden, daß die Abbildung dem abgebildeten in den kleinsten Theilen ähnlich wird" ("General solution of the problem: To map a surface so that the image is similar in the smallest parts to what is being mapped") published in 1825 in *Astronomische Abhandlungen von Altona*. It is now available in Gauss, *Werke* 4, 189, but in his student days Riemann looked for it in vain, for it was not available in the Göttingen library. It is not clear whether he became aware of the connection between conformal mappings and complex functions before or after reading Gauss' paper. If the latter, then it would make sense to ask why he tried so hard to find Gauss' paper. Of course, it is possible that he wanted to familiarize himself with all the publications of his revered teacher.

The paper must have immediately helped to clarify his thoughts. Gauss used complex numbers. He decomposed the line element of the surface, in particular a part of the plane given in terms of curvilinear coordinates p and q, into complex conjugate linear factors. Using the notation of his later surface

22: Augustin-Louis Cauchy

theory (see Section 3.1.2) we can spell this out as follows:

$$\begin{aligned} ds^2 &= Edp^2 + 2Fdpdq + Gdq^2 \\ &= \frac{1}{E}(Edp + (F + iW)dq)(Edp + (F - iW)dq) \end{aligned}$$

with $W = \sqrt{EG - F^2}$. Using a suitable complex integrating factor λ we make $\lambda(Edp + (F + iW)dq)$ an exact differential and pair off $\bar{\lambda}$ with the second factor in ds^2. In terms of the new real parameters P and Q we have

$$ds^2 = \frac{1}{E|\lambda|^2}(dP^2 + dQ^2).$$

From this we can read off the following result: The mapping of a surface into the plane with Cartesian coordinates P, Q and square of line element $dP^2 + dQ^2$, given by $(p, q) \mapsto (P, Q)$, is conformal. Moreover, Gauss noted that $w = P + iQ$ is a function either of $p + iq$ or of $p - iq$, and that in the former case the orientation of figures is preserved.

If we look at $w = f(z)$ as a mapping, then the connection between complex functions and conformal mappings is traced out in Gauss' paper. Today one can cavil at the absence of analytic assumptions in the paper — with hindsight we realize that E, F, and G must be assumed to be real-analytic (because $f(z)$ is holomorphic) and that the result is only local, i.e., it can hold only for sufficiently small parts of the surface. But this is irrelevant to the basic idea. What distinguishes Riemann's approach to the theory of functions from the approaches of Cauchy and of the Weierstrass school is the idea of a mapping. It is this idea that makes for the closeness between Riemann's view and our own.

The motto of Gauss' paper was "Ab his via sternitur ad maiora" ("These [results] pave the way to bigger things"). Riemann changed this slightly and used it as the motto for his Paris prize essay of 1861 (W. 391).

Suppose that a function u of x and y satisfies the differential equation (III). Then it can always be viewed as the real part of a holomorphic function, because there is a function v such that u and v satisfy the Cauchy-Riemann equations (I). There is a close connection between the differential equation (III) and almost all groups of problems of mathematical physics that have become tractable since the middle of the 18th century.

Given the many connections between holomorphic functions and geometric and physical problems, it would have been surprising if the differential equations (I) had first turned up as late as 1850. Hence the search for precursors.

Beginning in 1776 Euler was aware that (I) must hold if $f(z)$ is real for real z. A relevant hydrodynamical argument by d'Alembert goes back to 1752. Later came Euler's investigation of conformal mappings. Markushevich goes so far as to suggest that (I) be called the d'Alembert-Euler differential equations. But adoption of this suggestion would ignore their historical significance, for it was Cauchy and Riemann who first recognized that (I) is the central property of complex functions from which everything else is derivable.

Riemann's preference for compact formulas and brief descriptions is reflected in his use of the complex summary formula (II) as the definition of a complex function. Today we tend to follow Cauchy's usage and add an adjective. Cauchy used the adjective *monogenic*; this was displaced by the long-lived term *analytic*. The term used nowadays is *holomorphic*.

Cauchy's contributions to complex analysis began with a paper written in his youth, in 1814. In time he obtained the many results that are rightly named after him. Differential equations played a role in his papers from the beginning, and in the 1840s he spelled out explicitly the importance of the existence and continuity of $f'(z)$. The short *Comptes Rendus* notes published at the beginning of 1851, which Riemann referred to, point to these earlier investigations and make it rather clear that if the Cauchy-Riemann differential equations hold, then the differential quotient $\frac{dw}{dz}$ is independent of dz.

1.1.4 Integration

It is easy to explain the meaning of the integral of $w = u + vi$ along a path C in the complex plane. To do this we replace dz in $\int w\,dz$ by $dz = dx + dy \cdot i$

and carry out the multiplications formally. The result is two line integrals whose independence of the path is implied by the differential equations (I). This explanation can be obtained less formally by viewing the integral as the limit of an appropriate sum. The Cauchy Integral Theorem expresses the path-independence of the integral.

Cauchy proceeded in these ways in his works from 1846 on, and Riemann did much the same thing in his lectures. Both used the integral formulas of Gauss and Green. Incidentally, Riemann proved these formulas explicitly in his dissertation without referring to their originators (W. 12–14, 16).

The complex integral theorem, as well as other results on complex integration to be cited in the sequel, derive from Cauchy's earlier works. He obtained them by strenuous efforts over a number of decades. These efforts are discussed in considerable detail in Bottazzini 1992 and in Belhoste 1991, 126 ff., 208 ff.

Cauchy made things difficult for himself by failing to introduce the complex number plane. Gauss had obtained the complex integral theorem as early as 1811, presumably by using his integral theorem in the real plane. He communicated it to his friend Bessel, saying that it was very beautiful and not difficult to prove. (Gauss, *Werke* 8, 90–92). But nobody, including Riemann, could derive any profit from this.

Questions related to the computation of *real* definite integrals contributed in a vital way to the rise of complex analysis. In the introduction to one of his papers of 1818 Cauchy said that his investigations were an attempt to provide a rigorous justification of the methods of Euler (after 1759) and Laplace (after 1782) for the computation of real integrals using imaginaries.

We provide a simple example. The equality

$$\int_0^\infty e^{-ck} \cos kx \, dk = \frac{c}{c^2 + x^2} \quad \text{for} \quad c > 0$$

is obtained from $\int e^{bk}\, dk = \frac{e^{bk}}{b}$ for $b = ix - c$ by going over to the real part (Euler's formula) and by substituting the limits. In this case integration by parts yields the same result, and this proves its correctness. But the applicability of the known "real" methods was limited. Specifically, the use of Fourier's methods led very frequently to integrals whose evaluation by "real" methods either was impossible or led to seemingly paradoxical results. For example, Poisson tried to evaluate the integral

$$\frac{1}{\pi} \int_{-\infty}^{+\infty} \frac{\cos az}{b^2 + z^2} \, dz, \quad a, b > 0.$$

His first attempt, in 1815, ended in failure. In 1820 he put $z = t + ic$, c a positive constant, integrated over a complex path from $t = -\infty$ to $t = +\infty$ and obtained the value e^{-ab} for $0 < c < b$ and the value $\frac{e^{-ab} - e^{ab}}{2}$ for $c > b$.

Cauchy's method of residues was rooted in questions of this kind arising from mathematical physics around 1820.

The idea of a genetic construction of mathematics based on its historical evolution has attracted a certain amount of publicity. But this idea is without merit for complex analysis. Of course, Riemann's construction of his lectures, which is imitated to this day, a construction that proceeds from the complex number plane through differentiation to integration and puts at its center the Cauchy-Riemann differential equations, is genetic, but the historical evolution of complex analysis is very different from this construction. For more than a century mathematicians could go from the real to the complex without running into difficulties so long as they limited themselves to the operations of algebra and to differentiation. It was integration that showed that this formal transition, "le passage du réel à l'imaginaire," is not free from complications: the history of complex analysis begins with integration!

Riemann used the integral theorem to obtain a special case of the residue theorem (Neuenschwander 1987, 33). Let $f(z)$ be holomorphic in a region, with the exception of one of its interior points $z = a$, as well as on its boundary C consisting of a simple closed curve. Suppose that $f(z)(z-a) = c \neq \infty$ for $\lim z = a$; that is, on a small circle of radius r centered at a

$$f(z)(z-a) = c + \epsilon(z) \quad \text{with} \quad \max|\epsilon(z)| \to 0 \quad \text{for} \quad |z-a| = r \to 0.$$

(Here Riemann used words, for he had no symbols for the absolute value and for the maximum.) Using the integral theorem, he replaced the integral over C by the integral over the circle $z - a = re^{\varphi i}$. Using the stated assumptions and putting $dz = ire^{\varphi i}\,d\varphi = (z-a)i\,d\varphi$, he obtained

$$\int_C f(z)\,dz = i\int_0^{2\pi} f(z)(z-a)\,d\varphi = 2\pi ic + i\int_0^{2\pi} \epsilon(z)\,d\varphi \to 2\pi ic,$$

and finally

$$\int_C f(z)\,dz = 2\pi i \lim f(z)(z-a).$$

Replacement of $f(z)$ by $\frac{f(z)}{z-t}$, where $f(z)$ is holomorphic everywhere, yields the Cauchy integral formula (with z replaced by ζ and t replaced by z)

$$f(z) = \frac{1}{2\pi i}\int_C \frac{f(\zeta)}{\zeta - z}\,d\zeta.$$

Of course, Riemann did not use the term "holomorphic." Nor did he mention Cauchy's name or the term "residue." But he emphasized the importance of the formula repeatedly and forcefully: without an assumed expression for $f(z)$ it is possible to represent this function in a whole region by means of an integral.

After the inclusion of a few arguments on infinite series (to be discussed in the next section), Riemann obtained from the integral formula an expansion of $f(z)$ in a Taylor series or in a Laurent series in a manner we are well acquainted with today (Neuenschwander 1987, 37-40).

Riemann assumed that these results were well known to professionals. This is clear from the beginning of his paper on Abelian functions, published in 1857 (W. 88), where he wrote, with reference to (II):

> "According to a well-known theorem, this differential equation implies that the quantity w is representable by means of a series of integral powers of $z - a$, provided that it has a definite value everywhere in the neighborhood of a which changes continuously with z, and that this representability takes place up to a value of the distance from a, or up to a value of the modulus of $z - a$, for which a discontinuity sets in. But it follows from the considerations underlying the method of undetermined coefficients that the coefficients a_n are fully determined, provided that w is given on a finite, albeit arbitrarily small, line issuing from a.
>
> By combining the two arguments we can easily see the correctness of the following theorem:
>
> A function of $x + yi$, given in a part of the (x, y) plane, can be continued beyond it in a continuous manner in just one way ."
>
> ("Aus dieser Differentialgleichung folgt nach einem bekannten Satze, dass die Grösse w durch eine nach ganzen Potenzen von $z - a$ fortschreitende Reihe ... darstellbar ist, sobald sie in der Umgebung von a allenthalben einen bestimmten mit z stetig sich ändernden Werth hat, und dass diese Darstellbarkeit stattfindet bis zu einem Abstande von a oder Modul von $z - a$, für welchen eine Unstetigkeit eintritt. Es ergibt sich aber aus den Betrachtungen, welche der Methode der unbestimmten Coeffizienten zu Grunde liegen, dass die Coeffizienten a_n völlig bestimmt sind, wenn w in einer endlichen übrigens beliebig kleinen von a ausgehenden Linie gegeben ist.
>
> Beide Überlegungen verbindend, wird man sich leicht von der Richtigkeit des Satzes überzeugen:
>
> Eine Function von $x + yi$, die in einem Theile der (x, y)-Ebene gegeben ist, kann darüber hinaus nur auf Eine Weise stetig fortgesetzt werden.")

Without mentioning Cauchy, Riemann summarized his results, which he supposed to be known or easy to infer.

To properly appreciate the remarkableness of Riemann's transparent and lean account, one must compare it with the labored deliberations of others. It is true that all the building blocks were made available by Cauchy, Laurent, Liouville, and Puiseux, but the whole edifice remained to be erected. Readers interested in the prehistory of individual results should consult the previously mentioned works of Bottazzini and Belhoste, as well as Remmert 1991, I.

Riemann repeatedly stressed the importance of the assertion that everything follows from the differential equation (II). This is the Ariadne's thread which goes through his whole account of (complex) function theory. The French mathematicians did not attach comparable importance to it. Nor did Riemann in his dissertation. We can only regret that he did not write a textbook on function theory, and that we have had to wait until Neuenschwander brought forward (in 1987) this didactic achievement of Riemann. It remains for historians to clarify the effect of his lectures as transmitted by his listeners. Materials for such an investigation have been made available by Neuenschwander in a number of publications.

1.1.5 Power series

From the early history of analysis, power series have been an important aid in the study of transcendental functions. Newton went so far as to regard them as "literal expressions" that were generalizations of the decimal representations of the real numbers. It was easy to perform on them the most important operations, namely the rational operations as well as differentiation and integration, and a clever manipulator could use them to compute approximations. Power series were advanced by Taylor (1715), and even more by Maclaurin's *The Treatise of Fluxions* (1742), which summarized their use in these words: "When a fluent cannot be represented accurately in algebraic terms, it is then to be expressed by a converging series. In that case a few terms at the beginning of the series will be nearly equal to the whole." It took another half century for remainder estimates to be developed.

Euler worked with power series as if they were polynomials of infinite degree. He also had some success working with divergent series. Lagrange, who tried to remedy the deficient justification of infinitesimal and limit methods, decided to use formal power series, without regard to convergence, for a rigorous construction of analysis. He did this in a book, published in 1797, with the characteristic title *Théorie des fonctions analytiques, contenant des principes du Calcul différentiel, dégagés de toute considération d'infiniment petits, d'évanouissans, de limites et de fluxions, et reduit à l'analyse algébrique de quantités finies*. The only part of this book that turned out to be of lasting value is the remainder formula. One of the reasons for the limited success of this work was the growing importance of Fourier series, which made it abundantly clear that the important functions of *real* analysis are not limited to functions representable by power series. But in a sense Cauchy realized Lagrange's dream by showing that power series *do* suffice in *complex* analysis.

The sense of unease over some of the paradoxical results involving divergent series grew to the point of their being banned by Cauchy, who declared on p. iv of his book of 1821 that "une série divergente n'a pas de somme." In this "analyse algébrique" he did avoid divergent series and did not use either differentiation or integration, but he provided for the first time a complete and rigorous theory of convergent real and complex power series. At that time people began to take an interest in the conditions under which functions can be represented by series. The interest centered first on Fourier series, but later it extended to power series. Cauchy gave the famous example of the infinitely differentiable function e^{-1/x^2} unequal to its Taylor series. But this example "works" only in the real domain. Cauchy completely clarified the issue of representability of functions in the complex domain as early as 1831. Riemann adopted the essential part of Cauchy's methods. We proceed in accordance with Riemann's construction of his lectures.

The first application of Cauchy's integral formula was Riemann's proof of the theorem that a function holomorphic in a neighborhood of $z = a$ is representable as a series of powers of $z - a$, and that this series converges in the interior of the largest circle centered at a in which the function is holomorphic. The series diverges outside this circle. Cauchy proved this theorem in 1831 while in exile in Turin, but it was not until about 1840 that the "Turin theorem" became well known and was used effectively.

Riemann's proof of the theorem is preceded by general remarks on series with complex terms (Neuenschwander 1987, 34–35).

One must keep in mind that in the middle of the 19th century there were hardly any introductory texts intended for beginning students. The foundations of analysis were developed in lectures. Thus Riemann attended Stern's lectures on definite integrals and presented such a series of lectures in the summer of 1855. Dirichlet's lecture course with this title was first published 50 years after it was given (Arendt 1904). Riemann mentions occasionally (N. 657, Remark 1) that fundamental facts on linear algebra were stated in lectures dealing with other subjects, such as mechanics or Gauss' method of least squares.

The notes for Riemann's lectures contain the following necessary and sufficient condition for the convergence of a sequence of numbers S_n: as $n \to \infty$, $S_{n+m} - S_n$ tends to 0 for every m. The example of the harmonic series shows that this condition is not sufficient. This flawed formulation is not an accidental lapse of a copyist, for we find it in a note by Dedekind dated 1855 (see Dugac 1976) and in elaborations dated 1861, used by Neuenschwander (1987). It goes without saying that Riemann had a perfect knowledge of convergence. This is attested by his habilitation paper of 1854 (to be discussed in Chapter 2). It is not unusual to find the flawed version of Cauchy's convergence criterion in 19th-century texts, including Weierstrass 1886, 55. But in applications all is invariably done right.

It is possible that the formulation goes back to Cauchy 1821, 125. There it is stated that s_n converges to the limit s if and only if, for infinitely large values of the number n, the partial sums $s_n, s_{n+1}, s_{n+2}, \ldots$ differ from the limit s, and therefore also from one another, by infinitely small quantities. In other words, $s_{n+m} - s_n$ must be infinitely small for every infinitely large n and every m, *be it finite or infinitely large*. While Cauchy did not state this subordinate clause, he obviously meant it; indeed, $s_n - s$ is to be infinitely small for *every* infinitely large n, and therefore also for $n' = n + m$ with finite or *infinitely large* m. If one translates this into the language of limits, then one obtains the correct version of the convergence criterion: s_n converges if and only if $n_k \to \infty$ and $n'_k \to \infty$ imply that $s_{n_k} - s_{n'_k} \to 0$. One must write $n'_k = n_k + m_k$ rather than $n'_k = n_k + m$. It is conceivable that the reason for the flawed formulation was that the language of infinitesimals was still in use but was incorrectly interpreted.

The fact that the error was harmless in actual usage is illustrated by Riemann's proof, found in his lectures (Dugac, Neuenschwander), of termwise integrability of a series of (uniformly) convergent functions: Suppose that the series of functions $\sum u_k(x)$ converges in the interval from a to b. Put $p_n = \int_a^b u_n$, so that

$$p_n + p_{n+1} + \cdots + p_{n+m} = \int_a^b (u_n + u_{n+1} + \cdots + u_{n+m}).$$

In what follows the correct thing to do is to write n_k and m_k. It is clear that Riemann assumed uniform convergence and denoted by δ the maximum of the absolute value of $u_n + u_{n+1} + \cdots + u_{n+m}$. Since the integral on the right is in absolute value less than $\delta(b-a)$, $\delta \to 0$ implies the convergence of the series of integrals. Of course, the required uniformity of convergence, used by Riemann without acknowledgment, is justified in the context of power series, considered by him immediately thereafter.

We state the obvious: for a number of years, the otherwise careful Riemann was rather casual in his lectures when it came to questions of convergence of series. One can imagine that he considered these things to be of little importance and was therefore unwilling to devote much time to them. In fact, he repeatedly emphasized that it is not necessary to think of a function as given by means of an expression. Power series are expressions, and as such they should be avoided whenever possible.

Lending additional support to this conjecture is the steady decrease in the amount of attention Riemann seems to have paid to power series between 1856 and 1861. Thus we still find the formula, named after Cauchy and Hadamard, for the radius of convergence ρ of a power series $\sum a_k(z-z_0)^k$, namely $\rho^{-1} = \limsup \sqrt[k]{|a_k|}$, in a note in November 1856, but not in the notes from 1861. (Concerning the discovery of this formula in Riemann's notes see Laugwitz 1993 and Laugwitz and Neuenschwander 1994.) Until recently it was thought that this formula had been forgotten after Cauchy and had been rediscovered by Hadamard in 1890. It is of conceptual, rather than computational, interest. In each particular case more convenient tests are available. Moreover, the concept of lim sup involved conceptual considerations that were far from the minds of Riemann's contemporaries.

The first, and for a long time the only, book containing a complete theory of real and complex power series was Cauchy's work of 1821. Riemann borrowed it when he was still a student. (A facsimile version of the Göttingen borrowing list of January 1847 was reproduced by Neuenschwander 1981a, 227.) Cauchy dealt with increasingly more general versions of this formula in a number of places in his *Cours* and ended up with a theorem, stated on p.280, devoted to series with complex terms $u_n = p_n + q_n\sqrt{-1}$. Put $\rho_n^2 = p_n^2 + q_n^2$ and look for the limit or limits of $\rho_n^{1/n}$ as $n \to \infty$. The series converges if the largest of these limits is < 1 and diverges if it is > 1. (For $u_n = a_n z^n$ we obtain as a special case the earlier formula for the radius of convergence.) Cauchy's proof is a model of simplicity and clarity: in the first case we have as a majorant a convergent geometric series, and in the second case not every subsequence of the u_n converges to 0, i.e., the series diverges.

It is obvious that Riemann did not recall Cauchy's *Cours* when he was preparing his lectures. For one thing, the technical means he employed — the material was not intended for publication — are less general and less skillful than Cauchy's. Cauchy's account of series may have interested Riemann when he was a student attending Stern's lectures but did not accord with his later tendencies. When he decided, most likely for didactic considerations, to talk in his lectures about series, he worked everything out by himself.

We mention one more result, known in the textbook literature as *Riemann's theorem on removable singularities*: If $f(z)$ is bounded in a neighborhood of $z = a$ and holomorphic in that neighborhood with the possible exception of the point a, then one can change its value at a so that it is holomorphic there as well. Specifically, $g(z) = (z-a)f(z)$ is continuous and $h(z) = (z -$

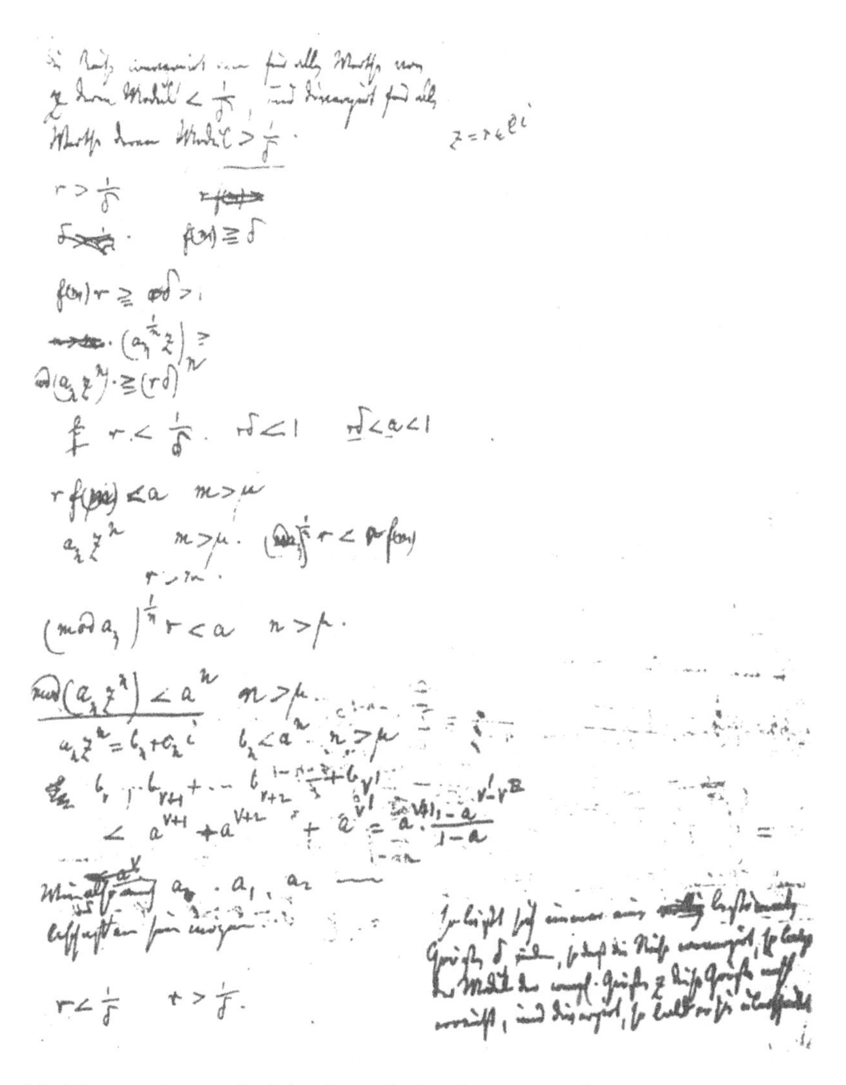

23: Riemann's proof of the formula for the radius of convergence $\rho = 1/\delta$. For a series with coefficients a_n, $f(m)$ denotes sup $\sqrt[n]{|a_n|}$ for $n > m$ and $\delta = \lim f(m)$. (Cod. Ms. Riemann 23.4, fol. 96^r, UB Göttingen)

$a)g(z) = h(a) + (z-a)g(z)$ is even differentiable with $h(a) = h'(a) = 0$. Thus $h(z) = (z-a)^2(c_2 + c_3(z-a)...)$. If we put $f(a) = c_2$, the required conclusion follows. This is a modern proof of the theorem.

Riemann stated this theorem in his dissertation (W. 23) and proved it in a more involved way, without using power series. It is possible to deduce it implicitly from his lecture notes. For Riemann the theorem was unimportant, or rather obvious. Its significance consists in the fact that a single point of discontinuity in the interior of a function's domain of holomorphy necessarily implies its unboundedness.

Incidentally, the preceding simple proof was not as obvious in the 19th century as it is for us today, and this is true of a great many theorems in a basic course in analysis. Around 1900 the *Bulletin of the American Mathematical Society* ran a number of discussions of difficulties in various proofs of our theorem [W.F. Osgood, "Some points in the elements of the theory of functions," 2 (1896), 296–302; E. Landau, "On a familiar theorem of the theory of functions," 12 (1905), 155–156; see also Remmert 1991, Bd.1, 167–168].

To get back to the history of power series. Cauchy's Turin theorem was used effectively in the 1840s in Paris, especially after having been generalized by Laurent in 1843: *Extension du Théorème de M. Cauchy relatif à la convergence du développement d'une fonction suivant les puissances ascendantes de la variable*. Cauchy himself developed his existence theorems for differential equations. After Newton, power series were firmly fixed in the consciousness of mathematicians as a fairly general class of functions. But now, given the mistaken notion that complex analysis encompasses real analysis, they were seen as the most general class of functions. This mistaken notion accommodated the penchant of contemporaries for the algorithmic, for operating with expressions.

During that decade, far from Paris, Karl Weierstrass was working in Eastern Prussia as a gymnasium teacher. While in Braunsberg, in the summer of 1853, he wrote a paper on Abelian functions which attracted a great deal of attention. He was brought to Berlin, made a member of the Academy in 1856, and appointed full professor in 1864.

Karl Weierstrass' (1815–1897) view of the centrality of power series was propagated by his many students. Riemann read Weierstrass' papers, published in 1856/57, in connection with the composition of his own paper on Abelian functions. Weierstrass defined an analytic function as the set of power series obtained from a single power series by analytic continuation along curves.

It is striking that in his first course of lectures in the winter term of 1855/56 Riemann introduced the Cauchy integral formula and, just as he had done in his dissertation, immediately went over to the Dirichlet principle, on which he based the remaining theory. A chapter on power series first turned up in his lectures in the winter term of 1856/57.

Neither Weierstrass nor his students seem to have noticed the Cauchy-Hadamard formula. Riemann did, and in this sense he may be said to have beaten the Weierstrass school at their own game, the exploration of power series.

Weierstrass clarified his principles once more in 1886, in a course of lectures titled *Ausgewählte Kapitel aus der Funktionenlehre (Selected topics in function theory)*. His summarizing statement was (Weierstrass 1886, 176): "The *ultimate* objective is always the representation of a function." ("Das *letzte* Ziel bildet immer die Darstellung einer Funktion"). At that time he made a concerted effort to apply this principle to real analysis as well. He thought of the approximation theorem he had stated in 1885 as a representation theorem: a function continuous on a compact set need not always be representable as a power series, but it is always representable as a uniformly convergent series of polynomials. For Weierstrass series were not just tools, they were the essence. He also made it clear in his lectures that he was aware of the closeness between his view of a function as an analytical expression and the view of a function that was dominant in the 18th century.

The contrast between the views of Weierstrass and Riemann could not have been sharper, and the well-intentioned attempts to reduce it, repeatedly undertaken from the time of Klein, were doomed to failure. For Riemann power series were just tools. They play virtually no role in his fundamental papers. He may have dealt with them in his lectures for pedagogical reasons: after all, power series supply examples students can appreciate and they help them to understand analytic continuaton. But otherwise they can be dispensed with.

A few years before the publication (in 1987) of Riemann's lecture notes, Neuenschwander (1980, 1981a,b) stated on a number of occasions that the lectures showed that Riemann had "advanced to a certain blending of 'Cauchy', 'Riemannian', and 'Weierstrassian' methods" (Neuenschwander 1980, 7). He based this assertion, which contradicts the view of mathematicians of the generation immediately following Riemann, on a few proofs in the lectures related to the identity theorem and to analytic continuation. But we think that conclusions about the scientific conceptions of a researcher cannot be based on the elementary parts of his or her lectures alone, just as one cannot conclude from a lecturer's occasional use of decimals that he or she attaches deep scientific value to the decimal representation of real numbers. In this book, and especially in Chapter 4, we can see that whenever possible Riemann eliminated special representations. For him, the essential thing was the differential equation and not representation by formulas, regardless of whether they involved series or the Cauchy integral formula. It was obvious to him how one continues solutions of linear differential equations, and so, for him, analytic continuation was likewise unproblematic. But he could not assume that his listeners had comparable knowledge and therefore made use of power series in accounts *ad usum Delphini*. Analytic continuation along a curve is a natural concept. Of course, it can not be ruled out that Riemann had read about it somewhere. As

24: Karl Weierstrass

we saw in the case of the Cauchy-Hadamard formula, and will see in other cases, Riemann frequently developed anew mathematical objects existing in the literature. He used to prepare his lectures during vacation time, far from Göttingen, and did not always have the necessary publications at hand.

Restriction of complex analysis to power series has conceptual as well as practical consequences. Important functions did not even enter the field of vision of Weierstrass and his students. One example is the zeta function. It has a pole at $z = 1$ and an essential singularity at $z = \infty$ and is holomorphic everywhere else. This means that it has a series representation

$$\zeta(z) = \frac{1}{z-1} + \sum_{n=0}^{\infty} c_n (z-1)^n$$

that is everywhere uniformly convergent. But there is not a word about any of this in the whole comprehensive theory of the zeta function. Indeed, this representation is of no use.

1.1.6 Further applications

Riemann dealt in considerable detail with applications of the integral formula. Suppose that $f(z)$ has singularities at finitely many $z = a_k$ and possibly at

$z = \infty$, or, as Riemann put it, is discontinuous or infinite there. He showed that $f(z) = \sum_0^\infty b_n z^n + \sum_1^\infty c_n (z - a_1)^{-n} +$ additional sums for finitely many a_k, with the first series converging for all finite z, the second for $z \neq a_1, \ldots$. The coefficients are

$$b_m = \frac{1}{2\pi} \int_0^{2\pi} \frac{f(z)}{z^m} \, d\varphi, \quad \text{and} \quad c_m = \frac{1}{2\pi} \int_0^{2\pi} f(z)(z - a_1)^m \, d\varphi, \ldots$$

respectively. In the first case the path of integration extends over a sufficiently large circle $z = R\, e^{\varphi i}$, in the second, over a sufficiently small circle $z - a_1 = r\, e^{\varphi i}, \ldots$. (The names "pole" and "essential singularity" are due to Briot-Bouquet and Weierstrass respectively. These singularities were of little importance to Riemann and he did not attach special names to them. The important things for him were branch points, and here his term was adopted).

Riemann stated a number of consequences:

1) $\lim\limits_{z=\infty} \left(\frac{f(z)}{z^{\mu+1}} \right) = 0$ implies $b_m = 0$ for $m \geq \mu + 1$.
2) If $\lim\limits_{z=a_k} (f(z)(z - a_k)^\mu)$ is not ∞, then this limit is 0 for $\mu = m$ and the series corresponding to a_k breaks off after the mth term.

He drew from all this the following conclusion, central to his viewpoint (Neuenschwander 1987, p. 47):

A single-valued function $f(z)$ is determined up to a constant if at each point at which it is discontinuous a function $\varphi(z)$ is given such that $f(z) - \varphi(z)$ remains bounded while $f(z)$ and $\varphi(z)$ become infinite.

In particular it follows that a function is rational if and only if it becomes infinite of finite order at, at most, finitely many points.

3) On the other hand, if $f(z)$ is finite for all finite values of z and infinite of finite order just at $z = \infty$, i.e., if

$$\lim_{z=\infty} \frac{f(z)}{z^{n+1}} = 0$$

for $n \geq m$, then $f(z)$ is an entire rational function of order m, for then only the first $m + 1$ terms of the first series remain.
4) If $f(z)$ is finite for all values of the argument including $z = \infty$, then $f(z)$ is constant.

The consequence 4 (Riemann derived it a second time directly from the integral formula) is usually named after Liouville, who proved it in his lectures around 1847 just for doubly periodic functions. Liouville's result was published only in 1879 (*Crelle's Journal* 88, 277), by Borchardt, who at one time sat in on his lectures. The theorem had been proved by Cauchy in full generality as early as 1844 (*Oeuvres* (1) 8, 378). For Riemann this result illustrated the fact that functions are basically determined by their singularities: a function without singularities is a constant.

The next application is the theorem that $\frac{1}{2\pi i}\int d\log f(z)$ is equal to the number of zeros of $f(z)$ (each counted with proper multiplicity) in the interior of the simple closed curve of integration supposed free of singularities of this function; if finitely many poles are present, then the value in question is equal to the number of zeros diminished by the number of poles (Neuenschwander 1987, 51).

From this result Riemann obtained the Fundamental Theorem of Algebra: an nth-order polynomial is infinite of order n at $z = \infty$, for its quotient by z^n is bounded. But then it must also be n times infinitely small of order one. Riemann called the attention of his listeners to Gauss' third proof (1819) of the Fundamental Theorem, which he wanted to relate to the function-theoretic proof.

It is possible that Riemann was unaware of the following simple indirect proof of the existence of a zero of a polynomial $f(z)$: if there were no zero, then $1/f(z)$ would be bounded and free of singularities, and thus a constant. (Incidentally, using this result and induction on the degree of a polynomial we can prove that it can be written as a product of linear factors.) It seems that, in spite of the fact that he had proved "Liouville's theorem," Cauchy too was unaware of this simple proof of the Fundamental Theorem of Algebra.

We continue our survey of Riemann's lectures, based on Neuenschwander 1987. The key theme of Riemann's conception is *the determination of a function from its singularities*. He clarifies it here for single-valued functions defined on $\mathbb{C}\cup\{\infty\}$ whose only singularities ("discontinuites" ("Unstetigkeiten")) are poles. It is immediately obvious that determination of the function also requires knowledge of its zeros. Now let $f(z)$ be a function whose zeros and poles are given in terms of position and order and let $g(z)$ be a rational function that shares the very same conditions. Then $f(z)/g(z)$ is bounded and therefore constant, and the constant is not 0.

Obviously, an essential assumption underlying the argument is the finiteness of the number of zeros and poles. Now Riemann goes over to the next question: to determine a function with infinitely many zeros whose only point of accumulation is ∞. What he is after is the product representation later named after Weierstrass. He uses a special case to explain the general procedure. He does it in such a way that by following his directions one could immediately give a proof of the Weierstrass product theorem.

We want to determine a function that has simple zeros at $z = 0, 1, 2, \ldots$ and is "finite for all finite z" ("für alle endlichen z endlich" ist). We cannot take the product of all $(z-n)$ because it violates the second condition. Consider $p_m(z) =$

$z(z-1)\dots(z-m)$ and $\log p_m(z) = \log z + \sum_1^m \log(1-\frac{z}{n}) + \sum_1^m \log(-n)$. Riemann notes that for $n > |z|$ the logarithmic series can be used, so that the following expansion is valid at least for sufficiently large n (but the expansion is set down for small n as well):

$$\sum_1^m \log(1-\frac{z}{n}) = -z\sum_1^m \frac{1}{n} - \frac{z^2}{2}\sum_1^m \frac{1}{n^2} - \frac{z^3}{3}\sum_1^m \frac{1}{n^3} - \cdots.$$

Here the first term on the right is "as infinite as" $-z\log m$. If in the expression for $\log p_m$ we drop the last sum on the right, then we obtain the function $g(z)$ such that

$$\log g(z) = \lim_{m=\infty}\left[\log z + \sum_1^m \log(1-\frac{z}{n}) + z\log m\right].$$

This function has the right zeros and "remains permanently finite" ("fortwährend endlich bleibt").

The road to this $g(z)$ is heuristic, but this is of no consequence to Riemann. All he wants is to find *some* function with the prescribed zeros. By contrast, Weierstrass always aims to obtain formula representations of *given* functions. Riemann also notes that if $f(z)$ is another function with the same zeros then $\log[f(z)/g(z)]$ is an entire function.

It seems that Riemann's listeners failed to pick up on his suggestion to investigate product expansions in general.

Another topic Riemann investigated in his lectures was the application of the method of residues to the evaluation of real definite integrals. He did this in more detail, and mentioned specifically that one can use this method to obtain the integrals computed earlier by Laplace. Here too he failed to mention the name of Cauchy, the inventor of the method.

This ends the introductory part of the lectures devoted to single-valued ("einwerthige") functions. Their study does not call for the concept of a Riemann surface. This material is now referred to, for good reason, as Cauchy's function theory. Actually, before Riemann nobody presented this material in this way, and the construction and content of his account can now be regarded as canonical. Konrad Knopp's little Göschen volume *Funktionentheorie. Erster Teil: Grundlagen der allgemeinen Theorie der analytischen Funktionen* (1. Aufl. 1913, 7. Aufl. 1949) virtually coincides with Riemann's introductory lectures [an English translation (by F. Bagemihl) of this material was published in two parts by Dover (in 1945 and in 1947) as *Theory of Functions*. (tr.)]. This book served for decades as the basis for lectures in German-speaking countries.

Very little had to be changed as time went on. As for concepts, it was found necessary to explicitly consider uniform convergence; moreover, notations were standardized, some proofs were simplified, and a few theorems — Casorati-Weierstrass, Rouché, Morera, the maximum modulus theorem — were added.

1.1.7 Multivalued functions and Riemann surfaces

Apart from their didactic presentation, the parts of the introductory lectures discussed so far fit within the framework of Cauchy's function theory. Of course, Riemann used consistently the Gaussian number plane and, more systematically than Cauchy, the differential equation $\frac{\partial w}{\partial y} = i\frac{\partial w}{\partial x}$. He had brought together existing materials and ideas in an elegant way.

We now come to the idea of a Riemann surface, the brilliant invention in Riemann's dissertation (W. 7–9) which was to make possible the further development and application of complex function theory. The historical context will be discussed in 1.2, but we insist, above all, that this was a flash of genius that cannot possibly be explained as a necessary consequence of the then available publications. The only way a modern mathematician, who has grown up with the idea of a function as a mapping, can properly appreciate the significance of Riemann's innovation is to realize that he was the first in the history of mathematics consistently to recognize the concept of a mapping as fundamental and to apply it immediately to the solution of classical open problems (on algebraic functions and their integrals; see 1.2). We are now ready to explain how the notion of a function as a mapping leads in a straightforward manner to that of a Riemann surface.

In the case of expressions such as $y^2 - x = 0$ we can put up for a while with the multivalued nature of $y = \pm\sqrt{x}$, and sometimes, say in the context of the solution formula for quadratic equations, such multivalued expressions are objectively justified. But in the case of problems such as integration it is mandatory that the integrand be uniquely determined at all points of the path of integration. To every point in the domain there must correspond just one functional value. As Riemann put it, even in the case of the single-valued integrand (bei der "einwertigen" Integrandenfunktion) $\frac{1}{z}$ the value $\log z$ of the integral is a multivalued function (ist der Integralwert eine "mehrwertige" Funktion). A multivalued function cannot be thought of as a mapping. On the other hand, functions like $\log z$ and $\sqrt{z}$ are needed. Hence the need for a new interpretation of multivalued functions.

Riemann's way out involves a change in the concept. In order to make $\sqrt{z}$ single-valued, we must, so to say, double the domain of definition at all points other than $z = 0$ and $z = \infty$, where the function is single-valued. Spread over the plane two infinitely thin surfaces — we will call them sheets — which are at first connected only at 0 and at ∞. Think of these two points, to be called branch points, as connected by a curve without double points. Slit each of the two sheets above the plane along the corresponding curve from 0 to ∞. Then join the "left edge" of the lower sheet to the "right edge" of the upper sheet and conversely. The result is a closed surface on which $\sqrt{z}$ can be defined as a single-valued function. This must be done in a continuous manner. If we begin at $z = re^{\varphi i}$ on the lower sheet with $w = \sqrt{r}e^{\varphi i/2}$ and traverse the complete circle of radius r centered at the origin, then we arrive at the point $z = re^{(\varphi+2\pi)i}$ on the upper sheet and associate with it the second square root $w = \sqrt{r}e^{\psi i/2}$ with $\psi = \varphi + 2\pi$. In the space model of this construction the two sheets penetrate each other along the cut-curve, but its points must be counted twice. Riemann saw in this no didactic difficulties. In the case of the logarithm we need infinitely many sheets, and the resulting surface involves no self-penetrations. Riemann wrote in 1857 (W. 90/91) (and practically the same text appears in the lecture notes (Neuenschwander 1987)):

> "A sheet of the surface continues around a branch point of the function into another [sheet], so that we can regard the surface in the neighborhood of such a point as a helicoidal surface with vertical axis at this point in the (x, y)-plane and infinitely small pitch. If after a number of traversals of z around the branch value the function takes on once more its former value ..., then, of course, we must assume that the topmost sheet of the surface continues through the other sheets into the lowest one,"
>
> ("Um einen Verzweigungspunkt der Function herum wird sich ein Blatt der Fläche in ein anderes fortsetzen, so dass in der Umgebung eines solchen Punktes die Fläche als eine Schraubenfläche mit einer in diesem Punkte auf der (x, y)-Ebene senkrechten Axe und unendlich kleiner Höhe des Schraubenganges betrachtet werden kann. Wenn die Function nach mehreren Umläufen des z um den Verzweigungswerth ihren vorigen Werth wieder erhält ..., muss man dann freilich annehmen, dass sich das oberste Blatt der Fläche durch die übrigen in das unterste fortsetzt,")

Clearly, this manner of forcing the single-valuedness of a function grew out of analysis — after all, one needs continuous curves for, besides other things, integration in the complex plane. One could expect algebraists to be less than pleased with the use of nonalgebraic tools for the treatment of algebraic functions. One such discontented algebraist was Dedekind, who tried to obtain the Riemann surfaces of algebraic functions in a purely algebraic way (see Section 1.3.8).

The first use Riemann made of his conceptual construct was to prove in the lectures a theorem that he barely hinted at in his dissertation (W. 39), namely, the characterization of the algebraic functions in terms of the compactness (closedness), to use a modern term, of their Riemann surfaces:

> "The common character of a species of functions, described in a similar manner in terms of operations on quantities, is reflected in the form of boundary and discontinuity conditions imposed on them. For example, if the region of variability of the quantity z extends simply or multiply over the whole infinite plane, and if, throughout the latter, the function is allowed to have discontinuities only at isolated points, where it can become infinite of finite order, then the function is necessarily algebraic, and, conversely, every algebraic function satisfies this requirement".
>
> ("Der gemeinsame Charakter einer Gattung von Functionen, welche auf ähnliche Art durch Grössenoperationen ausgedrückt werden, stellt sich dann dar in der Form der ihnen auferlegten Grenz- und Unstetigkeitsbedingungen. Wird z.B. das Gebiet der Veränderlichkeit der Grösse z über die ganze unendliche Ebene A einfach oder mehrfach erstreckt, und innerhalb derselben der Function nur in einzelnen Punkten eine Unstetigkeit, und zwar nur ein Unendlichwerden, dessen Ordnung endlich ist, gestattet, so ist die Function nothwendig algebraisch, und umgekehrt erfüllt diese Bedingung jede algebraische Function") (W. 39).

So much for the quotation from the dissertation. We recall that w is an algebraic function of z if $F(w, z) = 0$, F a polynomial. Suppose the degree of F as a polynomial in w is n and its degree in z is m. For every z there will be n roots w; the Riemann surface will have n sheets. The surface will be connected if and only if F is irreducible, i.e., if and only if F does not split into two (polynomial) factors.

Assume, conversely, that we are dealing with an n-sheeted Riemann surface. In his lectures Riemann assumes at first that w takes on for every z exactly n values $w_1, w_2, \dots, w_n$, which are finite at $z = \infty$, and that, invariably, just one of the w_k has a first-order pole at one of the $z = c_1, c_2, \dots, c_m$. The c_j must not be branch points. The claim is that the w_k are the roots of an algebraic equation $F(w, z) = 0$. (For this and for what follows see Neuenschwander 1987, 63–66).

Riemann forms the product $(s - w_1)(s - w_2) \cdots (s - w_n)(z - z_1) \cdots (z - z_m)$ and notes that this is a single-valued function of z in the whole plane which becomes infinite of order m at $z = \infty$, and so is an entire function of degree

m in z, and, at the same time, an entire function of degree n in s. He denotes it by $F(s, z)$. For $s = w_k$, $F = 0$. Riemann observes laconically that the initial restrictions (just first-order poles, boundedness at $z = \infty$, none of the c_j a branch point) can be dropped without invalidating the theorem. Nowadays we tend to avoid the use of the auxiliary variable s by exploiting the fact that the elementary symmetric functions of the w_k are single-valued, but this is just a modification of Riemann's idea of the proof.

At first the Riemann surface seemed to be just a means for making multi-valued functions, deprived of the status of functions by Cauchy's reform of the foundations of analysis, into legitimate objects: now to every value in the domain of definition there again corresponded just one functional value. One might have expected the followers of Cauchy to be the first to welcome the idea of a Riemann surface. But Briot and Bouquet rejected it as too difficult. Wherever necessary, they distinguished functional values belonging to different branches of the same function by means of indices, but this device was transparent only in the simplest cases. Weierstrass could never bring himself to approve of Riemann surfaces. In contrast to the French, who regarded them as a flash of genius, Weierstrass tried to deny completely their scientific justification by regarding them as mere "means of sensualization" ("Versinnlichungsmittel") (Weierstrass 1886, 144). While the well-intentioned Klein, with his tendency to illustrate by means of examples, was sometimes wide of the mark, he at least tried to publicize them in a popular way.

The old notion that geometry is an intruder in analysis had not yet died out in the second half of the 19th century. We saw that, in his lectures, Riemann managed immediately to demonstrate an astounding connection between algebra, analysis, and geometry: a function $w = f(z)$ is algebraic, i.e., an equation $\sum_{j,k=0}^{m,n} a_{jk} w^j z^k = 0$ holds, if and only if the corresponding Riemann surface is compact. (The notion of compactness was formulated later. In the present context it includes the one-point (∞) compactification of the plane, modeled by the number sphere as its stereographic image, and the finiteness of the number of sheets of the Riemann surface.)

The reader will find it useful to compare Riemann's proofs with the ones accepted today. As a rule, modern proofs of the theorem just stated are related to Riemann's idea, but include a great many details and assume that the reader is familiar with material that takes up hundreds of pages. By contrast, Riemann burdened his readers with this result after a few weeks of lectures.

The opposition to the mixing of geometry with algebra and analysis was fully justified as long as geometry was interpreted in the physical sense. But this practice was altered so radically by Hilbert that the great majority of mathematicians dropped their opposition. Riemann, as we explain in detail in

Chapter 3, anticipated this development. Once geometry was provided with a mathematical ("abstract," axiomatic, conceptual) basis, its connections with other branches of mathematics became particularly interesting and fruitful. Riemann's theorem is a case in point: the geometric-topological nature of a Riemann surface, its compactness, turned out to be equivalent to the algebraic property of its associated functions, including their analytic property of holomorphy. Hilbert's Fifth Problem of 1900 deals with another question of this kind: certain topological properties of groups are equivalent to the existence of an analytic structure.

One might be tempted to say that Riemann solved a problem that did not exist until the advent of the concept of a Riemann surface. But this is not so. Algebraic functions existed earlier, and so did the problem of changing their implicit form $F(w, z) = 0$ to an explicit form $w = f(z)$, and it is this problem that was solved by the concept of a Riemann surface.

An important subject that Riemann treated in his lectures very briefly was the expansion in powers of $(z - a)^{1/\nu}$ in the neighborhood of a branch point a, a result due to V. Puiseux (1822–1883). Riemann did not mention him in his papers but he did refer to him in the theses to his dissertation, as well as in his lectures of 1855/56. He was undoubtedly familiar with Cauchy's discussion of Puiseux' results in the *Comptes Rendus* of early 1851. As Riemann put it, w becomes (at least for $z \neq a$) a monodromic (holomorphic) function of the variable $\zeta = (z - a)^{1/\nu}$. One can state this as follows: locally, a new parameter ζ is introduced on the Riemann surface so that w is a holomorphic function of ζ in the neighborhood of $\zeta = 0$ (corresponding to $z = a$); 0 itself can be a singularity. Later such a parameter came to be known as a local uniformizing variable.

Puiseux dealt with algebraic functions defined by $F(w, z) = 0$ and showed, besides other things, that single-valued algebraic functions are rational and that — to use Riemann's language — a polynomial F is irreducible if and only if its Riemann surface is connected.

1.1.8 Doubly periodic functions

We insert here a short section on a class of functions that have played an important role in the history of mathematics. Of course, the actual road that led to these functions is very different from the one that we regard as natural today and which we shall sketch here. Periodic functions are especially important in real analysis of a single variable. Since $\mathbb{C}$ is a two-dimensional vector space over $\mathbb{R}$, it makes sense to look for functions with two linearly independent periods k_1 and $k_2 \in \mathbb{C}$, i.e. for functions $f(z)$ such that

$$f(z + mk_1 + nk_2) = f(z)$$

for all integral m and n. The account that follows is based on Stahl 1899, a version of Riemann's lectures. We choose a "period parallelogram" P with vertices $z_0, z_0 + k_1, z_0 + k_2$ and $z_0 + k_1 + k_2$ whose boundary is free of zeros and singularities of $f(z)$. The periodicity of $f(z)$ implies the sameness of the functional values on each pair of parallel sides of P. In turn, this implies that the contour integral $\int_P f(z\,dz)$ is equal to 0, and so too is the sum of the residues. If our function were free of singularities inside P as well as on its boundary, then it would be bounded in the whole plane and therefore constant. A nonconstant function with residue sum 0 could not have just one first-order pole inside P. We consider the simplest case: $f(z)$ has inside P exactly two first-order poles at $z_1 \neq z_2$ with respective residues A and $-A$ (bear in mind that $\int_P f(z\,dz) = 0$). Incidentally, since the contour integral $\int_P d\log f(z)$ is also 0, it follows that $f(z)$ has inside P exactly two simple zeros or just one double zero.

$f(z)$ behaves at z_1 like $\frac{A}{z-z_1}$ and at z_2 like $\frac{-A}{z-z_2}$. But then $f'(z)$ behaves at z_1 like $\frac{-A}{(z-z_1)^2}$ and at z_2 like $\frac{A}{(z-z_2)^2}$. Since, again by double periodicity, $\int d\log f'(z)$ taken over P also vanishes, it follows that $f'(z)$ has exactly four zeros in P. We are about to find them. Put $g(z) = f(z_1 + z_2 - z)$. $g(z)$ behaves at z_1 like $\frac{A}{z_2-z}$ and at z_2 like $\frac{-A}{z_1-z}$. Since the difference $f(z) - g(z)$ has no singularities, it is bounded, and therefore constant. Putting $z = (z_1+z_2)/2$ yields $f(z) = g(z)$, and thus $f(z_1+z_2-z) = f(z)$ and $f'(z_1+z_2-z) = -f'(z)$. But then $f'(z) = 0$ for $z_1 + z_2 - z = z + mk_1 + nk_2$. It follows that the four zeros in P are

$$\zeta_1 = \frac{z_1 + z_2}{2}, \quad \zeta_2 = \zeta_1 \pm \frac{k_1}{2}, \quad \zeta_3 = \zeta_1 \pm \frac{k_2}{2}, \quad \zeta_4 = \zeta_1 \pm \frac{k_1}{2} \pm \frac{k_2}{2}.$$

Incidentally, these zeros are the vertices of a parallelogram with sides half the length of the sides of P and parallel to the latter. Now $(f'(z))^2$ has fourth-order poles at z_1 and z_2 and double zeros at ζ_j, $j = 1, \ldots, 4$. Put $w = f(z)$ and $f(\zeta_j) = a_j$. Note that

$$w - a_1 = f(z) - f(\zeta_1) = f'(\zeta_1) + f(\zeta_1)^2(\ldots) \text{ etc.}$$

But then $w = f(z)$ and the function

$$(w - a_1)(w - a_2)(w - a_3)(w - a_4)$$

have the same poles and zeros in P. Hence the quotient of the two functions is constant and we have

$$f'(z) = \frac{dw}{dz} = c\sqrt{(w - a_1)(w - a_2)(w - a_3)(w - a_4)},$$

or

$$z = C \int^{w} \frac{dw}{\sqrt{(w-a_1)(w-a_2)(w-a_3)(w-a_4)}}. \tag{E}$$

On the right side is a so-called elliptic integral. Surprisingly enough, it is obtained as the inverse function of a doubly periodic function $w = f(z)$ of the simplest nontrivial type. These integrals turned up very early in connection with the computation of the circumference of an ellipse and with related problems, which explains their name. The inverse functions of elliptic integrals, and more generally the doubly periodic functions, were then called elliptic functions.

This argument, which used only Cauchy's function theory, led us to integrals of the form $\int dx/\sqrt{P(x)}$, $P(x)$ a polynomial. Historically it was such integrals that provided the starting point for the theory.

Incidentally, it makes little difference whether the polynomial is a cubic or a quartic. Indeed, putting $x = 1/t$ we obtain

$$\frac{dx}{\sqrt{x(x-a)(x-b)(x-c)}} = \frac{-dt}{\sqrt{(1-at)(1-bt)(1-ct)}}.$$

From the very beginning of the calculus mathematicians were aware that in the case of quadratic polynomials all one needs is logarithms and trigonometric functions. As early as 1694 Jacob Bernoulli encountered $P(x) = 1 - x^4$ in connection with a mechanical problem, the so-called elastica, and immediately provided a geometric interpretation. In rectangular coordinates, the equation of a lemniscate, a "reclining figure eight," is $(x^2+y^2)^2 = 2(x^2-y^2)$. In polar coordinates it is $r^2 = 2\cos 2\varphi$. For its arclength Bernoulli obtained

$$u = b(x) = \int_0^x \frac{dt}{\sqrt{1-t^4}}.$$

Here there are formal analogies with the unit circle, for which we have

$$\arcsin x = \int_0^x \frac{dt}{\sqrt{1-t^2}}.$$

For $x = 1$ we get the quarter circle, and so the inverse function, the sine, is periodic with period

$$2\pi = 4\int_0^1 \frac{dt}{\sqrt{1-t^2}}.$$

Similarly, the inverse function of $b(x)$, the lemniscatic sine $\operatorname{sl} u$, has period

$$2\omega = 4\int_0^1 \frac{dt}{\sqrt{1-t^4}}.$$

If we replace t in the integral by it, then we see immediately that the lemniscatic sine has the additional period $2i\omega$. This was first observed by the young Gauss, a century after the first appearance of the integral.

A splendid illustration of Riemann's mode of thought is that we can see "practically without computing" ("fast ohne Rechnung") that if $P(z)$ is a quartic polynomial with four different zeros $z_1, \ldots, z_4$, then the inverse function of the elliptic integral $\int dz/\sqrt{P(z)}$ is necessarily doubly periodic.

This is so because, due to the square root, the Riemann surface of the integrand is two-sheeted over the plane compactified by the addition of ∞ and its only singularities are branch points at $z_1, \ldots, z_4$. Make nonintersecting cuts from z_1 to z_2 and from z_3 to z_4 and join the sheets crosswise along these cuts. The result is a torus. Indeed, since each of the slit sheets has two holes and is closed at ∞ it is topologically a piece of a cylinder, and joining the edges of these two cylinder pieces yields a torus. Let one crosscut be a closed meridian curve C_1 and another a parallel C_2. Let the values of the integral over these two curves be ω_1 and ω_2 respectively. Since traversal of each of these curves brings us back to the starting point, it follows that the inverse function must have the two periods ω_1 and ω_2.

For higher-order polynomials everything gets to be much more complicated. Riemann envisaged the more complicated Abelian integrals in his dissertation (§20; see Section 1.2.1 below).

The following is another fairly natural way of obtaining doubly periodic functions. The series $\sum_{-\infty}^{+\infty} a_n e^{2\pi inz}$ are functions with period 1. Can one choose the coefficients a_n so that another, complex, period appears? Following Jacobi, Riemann pursued this question in his lectures (Stahl 1899, third section). This line of thought leads to the theta functions; the name is due to Jacobi's denoting them by the letter theta. While the function

$$\vartheta(z) = \sum_{n=-\infty}^{+\infty} (-1)^n q^{n^2} e^{2\pi inz}$$

is not itself doubly periodic, one can obtain from it such functions by forming suitable quotients. Thus for $q = e^{i\pi\tau}$, $\vartheta(z + \frac{1}{2})/\vartheta(z)$ has the periods 1 and τ.

Klein (1926, 43) made fun of the subsequent development: "There is a whole science of so-called theta-relations, i.e., of identities connecting theta functions with suitably chosen arguments, which were for some time a prime hunting ground for all mathematicians." ("Es gibt nun eine ganze Lehre von den sog. Thetarelationen, d.h. von den Identitäten, die zwischen den Thetafunktionen geeignet ausgewählter Argumente statthaben, die eine Zeit lang ein Hauptjagdgebiet für alle Mathematiker bildete"), and regarded it as a mere fashion. Of course, there are arguments in favor of these functions: they converge very rapidly for im $\tau > 0$ and are solutions of the heat equation. Riemann, who used series only when it was unavoidable, used them in his papers, beginning in 1857, in the case of $p > 1$ arguments. What had given Jacobi great pleasure was that these functions have number-theoretic significance as well.

We conclude with a look at the state of affairs at the time when Riemann wrote his dissertation. In 1851 the Paris Academy awarded a prize to a paper by J.G. Rosenhain (1816–1887) "Sur les fonctions de deux variables et à quatre périodes, qui sont les intégrales ultra-elliptiques de la première classe." The competition, involving the problem of representation of Abelian integrals of the first kind by a formula, had been announced as early as 1846, and G.A. Göpel's (1812–1847) complete solution appeared in 1847 in *Crelle* 35, 277–312. Göpel died shortly before the publication of his paper, and Jacobi wrote in his honor an obituary note which was printed immediately following the paper, in *Crelle* 35, 313–318. In it Jacobi said that he had never met Göpel, who studied in Berlin when Jacobi was in Königsberg and who later worked modestly as a librarian in Berlin, and he went on to praise Göpel's achievement in these words: "Masterly is the manner in which he found the differential equations, and, undaunted by their complexity, used a suitable substitution to put them in the required form of systems of hyperelliptic differential equations set down by me, and thereby completely settled the posed problem." ("Meisterhaft ist die Art, wie er die Differentialgleichungen welche er findet, ungeschreckt von ihrer Complication, durch eine passende Substitution in die verlangte Form der von mir aufgestellten Systeme der hyperelliptischen Differentialgleichungen bringt und dadurch das gestellte Problem vollständig erledigt.")

This paper appeared in Crelle's Berlin journal when Riemann was a student in Berlin. It is possible that it too provided a motive for his attempt to escape from this huge network of formulas and to find a more conceptual and transparent method of handling the problem. We will see, however, that in his 1851 dissertation he did not solve any concrete problems of the kind just mentioned.

1.2 The dissertation of 1851

1.2.1 Riemann's view of the motives for the paper: Article 20 of the dissertation, Part I

We are indeed very lucky that in his very first publication, the dissertation of 1851, Riemann himself explains his view of how his paper ties into the historical tradition and what role he assigns to his own contributions.

We begin by quoting the first half of Article 20 (W. 37/38); the letters in parentheses refer to the explanations that follow the quotation. Riemann's footnotes are included in the quoted material.

"The origin and immediate cause of the introduction of complex quantities into mathematics is the theory of simple* laws of dependence among variable quantities, expressed by operations on quantities. [* Here we regard as elementary operations addition and subtraction, multiplication and division, integration and differentiation, and view a law of dependence as being the simpler, the fewer elementary operations the dependence requires. In fact, all the functions that have been used until now in analysis can be defined by means of a finite number of these operations.] *(A)*

If we use these laws of dependence on an extended scale by assigning complex values to the variable quantities to which they apply, then a harmony and regularity emerge that otherwise remain hidden. *(B)*

Thus far, the cases in which this has happened encompass a small domain — virtually all of them can be reduced to laws of dependence between two variables in which one is an algebraic** function of the other [** i.e., the two are connected by an algebraic equation] or a function whose differential quotient is an algebraic function *(C)* — but almost every step taken here has not only given a simpler and more unified form to results obtained without complex quantities *(D)*, but has also paved the way for new discoveries, a fact proved by the history of investigations of algebraic functions, trigonometric or exponential functions *(E)*, elliptic and Abelian functions *(F)*."

("Die Einführung der complexen Grössen in die Mathematik hat ihren Ursprung und nächsten Zweck in der Theorie einfacher* durch Grössenoperationen ausgedrückter Abhängigkeitsgesetze zwischen veränderlich-en Grössen. [* Wir betrachten hier als Elementaroperationen Addition und Subtraction, Multiplication und Division, Integration und Differentiation, und ein Abhängigkeitsgesetz als desto einfacher, durch je weniger Elementaroperationen die Abhängigkeit bedingt wird. In der That lassen sich durch eine endliche Anzahl dieser Operationen alle bis jetzt in der Analysis benutzten Functionen definiren.] *(A)*

Wendet man nämlich diese Abhängigkeitsgesetze in einem erweiterten Umfange an, indem man den veränderlichen Grössen, auf welche sie sich beziehen, complexe Werthe gibt, so tritt eine sonst versteckt bleibende Harmonie und Regelmässigkeit hervor. *(B)*

Die Fälle in denen dies geschehen ist, umfassen zwar bis jetzt erst ein kleines Gebiet - sie lassen sich fast sämtlich auf diejenigen Abhängigkeitsgesetze zwischen zwei veränderlichen Grössen zurückführen, wo die eine entweder eine algebraische** Function der anderen ist [** d.h. wo zwischen beiden eine algebraische Gleichung Statt findet] oder eine solche Function, deren Differentialquotient eine algebraische Function ist *(C)* -, aber beinahe jeder Schritt, der hier gethan ist, hat nicht bloss den ohne Hülfe der complexen Grössen gewonnenen Resultaten eine einfachere, geschlossenere Gestalt gegeben *(D)*, sondern auch zu neuen Entdeckungen die Bahn gebrochen, wozu die Geschichte der Untersuchungen über algebraische Functionen, Kreis- oder Exponentialfunctionen *(E)*, elliptische oder Abel'sche Functionen *(F)* den Beleg liefert.")

(A) Let us add to the four rational operations the definition $i^2 = -1$, which is based on them. Adjunction of i to the field $\mathbb{R}$ yields the field $\mathbb{C}$ in which every polynomial can be written as a product of linear factors. By including integration and differentiation among the elementary operations, Riemann goes immediately beyond the purely algebraic fact stated by us in modern terms. His key interest is *analysis*, which is primarily concerned with *functions*. His view of the constructive or recursive buildup of the supply of functions in analysis by means of single or multiple applications of these operations is noteworthy in that it leaves out infinite series, in particular power series, as a way of defining functions. Nor are functional equations mentioned as a tool for generating functions.

(B) Here complex analysis is seen as a means to a better understanding of real analysis. An oft-cited example is the dependence of the radius of convergence of a series expansion on the complex singularities of the represented function, but it soon becomes clear that Riemann has other things in mind as well.

(C) Algebraic relations such as $F(x, y) = 0$, $G(x, y, y') = 0$, and perhaps also $H(x, y, y', y'') = 0$, lead far beyond the obvious examples of the logarithm, the exponential function, and the trigonometric functions. On the other hand, it is preferable to obtain, say, the gamma function using the generating principle in (A) rather than the Euler integral.

(D) This begins with the Fundamental Theorem of Algebra. To mention examples that are not part and parcel of Riemann's function theory we should perhaps point to the substantial simplification resulting from the use of complex notation in Fourier series and integrals.

(E) The most conspicuous early example is the Euler formula $e^{it} = \cos t + i \sin t$.

(F) *Elliptic and Abelian functions* bring Riemann to the circle of topics that is the actual starting point for his complex analysis, and that is basically different from Cauchy's entrance into the subject. We saw that Cauchy began with the computation of definite integrals required in mathematical physics at the beginning of the 19th century. On the other hand, the history of elliptic functions and of their generalizations by Abel transpired largely within "pure" mathematics. We limit our discussion of this issue to the relevant hints in the previous section and recommend the account of salient points of the historical development in Stillwell 1989, 152–166. Of course, we will devote more attention (in 1.3.3) to a high point of the development known as Abel's Theorem, because it is of decisive importance for the approach to Riemann's paper of 1857.

1.2.2 A short account of the contents of the dissertation

The doctoral dissertation of 1851 is titled *Grundlagen für eine allgemeine Theorie der Functionen einer veränderlichen complexen Grösse (Foundations of a general theory of functions of a single variable complex quantity)* (W. 3–43). It is of modest size. Having given an account of Riemann's terminology in the introductory sections in connection with his lectures, and having compared it with its modern version, we can afford from now on to use the modern terms with which the reader is more familiar. We do not go here into the part of the dissertation discussed in 1.1.

Riemann defines holomorphic functions as complex single-valued functions on Riemann surfaces satisfying the Cauchy-Riemann differential equations. There can be isolated singularities. If they appear at finite points they must be poles. Such meromorphic functions are viewed as conformal mappings between two Riemann surfaces. We must always think of the complex plane as extended by the addition of the point ∞ (as the Riemann complex number sphere or as a complex projective straight line).

Functions must be thought of not as given by expressions but as determined (to within arbitrary constants) by the positions and nature of their singularities. This leads to the question of the construction of functions with prescribed properties on a given Riemann surface. Here the topology of the surface is of decisive importance. The surface T is decomposed by means of n crosscuts into a system of m simply connected surface pieces. The number $n - m$, which is independent of the manner of decomposition, is called the order of connectivity of T (W. 10–11); incidentally, in modern terms, this number is equal to the negative of the Euler characteristic of T.

In order to construct appropriate functions on T, Riemann uses a variational principle. (He called it later the Dirichlet principle because he came to know similar procedures in Dirichlet's lectures, and the historically unjustified name stuck.) First T is made into a simply connected surface T^* by means of crosscuts. Then, subject to suitable boundary conditions, the integral

$$\int [(u_x - v_y)^2 + (u_y + v_x)^2]\, dx\, dy$$

is minimized on this surface. If there are singularities to be taken into consideration, then the integral is somewhat modified. With the possible exception of the boundary of T^*, the pair of functions u, v associated with the minimum is a holomorphic function $f = u + iv$. It should be noted that the functional values on the two edges of a crosscut need not coincide; jumps ("periods") may occur.

The paper ends with an application of these methods to the Riemann mapping theorem. This theorem asserts that the topological equivalence of two surfaces or regions implies their conformal equivalence, i.e., the existence of a conformal mapping between them. Here the theorem is first stated for regions in the complex plane that are homeomorphic to a circular disk.

We will examine the individual key words while considering further developments in the work of Riemann and others.

We explain briefly, in modern terms, the form of inference Riemann learned from Dirichlet. Let $I(\varphi, \psi)$ be the integral of $\varphi_x \psi_x + \varphi_y \psi_y$ over a region G and let $J(\varphi) = I(\varphi, \varphi)$. Let η be a function that vanishes on the boundary ∂G of G.

$$J(\varphi + t\eta) = J(\varphi) + 2tI(\varphi, \eta) + t^2 J(\eta)$$

implies that if $J(\varphi) \leq J(\varphi + t\eta)$ is to hold for all t, then we must have $I(\varphi, \eta) = 0$. Put $\Delta\varphi = \varphi_{xx} + \varphi_{yy}$. Our last result, the vanishing of η on ∂G, and the Gauss integral formula (Gauss' theorem) imply that

$$0 = \int_{\partial G} (\varphi_x \eta\, dy - \varphi_y \eta\, dx) = \int_G (\Delta\varphi)\eta\, dF + I(\varphi, \eta) = \int_G (\Delta\varphi)\eta\, dF\,.$$

Since this holds for every η, it follows that $\Delta\varphi = 0$. In other words, a function that minimizes $J(\varphi)$ is a solution of $\Delta\varphi = 0$. To be sure, the argument does not prove the *existence* of such a function, and this elicited justified criticism.

It is relatively easy to prove the uniqueness of the solution of the boundary-value problem. If ψ were another solution, then $\eta = \varphi - \psi$ would vanish on

∂G. Moreover,

$$J(\varphi) = J(\psi) + 2I(\psi, \eta) + J(\eta)$$

and

$$I(\psi, \eta) = \int_{\partial G} \eta(\psi_x \, dy - \psi_y \, dx) - \int_G (\Delta\psi)\eta \, dF = 0 .$$

But then

$$J(\varphi) = J(\psi) + J(\eta) \geq J(\psi) .$$

In view of the minimality of $J(\varphi)$, the inequality sign in the last expression must be replaced by an equality sign. But then $J(\eta) = 0$, i.e., $\eta_x = \eta_y = 0$. Since $\eta = 0$ on ∂G, it follows that $\eta = 0$, and therefore $\psi = \varphi$ throughout G.

1.2.3 Riemann's summary of the dissertation. The program: Article 20, and Part II, Article 22

Riemann continues (W. 38; as before, the letters in parentheses refer to our subsequent comments):

> "We are about to indicate briefly what the theory of such functions gains from our investigations.
>
> The previous methods of treating such functions always set down as a definition an *expression* for the function, whereby its value was given for *every* value of its argument *(A)*; our investigation shows that, as a result of the general character of a function of a variable complex quantity, in such a definition some of the data are a consequence of the remaining ones, namely, the proportion of data has been reduced to those indispensable for the determination. This simplifies significantly the treatment of the latter. For example, in order to show the equality of two expressions for the same function it was usually necessary to transform one of them into the other, i.e., one had to show that they agreed for every value of the variable quantity; now it is enough to establish their agreement for a much smaller proportion [of values]. *(B)*
>
> A theory of these functions on the basis provided here would determine the presentation of a function (i.e., its value for every value of its argument) independently of its mode of determination by operations on quantities, because one would add to the general concept of a function of a variable complex quantity just the attributes necessary for the determination of the function, and only then would one

go over to the different expressions the function is fit for. *(C)* The common character of a type of functions expressed in a similar way by operations on quantities shows itself in the form of boundary and discontinuity conditions imposed on them. *(D)* For example, if the region of variability of the quantity z extends simply or multiply over the whole infinite plane A, and if, within it, the function can have a discontinuity only at several points, namely, just an infinity of finite order (we add that for an infinite z this quantity itself is infinite of the first order, but for every finite value z' of this quantity, $\frac{1}{z-z'}$ is infinite of the first order), then the function is necessarily algebraic, and conversely, every algebraic function satisfies this condition.

However, we now refrain from the realization of this theory, intended, as noted, to bring to the fore simple laws of dependence conditioned by operations on quantities, for we rule out, at present, consideration of an expression of a function. *(E)*

For the same reason, it is not our concern here to set forth the utility of our theorems as a foundation for a *general* theory of these laws of dependence, for this would require a proof that the concept of a function of a variable complex quantity adopted here as a basis coincides completely with that of a dependence* expressible by operations on quantities. [* By this is meant every dependence that can be expressed by means of a finite or infinite number of the four simplest computing operations, addition and subtraction, multiplication and division. The expression "operations on quantities" (in contrast to "operations on numbers") is to refer to computing operations in which the commensurability of the quantities is irrelevant.]" *(F)*

("Es soll kurz angedeutet werden, was durch unsere Untersuchungen für die Theorie solcher Functionen gewonnen ist.

Die bisherigen Methoden, solche Functionen zu behandeln, legten stets als Definition einen *Ausdruck* der Function zu Grunde, wodurch ihr Werth für *jeden* Werth ihres Arguments gegeben wurde *(A)*; durch unsere Untersuchung ist gezeigt, dass, in Folge des allgemeinen Charakters einer Function einer veränderlichen complexen Grösse, in einer Definition dieser Art ein Theil der Bestimmungsstücke eine Folge der übrigen ist, und zwar ist der Umfang der Bestimmungsstücke auf die zur Bestimmung nothwendigen zuruckgeführt worden. Dies vereinfacht die Behandlung derselben wesentlich. Um z.B. die Gleichheit zweier Ausdrücke derselben Function zu beweisen, musste man sonst den einen in den andern transformiren, d.h. zeigen, dass beide für jeden Werth der veränderlichen Grösse übereinstimmen; jetzt genügt der Nachweis ihrer Übereinstimmung in einem weit geringern Umfange. *(B)*

Eine Theorie dieser Functionen auf den hier gelieferten Grundlagen würde die Gestaltung der Function (d.h. ihren Werth für jeden Werth ihres Arguments) unabhängig von einer Bestimmungsweise derselben durch Grössenoperationen festlegen, indem zu dem allgemeinen Begriffe einer Function einer veränderlichen complexen Grösse nur die zur Bestimmung der Function nothwendigen Merkmale hinzugefügt würden, und dann erst zu den verschiedenen Ausdrücken deren die Function fähig ist übergehen. *(C)* Der gemeinsame Charakter einer Gattung von Functionen, welche auf ähnliche Art durch Grössenoperationen ausgedrückt werden, stellt sich dann dar in der Form der ihnen auferlegten Grenz- und Unstetigkeitsbedingungen. *(D)* Wird z.B. das Gebiet der Veränderlichkeit der Grösse z über die ganze unendliche Ebene A einfach oder mehrfach erstreckt, und innerhalb derselben der Function nur in einzelnen Punkten eine Unstetigkeit, und zwar nur ein Unendlichwerden, dessen Ordnung endlich ist, gestattet (wobei für ein unendliches z diese Grösse selbst, für jeden endlichen Werth z' derselben aber $\frac{1}{z-z'}$ als ein unendlich Grosses erster Ordnung gilt), so ist die Function nothwendig algebraisch, und umgekehrt erfüllt diese Bedingung jede algebraische Function.

Die Ausführung dieser Theorie, welche, wie bemerkt, einfache durch Grössenoperationen bedingte Abhängigkeitsgesetze ins Licht zu setzen bestimmt ist, unterlassen wir indes jetzt, da wir die Betrachtung eines Ausdruckes einer Function gegenwärtig ausschliessen. *(E)*

Aus demselben Grunde befassen wir uns hier auch nicht damit, die Brauchbarkeit unserer Sätze als Grundlagen einer *allgemeinen* Theorie dieser Abhängigkeitsgesetze darzuthun, wozu der Beweis erfordert wird, dass der hier zu Grunde gelegte Begriff einer Function einer veränderlichen complexen Grösse mit dem einer durch Grössenoperationen ausdrückbaren Abhängigkeit* völlig zusammenfällt. [* Es wird darunter jede durch eine endliche oder unendliche Anzahl der vier einfachsten Rechnungsoperationen, Addition und Subtraction, Multiplication und Division, ausdrückbare Abhängigkeit begriffen. Der Ausdruck Grössenoperationen soll (im Gegensatze zu Zahlenoperationen) solche Rechnungsoperationen andeuten, bei denen die Commensurabilität der Grössen nicht in Betracht kommt.]" *(F)*)

We now quote Article 22 because it too is programmatic in nature; it is related to the special version of the Riemann mapping theorem presented in Article 21.

> "We omit here the complete implementation of the investigation in the previous Article, which is devoted to the more general case in which to one point of one surface there correspond several points of the other and the surfaces are not assumed to be simply connected. The reason for this is that, from a geometrical viewpoint, our whole investigation could have been conducted in a more general form. The restriction to plane surfaces which are schlicht with the exception of several points is not essential; what is more, the problem of mapping one arbitrarily given surface on another, subject to similarity in the smallest parts, can be dealt with in quite an analogous way. In this matter we content ourselves with referring to two of Gauss' papers, namely the one quoted in connection with Article 3 and the disquis. gen. circa superf. art. 13." *(G)*.
>
> ("Die vollständige Durchführung der Untersuchung des vorigen Artikels für den allgemeinern Fall, wo Einem Punkte der einen Fläche mehrere Punkte der andern entsprechen sollen, und ein einfacher Zusammenhang für dieselben nicht vorausgesetzt wird, unterlassen wir hier, zumal da, aus geometrischem Gesichtspunkte aufgefasst, unsere ganze Untersuchung sich in einer allgemeinern Gestalt hätte führen lassen. Die Beschränkung auf ebene, einzelne Punkte ausgenommen, schlichte Flächen, ist nämlich für dieselbe nicht wesentlich; vielmehr gestattet die Aufgabe, eine beliebig gegebene Fläche auf einer andern beliebig gegebenen in den kleinsten Theilen ähnlich abzubilden, eine ganz ähnliche Behandlung. Wir begnügen uns, hierüber auf zwei *Gauss'sche* Abhandlungen, die zu Art. 3 citirte und die disquis. gen. circa superf. art.13, zu verweisen.")

(A) One is inclined to refer to Cauchy and Laurent, who also drew conclusions not from expressions but from properties of a function, namely the continuity of f and f'. The example of the Cauchy integral formula is relevant for (B), for it implies that in order to determine a function in a simply connected region it suffices to prescribe its values on the boundary of that region. It is also conceivable that Riemann had in mind his leitmotiv, the determination of a function by its singularities.

We supplement Riemann's heading for Article 20 (in the table of contents (W. 45)), in which again only the reduction of the number of determining components is taken into consideration, but neither the subsequent program nor the preceding historical summary:

> "20. The previous mode of determination of a function by means of operations on quantities contains superfluous components. As a result of the considerations carried out here, the number of determining components of a function has been reduced to the necessary size."
>
> ("20. Die frühere Bestimmungsweise einer Function durch Grössenoperationen enthält überflüssige Bestandtheile. Durch die hier durchgeführten Betrachtungen ist der Umfang der Bestimmungsstücke einer Function auf das nothwendige Mass zuruckgeführt.")

(C) Here we also find precursors. It suffices to recall the determination of a function by functional equations (say the gamma function) or by differential equations that give rise to series expansions for the solutions. But what Riemann has in mind becomes clear in (D), where he tries to characterize whole classes of functions by boundary conditions and singularities. The subsequent example, the characterization of algebraic functions by the compactness or closedness of the corresponding Riemann surfaces, is illuminating. It is not established in the paper. This omission is justified in a strange way (E).

(E) and (F) are programmatic and, strictly speaking, can be understood only in retrospect, on the basis of the knowledge of Riemann's later papers. The second half of the sentence in (F) could be interpreted as a reference to the expandability of functions in infinite series (power series, Laurent series), in infinite products, and in continued fractions; but Riemann did not return to the two latter representations in general form. The editor's footnote to this clause shows that in his time too it was not clear what Riemann meant.

Riemann must have known that, in general, the sum of a convergent series of rational functions need not even be continuous. Thus the series whose partial sums are $\frac{1-z^n}{1+z^n}$ has the sum 1 for $|z| < 1$, the sum -1 for $|z| > 1$, and the sum 0 for $z = 1$. On the other hand, using Cauchy's function theory one can easily prove that Riemann's assertion of the equivalence of holomorphy and representability by "operations on quantities" ("Grössenoperationen") is at least locally correct (and it cannot possibly hold on multi-sheeted Riemann surfaces). To prove it in one direction we can use the local expandibility of a holomorphic function in a power series. To prove it in the other direction we can use the following immediate consequence of the Cauchy integral formula: the limit of a sequence of holomorphic functions which converges uniformly in a region and on its boundary is holomorphic. We saw in the account of Riemann's lectures that, while he did not explicitly formulate the notion of uniform convergence, he used it implicitly.

Regardless of what Riemann had in mind when he stated the vague formulation, the following is of interest: while rejecting the assumption that functions are given by expressions, he claims that it is possible to establish the equivalence of the concept of holomorphy with that of functions expressible by operations on quantities! We can compare this with the attitude of Weierstrass, for whom expressions for functions were invariably the starting point. Even when he drew conclusions from other properties of functions, such as continuity, he adhered to the credo: the ultimate aim is always the representation of a function (Laugwitz 1992).

Since Riemann returned stubbornly and repeatedly to the topic of representation of a function by an expression, it is relevant to mention the remarkable view which he expressed at the very beginning of the dissertation (W. 3). With reference to real-valued functions he said:

> "Formerly one attributed the property of being determinable by a single law of dependence for all values of z lying within a given interval to just a certain type of functions (functiones continuae, in Euler's terminology); but recent investigations have shown that there are analytic expressions by which every continuous function can be represented on a given interval."
>
> ("Die Fähigkeit, für alle innerhalb eines gegebenen Intervalles liegenden Werthe von z durch dasselbe Abhängigkeitsgesetz bestimmt zu werden, schrieb man früher nur einer gewissen Gattung von Functionen zu (functiones continuae, in Eulers Sprachgebrauch); neue Untersuchungen haben indess gezeigt, dass es analytische Ausdrücke giebt, durch welche eine jede stetige Function für ein gegebenes Intervall dargestellt werden kann.")

He probably had in mind representations by Fourier series. Continuity implies representability by an analytic expression!

Incidentally, continuity is defined as follows (W. 46):

> "By the expression 'the quantity w varies continuously with z between the limits $z = a$ and $z = b$' we mean the following: in this interval, to every infinitely small change in z there corresponds an infinitely small change in w, or to put it more tangibly: for an arbitrarily given quantity ϵ one can always choose a quantity α so that, within an interval for z smaller than α, the difference between two values of w is never greater than ϵ."
>
> ("Unter dem Ausdruck die Grösse w ändert sich stetig mit z zwischen den Grenzen $z = a$ und $z = b$ verstehen wir: in diesem Intervall

> entspricht jeder unendlich kleinen Änderung von z eine unendlich kleine Änderung von w, oder, greiflicher ausgedrückt: für eine beliebig gegebene Grösse ϵ lässt sich stets die Grösse α so annehmen, dass innerhalb eines Intervalles für z, welches kleiner als α ist, der Unterschied zweier Werthe von w nie grösser als ϵ ist.")

We would call this uniform continuity.

In 1851 Riemann still thinks (W. 3–4) that

> "It therefore makes no difference whether one defines the dependence of the quantity w on the quantity z as arbitrarily given or as subject to definite operations on quantities. The two concepts are congruent as a result of the theorems mentioned."
>
> ("Es ist daher einerlei, ob man die Abhängigkeit der Grösse w von der Grösse z als eine willkürlich gegebene oder als eine durch bestimmte Grössenoperationen bedingte definirt. Beide Begriffe sind in Folge der erwähnten Theoreme congruent.")

(G) Riemann says that his mapping theorem holds in greater generality, and that his proof can be carried over by means of the Dirichlet principle. The reference to Article 13 of Gauss' surface theory of 1827 is very strange, for that article has nothing to do with conformal mapping. Rather, it deals with the intrinsic geometry of surfaces and thus belongs to the prehistory of Riemann's differential geometry but not to function theory. The only reason for regarding it as relevant here is that the requirement of embedding the surface in three-dimensional space is waived in it.

1.2.4 On the prehistory of the dissertation

Nowadays a doctoral dissertation must include a complete list of all source materials and show the author's familiarity with the relevant previous works. It was different in Riemann's time, in spite of the relative paucity of publications to be consulted. But even if we take into consideration the then usual bibliographical incompleteness, it is strange that Riemann cites no sources other than Gauss, and even some of these have no recognizable connection with his own paper.

We discussed the influence of Gauss' publications on Riemann in Sections 1.1.2 and 1.1.3. Now we wish to consider the influence on Riemann of the works of Cauchy and Dirichlet, and of mathematical physics.

Is it possible that when he wrote his dissertation Riemann did not know of the comprehensive works of Cauchy and of the younger Frenchmen? Did he rediscover all of their results? These are questions we propose to investigate carefully.

We know (see Neuenschwander 1981, 91) that already as a first-year student Riemann borrowed from the Göttingen university library (besides other works) Cauchy's *Cours d'analyse* and various issues of his *Exercices de mathématiques*, Legendre's *Elliptic Functions*, and Moigno's Cauchy-based *Differential Calculus*. At the very least, he must have gathered from the *Exercices* that Cauchy had greatly advanced complex analysis, and from the *Cours* that he had dealt with complex power series. It is safe to assume that during the two years 1847/49 in Berlin Riemann continued to watch for Cauchy's papers.

We know so little about his Berlin years that every hint, however questionable, must be followed up. For example, there is the well-known anecdote in E.T. Bell's *Men of Mathematics* (first published in 1936 and reprinted many times since), in the chapter on Riemann titled "Anima candida," that involves J.J. Sylvester (1814–1897). What supposedly happened (in 1896) is this: Sylvester lived "in a hotel on the river at Nuremberg, where I conversed outside with a Berlin bookseller, bound, like myself, for Prague ... He told me he was formerly a fellow pupil of Riemann, at the University, and that, one day, after receipt of some numbers of the *Comptes Rendus* from Paris, the latter shut himself up for some weeks, and when he returned to the society of his friends said (referring to the newly published papers of Cauchy), 'This is a new mathematic'."

This seemingly obscure source is not necessarily unreliable. The old Sylvester must have immediately jotted down the incident; he died soon after (on 15 March 1897). An examination of the relevant numbers of the *Comptes Rendus* shows that they actually contain very important results of Cauchy. They are all found in Cauchy, *Oeuvres* (1) 10, and the page numbers below refer to this material.

Of course, finding the essential in this material is tiresome and may have taken Riemann a great deal of time. Cauchy's style was no longer as systematic and clear as in his textbooks from the 1820s. He gives in detail a great deal of material which strikes us as unimportant, holds forth on his past accomplishments, and, on the other hand, sometimes takes it for granted that the reader is familiar with certain of his former considerations.

It is a far from simple task to decide what assumptions Cauchy made about his complex functions, i.e., how he defined at that time what later came to be known as holomorphic functions. On p. 137, in a note of 21 September 1846, we see that he assumes the existence of an antiderivative $F(z)$ of a given complex function $f(z)$. Today every student knows that this is a correct local characterization of the holomorphy of $f(z)$, now known as Morera's theorem.

Everything else follows from it with the help of theorems obtained by Cauchy in the early 1830s during his Turin exile. Apart from all this, an expert on the *Cours d'analyse* could infer from the Cauchy(-Hadamard) formula for the coefficients that a convergent power series has an antiderivative in the interior of the circle of convergence that can be obtained from the series by termwise integration.

Around that time Cauchy had brought himself to use, temporarily, the complex number plane (pp. 136/137) and the notations $z = x + y\sqrt{-1}$ and $f(z) = u + v\sqrt{-1}$. He mentions later, in connection with Gauss and Kummer, that the German mathematicians use the letter i for the symbol $\sqrt{-1}$ (p. 313, note of 28 June 1847). In contrast to his earlier manner, he now presents his arguments geometrically, and he discusses the logarithm function as a "very simple application." He notes as early as 10 August (p. 76) that it is useful to fix the principal value of the logarithm by restricting the imaginary part to the interval from $-\pi$ (exclusive) to $+\pi$ (inclusive).

What prompted Cauchy to resume work in complex analysis was a paper by A. Lamarle that appeared in the April 1846 number of *Liouville's Journal* (see Belhoste 1991, 207). Cauchy acknowledges this on p. 79. Lamarle had tried to re-prove Cauchy's Turin theorems and cut the plane along the positive real axis, whereas we, like Cauchy, prefer to cut it along the negative real axis. Of course, it is a far cry from the use of a single sheet on which a function is single-valued to splicing several sheets along cuts and to constructing a Riemann surface, and neither Cauchy nor his successors dared to take this step. But it is possible that Cauchy inspired Riemann's idea.

The first of a series of Cauchy's communications appeared on 3 August 1846 (p. 70) and included careful arguments on line integrals. Cauchy considers boundary curves of pieces of surfaces and discusses the significance of orientation. He states Green's integral formula and includes the result that if a surface is decomposed by means of properly oriented curves, then the integral over the boundary is the sum of the integrals over the curves used in the decomposition. He applies this to complex functions on 10 August (p. 75 ff.).

In view of the Cauchy-Riemann differential equations, the Cauchy integral theorem is an immediate consequence of Green's formula. Cauchy does not mention this explicitly, but it must have been obvious to him, especially after the assumption of the existence of an antiderivative. He notes immediately that due to the achieved single-valuedness of the logarithm one can make new and more extensive (nouvelles et plus étendues) applications of, besides other things, the residue calculus (p. 76); he states the residue theorem, and he computes improper integrals with integrands of the form $(f(x))^{\mu} g(x)$ for which one can actually use the now single-valued logarithm. He can now replace an integral over a closed path by the sum of integrals over infinitely small circles centered at the finitely many singularities in the interior, and can thus use his beloved singular integrals (p. 133; 14 September). He immediately announces an extension of this concept: having joined singular points by means of a line one should think of the latter as enclosed by an infinitely narrow curve that can be used as a path of integration. Obviously, changes of the path of integration that do not involve crossing a singularity do not change the value of the integral.

Integrands of the kind just considered occur in elliptic integrals, and Cauchy is aware of their relevance for such integrals (pp. 166/167; 12 October). This is essential for Riemann's papers. Cauchy says that he has completely surmounted the serious difficulties in the theory of elliptic functions and Abelian integrals publicized by Eisenstein in *Crelle* 17 and in *Liouville's Journal* Vol.10. Using his, i.e., Cauchy's principles, one can now correct the erroneous concepts, hitherto thoughtlessly tolerated by mathematicians. In the case of such integrals it is inadmissible to speak carelessly about inverse functions. Rather, says Cauchy, one must speak of the complete integral of the relevant differential equation $dt = f(x)dx$, which must be considered in the complex plane and integrated along paths, and he mentions the different "periodicity indices." He can explain the periodicity of the inverse function of $\int_a^t f(z)\,dz$ by the fact that after traversing the closed path several times $f(z)$ takes on once more the same value.

While this program was not carried out by Cauchy in detail, we can, with hindsight, discern in it essential thoughts related to Riemann's subsequent considerations.

We have Riemann's own evidence that he knew the relevant papers of Cauchy and of other French mathematicians before he was awarded the doctoral degree. Neuenschwander (1980, 9) notes in an appendix that Riemann wrote the following in a sketch of the defense of his dissertation:

> "This view was expressed and elaborated on in several subsequent lectures by Cauchy, who had concerned himself with the theory of complex quantities before, and to a greater extent, than other Frenchmen, at the meeting of the Paris Academy that took place on 31 March of this year on the occasion of a report on a paper by Puiseux" (see also Neuenschwander 1981a, 226).
>
> ("Diese Ansicht ist von Cauchy, welcher sich unter den Franzosen zuerst und am meisten mit der Theorie der complexen Grössen beschäftigt hat, in der Sitzung der Par[iser] Ak[ademie] v[om] 31. März dieses Jahres bei Gelegenheit eines Berichts über eine Arbeit von

> Puiseux ausgesprochen worden und in mehreren folgenden Vorträgen weiter ausgeführt")

Unfortunately this is the only excerpt published by Neuenschwander, who just wanted to prove the undisputed fact that Riemann knew Cauchy's papers at the time when he wrote his dissertation. The unanswered question is, which of Cauchy's views Riemann is referring to. It is possible that he has in mind the difference between their views on complex functions, a point he was to make in his lectures ten years later (see 1.1.3).

It is safe to guess that Riemann had not read Cauchy carefully enough. Indeed, in the previously mentioned note Cauchy immediately states a definition of a derived function and gives the Cauchy-Riemann equations as conditions for the differentiability of a function. Two subsequent notes, dated respectively 7 April and 12 May 1851, make it clear that Cauchy uses this condition to reprove the expandability in a Laurent series and the residue theorem. Even if we ignore all of Cauchy's previous papers it is obvious that he had clearly elaborated the fundamental significance of the Cauchy-Riemann differential equations before Riemann was awarded the doctorate and that Riemann knew these notes at that time.

We pointed out in Section 1.1.3 that ten years later Riemann stated in his lectures that his definition differed from that of Cauchy, who presumably regarded every function $u(x, y)+iv(x, y)$, u, v real, as a function of the complex variable $x+iy$, and then proceeded to formulate the restriction to "monogenic functions." Actually, Cauchy never *worked* with any other functions. Moreover, as we saw, in previous years he used conditions equivalent to the concept of a monogenic function, such as the existence of an antiderivative.

Another safe conclusion is the following: since Riemann knew at the end of 1851 that the *Comptes Rendus* contained relevant papers — after all, he speaks also of other Frenchmen — he must by then, at the latest, have consulted other annual volumes. At any rate, one can reasonably expect this of a doctoral candidate, all the more so because the journal literature was at that time surveyable virtually at a glance. The whole thing is strange.

We note in passing that we do not know whether the disputation set for 16 December 1851 (Dedekind, W. 545) took place, and if so, in what form. On 24 November Riemann asked the faculty

> "to dispense with the disputation required for the attainment of the doctorate"
>
> ("um Dispensation von der zur Erlangung der Doctorwürde erforderlichen Disputation").

Weber supported him. "In the case under consideration I have misgivings about the disputation if only because the faculty will definitely not be in a position to produce an opponent if the candidate himself does not find one among his friends. Such a predicament may very easily arise." ("In dem vorliegenden Falle finde ich gegen die Disputation schon darum ein Bedenken, weil wenn der Candidat nicht selbst unter seinen Freunden einen Opponenten findet, die Facultät gar nicht im Stande sein würde, einen Opponenten zu stellen. Die Verlegenheit dürfte sehr leicht eintreten.") But Gauss made a terse pronouncement: "Without a disputation a Doctor is not *rite promotus*." ("Ohne Disputation ist ein Doctor kein rite promotus.") After all, Riemann wanted to become a university instructor (Privatdozent). (All this is based on Remmert 1991, 47. On p. 48 there is a complaint that the theses had not been announced.) Incidentally, the outlines of the theses mentioned by Neuenschwander 1980, 9 are completely different from the six theses published by Noether as early as 1902 (N. 710/711). We will discuss the latter in 3.2 because their content is primarily physical. These theses are so constructed that it would have been easy to find an opponent. They have nothing in common with the content of the dissertation.

Another strange story is associated with the "Dirichlet Principle." Riemann used it for the first time in his lectures, specifically at the time when Gauss died and Dirichlet was called to Göttingen to take his place. In this connection Neuenschwander (1981a, 225) cites a draft of a letter by Riemann in which he acknowledges that Gauss, Thomson, and Kirchhoff had also used this method, and that he named it after Dirichlet when the latter told him that he had dealt with it in his lectures already in the early 1840s.

On 6 February 1882 Prym, a former student of Riemann, wrote a letter to Felix Klein (Neuenschwander 1981a, 228, footnote) which bears not only on this issue but also on the more general issue of the origin of Riemann's function theory.

Prym reports on conversations with Riemann in Pisa early in 1865. Riemann was presumably led to his theory of complex functions by the following observation: it can happen that relations obtained from series expansions of functions retain their validity outside the regions of convergence of these series; in particular, like Euler, one can obtain correct results by operating with divergent series. In this connection Riemann presumably asked himself

> "what actually continues functions from region to region."
>
> ("was denn eigentlich die Functionen aus dem einen Gebiete in das andere fortsetze.")

He arrived at the insight that it was the partial differential equation, and Dirichlet agreed with him.

Nothing in Prym's letter tells us how Riemann arrived at this far-from-obvious insight, which had escaped all his predecessors. It was, in Prym's view, a decisive flash of Riemann's genius, and it occurred during his Berlin student years. Again according to Prym, a necessary consequence of this insight into the significance of the partial differential equation was that he availed himself of the tools of his teacher Dirichlet's potential theory. But Riemann's dissertation contains no reference whatsoever to Dirichlet.

Klein wanted Prym to confirm his oft-stated guess that Riemann had come to his function theory through mathematical physics, but this Prym would not do. The insight into the centrality of the differential equation led naturally to the application of the methods of potential theory, in which it also plays the key role.

Reconstruction of the prehistory of the issue is complicated by additional facts. While in Berlin, Riemann took Dirichlet's course in partial differential equations but not the one in potential theory (see Neuenschwander 1981a, 225), and in the note previously mentioned he refers not to lectures but to Dirichlet's personal communications. Another curious fact is that in spite of the obvious relevance of Gauss' paper on forces that behave like r^{-2}, Riemann does not mention him even once in this connection.

Incidentally, Riemann must be blamed for the historically completely inappropriate naming after Dirichlet of a certain class of boundary value problems associated with partial differential equations.

More important for us is the part of Prym's letter that names divergent series of functions as another source of Riemann's function theory, though they are not mentioned as such in his dissertation. Euler was one of those who dealt with such series. (We will find it necessary to consider this subject from a different viewpoint in 1.5 and in the chapter on real analysis.) Viewed in retrospect, Euler's formulation is tantamount to the principle of analytic continuation: if an expression $f(z)$ is equal to (the sum of) a series in some region of the variable z, then it can be regarded as its value outside its region of convergence. Historically, this issue could have given rise to two groups of questions, one associated with the identity theorem and the other with the discussion of manyvalued functions. Both belong to the question of whether one can assign a unique value to a divergent series — a possibility of which Euler was convinced on the basis of his experience with such series. While the course of history was different, it is nevertheless possible to establish a connection between the respective methodologies of Euler and Riemann, namely the recourse to differential equations; of course, Euler and others, for example Lagrange, used ordinary rather than partial differential equations in special cases of continuation.

In Riemann's case this view of continuation predated that involving chains of circles and power series expansions, which he may have introduced in his later lectures because they were more easily understood by students. As mentioned earlier, Neuenschwander (1980, 1981a,b) attributes the more frequent appearance of power series in Riemann's work in later years to the influence of Weierstrass. But Riemann regarded as important the "principal parts" of Laurent and Puiseux series, i.e., the parts with negative exponents, for these determined the nature of the singularities.

While there are no references to mathematical physics in the dissertation and in the 1857 papers based on it, it is certain that Riemann's methods, and possibly also his formulations of questions, were strongly influenced by the mathematical procedures of the physics of his time. This is borne out by a fragment (Nr. XXVI, W. 440–446) (unfortunately undated) in which he enters into two physical formulations of questions derived from electricity and heat. These questions have the following background.

Beginning in the middle of the 18th century, the importance of partial differential equations, enhanced by the requirements of physics, became ever clearer. As a result, they became one of the principal topics of research in the first decades of the 19th century. Paris was then a leading research center whose most active members included Poisson, Cauchy, and above all Fourier. The young Dirichlet was strongly influenced by Fourier.

We know of Riemann's early interest in physics. Already during his first term at the university he took an interest in Gauss' papers on terrestrial magnetism. This is shown by the items he borrowed from the library. Between January 1850 and April 1851 he took out four times the 1840 yearbook of the magnetism society containing Gauss' paper on forces that behave like r^{-2}, although he did not quote it in the dissertation (Neuenschwander 1981b, 90).

The mathematical object that played a role everywhere was the Laplacian

$$\Delta u = u_{xx} + u_{yy} + u_{zz}.$$

Time-dependent problems appeared largely in mathematical modeling exemplified by $u_t = c^2 \Delta u$ (the heat equation) and by $u_{tt} = c^2 \Delta u$ (the wave equation). For time-independent problems in two variables they reduce to $u_{xx} + u_{yy} = 0$. Hindsight makes obvious the connection with complex functions, but the fact is that Riemann was the first to use it systematically.

In Riemann's note, elaborated by Heinrich Weber, we read the following (W. 440):

"Consider the problem of determining the steady-state distribution of static electricity or of temperature in infinitely long cylindrical conductors with parallel generators, subject to the assumption that in the former case it is the distribution forces and in the latter case the surface temperatures which are constant along straight lines parallel to the generators. This problem is solved as soon as we find a solution of the following mathematical problem:

Let S be a plane, connected, and simply extended surface bounded by arbitrary curves. Determine a function u of rectangular coordinates x, y that satisfies the differential equation

$$\frac{\partial^2 u}{\partial^2 x} + \frac{\partial^2 u}{\partial^2 y} = 0$$

in the interior of the surface S and takes on arbitrary prescribed values on its boundaries."

("Das Problem, die Vertheilung der statischen Elektricität oder der Temperatur im stationären Zustand in unendlichen cylindrischen Leitern mit parallelen Erzeugenden zu bestimmen, vorausgesetzt, dass im ersteren Fall die vertheilenden Kräfte, im letzteren die Temperaturen der Oberflächen constant sind längs geraden Linien, die zu den Erzeugenden parallel sind, ist gelöst, sobald eine Lösung der folgenden mathematischen Aufgabe gefunden ist:

In einer ebenen, zusammenhängenden, einfach ausgebreiteten, aber von beliebigen Curven begrenzten Fläche S eine Function u der rechtwinkligen Coordinaten x, y so zu bestimmen, dass sie im Innern der Fläche S der Differentialgleichung genügt:

$$\frac{\partial^2 u}{\partial^2 x} + \frac{\partial^2 u}{\partial^2 y} = 0$$

und an den Grenzen beliebige vorgeschriebene Werthe annimmt.")

This is known as the first problem of potential theory. The Paris solution methods were based on Fourier series and integrals, and one obtained two representations of the solution. Specifically, let S be the unit disk centered at the origin, let r, φ be polar coordinates, and let $U(\varphi)$ be the prescribed boundary values for $r = 1$. Then one solution is given by the series

$$u(r, \varphi) = \frac{a_0}{2} + \sum_{n=1}^{\infty} a_n \cos n\varphi + b_n \sin n\varphi = \sum_{n=-\infty}^{\infty} c_n e^{in\varphi},$$

where a_n, b_n, c_n are the Fourier coefficients of the boundary values, and the other by the Poisson integral

$$u(r, \varphi) = \int_0^{2\pi} P(r, \varphi - \theta)\, U(\theta)\, d\theta$$

with kernel $P(r, \alpha)$. The Cauchy integral formula tells us that the kernel is the real part of

$$\frac{1}{2\pi i} \int \frac{f(\varphi)}{\zeta - z}\, d\zeta$$

taken over the unit circle $\zeta = e^{i\theta}$ with $z = re^{i\varphi}$. As for $f(\zeta)$, it, or at least its real part, must be equal to the prescribed boundary function $U(\theta)$. (The fact that we are using here the integral formula in a somewhat more general way than warranted by its initial derivation is of no importance because the boundary values can be rather arbitrary.) The series representation can also be viewed as the real part of a complex power series, in which case u is the real part of $f(z)$.

The Paris mathematicians were interested in the solution of as many problems as possible, including three-dimensional ones, and they looked for series and integral representations. Since the connection with the complex realm is helpful only for special problems, it was of marginal interest.

The picture is very different if, like Riemann, one knows the role of conformal mappings, for then it is clear that the solution of the boundary value problem for the unit disk can be carried over to all simply connected regions with boundary by using the Riemann mapping theorem. But to prove the latter one had to rely on the Dirichlet principle.

All these considerations were certainly well known to Riemann but none is found in his notes. He immediately considers multiply connected regions and later assumes that all boundary curves are circles. He is looking for a function

> "that is real on all boundary curves of S, becomes infinite of first order at one point of each of these curves, but is otherwise finite and continuous throughout the surface S."
>
> ("welche an sämtlichen Grenzcurven von S nur reell ist, in je einem Punkt einer jeden dieser Grenzcurven unendlich von der ersten Ordnung wird, übrigens aber in der ganzen Fläche S endlich und stetig bleibt.")

With Poisson in mind, one is likely to think of the real part of $\sum_k \frac{1}{z-z_k}$, where z_k is on the kth boundary curve. But Riemann asks for a complex function. If one of the circles is the real axis, then $\frac{1}{z-z_k}$ yields what is needed. For each of the other circles one must bring in a conformal mapping; more specifically, a linear fractional transformation. If there are n boundary curves then the required function is a quotient of two polynomials of degree n. If z_k is introduced as a function of the parameter, then we obtain an analogue of the Poisson kernel. But Riemann argues differently. After all, our simple construction does not satisfy his requirement that the function should be real-valued on all circles; as it is, each summand is real-valued just on its circle.

In summary, we have by now the following picture of the origin of the dissertation. Already after his first term at Berlin, during the summer vacation of 1847, Riemann was so thoroughly immersed in the task of introducing complex numbers into function theory that in a letter to his father, dated 29 November 1847, he wrote about a virtually completed paper devoted to the solution of a problem formulated by Eisenstein and about indications of its possible publication with the help of Jacobi (Neuenschwander 1981c). If all he had wanted to do was to convince his father of the fruitfulness of his Berlin studies he could have been far more vague. We know from Dedekind's recollection (W. 544) and from Prym's letter that he wanted the partial differential equation to be the new foundation that would replace the formula-based treatment. It is possible that this was the time when he talked to Dirichlet about the connection between the values of divergent series and continuation determined by the partial differential equation. He must have known that the circle of problems he was preoccupied with included conformal mappings, for he tried resolutely to find Gauss' paper of 1825 (and eventually succeeded, at the beginning of 1849). In spite of the anecdotal character of the source in E.T. Bell we can take it for granted that he studied intensively the *Comptes Rendus*. Of course, it is beyond doubt that he attended Jacobi's lectures on elliptic functions and that he discussed this topic with Eisenstein. The building blocks for his future dissertation were available before his return to Göttingen for the summer term of 1849. There he attended no mathematical lectures.

This was the second time the young Riemann wrote home about publication possibilities. The first such communication dates from 1847 and involves his paper W. 353–366. We do not know why the first possibility came to naught and we do not know why he declined Jacobi's offer of publication of his new deliberations. Had he become less self-confident? Did he think he had too little to offer? Did he want to use his method to find further results himself? The latter is plausible, for he did not want to talk to Eisenstein about his approach. Thus the announcement of his methods and results was delayed until the end of 1851. The only person in Göttingen who could have encouraged him to publish without delay was Gauss. We know nothing about how much Gauss knew about Riemann's dissertation before its submission.

We have added this material on the evolution of Riemann's dissertation in order to supplement his scant remarks about the motives behind his work in complex analysis.

We postpone discussion of the topological component in the dissertation, but mention now that Riemann got the idea of using cuts to make a surface into a simply connected surface during a private conversation with Gauss that focussed on other matters. This vague hint is due to Betti (A. Weil, 1979).

1.2.5 The effect of the dissertation

Today we are inclined to regard Riemann's dissertation as one of the most important achievements of 19th-century mathematics, but its immediate effect was rather slight. We saw that, in the second part of Article 20, Riemann himself emphasized just one principle, namely the determination of a function by as few data as possible and the elimination of expressions as definitions of functions. Given its vague formulation, this principle must have struck his contemporaries as neither new nor interesting. Riemann was as restrained in his statement as he was in the specification of his sources.

The first person who had to read the paper carefully was the referee for the Göttingen faculty, that is, Gauss. His report read as follows: "The paper submitted by Herr Riemann is a concise testimony to its author's thorough and penetrating studies of the area to which the subject treated therein belongs; of a diligent and ambitious, truly mathematical spirit of investigation, and of praiseworthy and fertile independence. The report is prudent and concise, and in places even elegant; nevertheless, most readers might well wish for even greater transparency of arrangement in some of the parts. Taken in its entirety, it is a solid and valuable work which not only meets the requirements usually set for test papers for the attainment of the doctorate but exceeds them by far." ("Die von Herrn Riemann eingereichte Schrift legt ein bündiges Zeugniß ab von den gründlichen und tief eindringenden Studien des Verf. in demjenigen Gebiete, welchem der darin behandelte Gegenstand angehört; von einem strebsamen echt mathematischen Forschungsgeiste, und von einer rühmlichen productiven Selbstthätigkeit. Der Vortrag ist umsichtig und concis, theilweise selbst elegant: der größte Theil der Leser möchte indeß wohl in einigen Theilen noch eine größere Durchsichtigkeit der Anordnung wünschen. Das Ganze ist eine gediegene werthvolle Arbeit, das Maaß der Anforderungen, welche man gewöhnlich an Probeschriften zur Erlangung der Doctorwürde stellt, nicht bloß erfüllend sondern weit überragend.")

If one has a certain amount of experience with evaluations and forgets for a moment that here the *princeps mathematicorum* is writing about a person des-

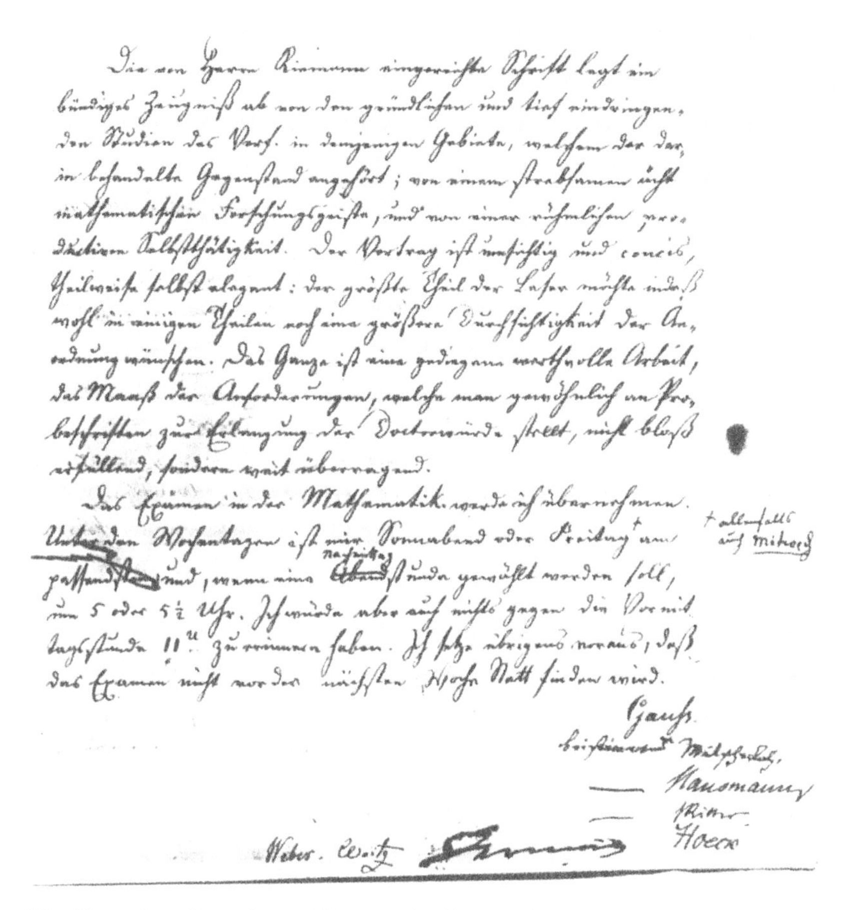

Die von Herrn Riemann eingereichte Schrift legt ein bündiges Zeugniß ab von den gründlichen und tief eindringenden Studien des Verf. in demjenigen Gebiete, welchem der darin behandelte Gegenstand angehört; von einem strebsamen ächt mathematischen Forschungsgeiste, und von einer rühmlichen productiven Selbstthätigkeit. Der Vortrag ist umsichtig und concis, theilweise selbst elegant: der grösste Theil der Leser möchte indeß wohl in einigen Theilen noch eine grössere Durchsichtigkeit der Anordnung wünschen. Das Ganze ist eine gediegene werthvolle Arbeit, das Maaß der Anforderungen, welche man gewöhnlich an Probeschriften zur Erlangung der Doctorwürde stellt, nicht bloß erfüllend, sondern weit überragend.

Das Examen in der Mathematik werde ich übernehmen. Unter den Wochentagen ist mir Sonnabend oder Freitag[+] am passendsten, und, wenn eine Abendstunde gewählt werden soll, um 5 oder 5½ Uhr. Ich würde aber auch nichts gegen die Vormittagsstunde 11 U. zu erinnern haben. Ich setze übrigens voraus, daß das Examen nicht vor der nächsten Woche statt finden wird.

+ allenfalls auch Mittwoch

Gauß

Hausmann

Ritter

Hoeck

Weber. Waitz

25: Gauss' testimonial on Riemann's dissertation

tined to become probably the most distinguished of his students, then one gets the following impression. The referee recognizes that the author has penetrated deep into a highly specialized field and has done this with great diligence, independently, and without the referee having to suggest the topic to him. There is no mention of the author's new ideas, of the solution of problems, or of new methods, but it is recognized that he may well be showing signs of independent research activity. The presentation is terse, elegant only in spots, and on the whole not clear enough. An objective reader must wonder what was the basis for the "Doktorvater's" (doctoral adviser) very positive overall evaluation stated in the last sentence. Riemann wrote to his brother:

> "When I visited Gauss he had not yet read my paper, but he told me that for years he had been preparing a paper (and is occupied with

> this right now) whose subject is the same, or partly the same, as the one I am treating."
>
> ("Als ich bei Gauss war, hatte er meine Abhandlung noch nicht gelesen, sagte mir aber, dass er seit Jahren eine Schrift vorbereite (und gerade jetzt damit beschäftigt sei), deren Gegenstand derselbe, oder doch zum Theil derselbe sei, wie der von mir behandelte.")

(Incidentally, this passage was quoted by Schering in his memorial address in 1866 (N. 835).) So far, no one has been able to find any indication that Gauss had discussed with Riemann the contents of his paper or had given him any hints or suggestions. Riemann would have reported such things. After all, he mentioned the rather disappointing conversation with Gauss which comes down, more or less, to this: right now I happen to be writing on a related topic, but your paper has not interested me enough that I should immediately and eagerly plunge into it.

Some (e.g., Remmert (1991, Band 2, 158)) think that the old Gauss was "chary of praise" ("lobkarg"). But against this is the fact that a few years earlier he had praised young Eisenstein to the skies. We will make no guesses about the great Gauss' admittedly baffling behavior toward Riemann.

We summarize the essential mathematical concerns that originated in Riemann's dissertation.

(1) The idea of a Riemann surface. Here, for the first time, the domain of definition of a function becomes one of the data that determine it. The complex plane is compactified by the addition of a single point ∞, the Riemann surfaces over it are precisely defined, the connectivity number is introduced and recognized as a topological invariant. (Complex) analysis is carried out not locally but on manifolds, which are compact in the case of algebraic functions. Local representability (by power series) is proved but is of secondary importance.

(2) In addition to poles, branch points are recognized as characteristic types of singularities, and the local series expansions in terms of (negative or fractional) powers are rigorously justified (Article 13/14, W. 24–27).

(3) The existence (together with the continuity) of $f'(z)$ is equivalent to the Cauchy–Riemann differential equations (together with the continuity of the occurring partial derivatives) and to the conformal character of f. It is also equivalent to the local expandibility, which implies the existence of all derivatives. (Holomorphic or analytic functions.)

(4) The transformation of surface integrals into line integrals is a tool for proving theorems (Articles 7–12, W. 12–24) of the "Cauchy type."

(5) The ("Dirichlet") principle of the existence of a function that minimizes a surface integral is used to solve boundary-value problems by means of holomorphic functions.

(6) The Riemann mapping theorem is a consequence of (5).

The response of contemporaries was amazingly slight; hardly any of the more than 500 titles in Purkert's list covering the period from 1851 to 1891 (N. 869–895) and relevant to Riemann's dissertation appeared before his death. This is all the more surprising if we keep in mind that two of Riemann's papers that presented the ideas of his dissertation in greater detail and applied them to the solution of problems appeared in 1857. Things were no different when it comes to textooks. For example, Heinrich Weber's *Elliptische Functionen* of 1891 contains nothing relating to Riemann. Thus one can hardly speak of a significant impact of Riemann's ideas during his lifetime and in the first 25 years after his death. In the subsequent sections we will examine the question of the very special directions in which Riemann influenced research and the question of which elements of his essential ideas failed initially to attract attention.

Let us return to the year of the composition of the dissertation. Jacobi died on 18 February 1851. Dirichlet pushed Riemann in another direction, which led to his habilitation paper on trigonometric series. Representatives of the algorithmic direction could hardly be expected to approve of Riemann's dissertation. Eisenstein died on 11 October 1852 and Weierstrass had not yet appeared on the scene. The French mathematicians, whose contributions were not explicitly acknowledged in the dissertation, could at best be expected to recognize the concept of a Riemann surface as new. At the same time, they viewed it as too complicated and superfluous. Moreover, Cauchy's students soon got used to working with complex functions in the complex plane in much the same way as Cauchy, who had used complex formulations for his integral theorems and for his method of residues as early as 1831. They must have regarded the method of real partial differential equations as a backward step. At the time doubly periodic functions were in fashion, and they could be dealt with without the use of Riemann surfaces.

Of course, in time the six previously listed key issues associated with the dissertation exerted a powerful effect. What follows is a survey describing this effect.

The effect of (6) was later especially notable in applied mathematics. For a disk, the first boundary value problem for the potential equation $u_{xx} + u_{yy} = 0$ is solved by the Poisson integral which expresses the function u in terms of its boundary values. Since the differential equation is invariant under conformal mappings, we obtain a solution of this problem for any simply connected region bounded by a curve by mapping the disk conformally onto this region. But this is just an existence statement, and Riemann's theorem does not directly yield a formula representing the solution. Such representations were eventually obtained for regions of practical importance by H.A. Schwarz, E.B. Christoffel, and others.

The mapping theorem became effective in many respects independently of applications and of the other objectives and contents of the dissertation. It is an instance of Riemann's novel view of mathematics. For one thing, it illustrates the fruitfulness of the notion that functions are simply mappings. For another, it is a global proposition; all Gauss could prove was the conformal equivalence of small pieces of surfaces. Finally it was one of the deeper existence theorems to emerge after Cauchy's existence theorems about solutions of differential equations. For adherents of algorithms this was an unusual type of proposition; indeed, *they* took note of transformations only if they were associated with effective formulas. It is also noteworthy that the theorem shows that the theory of functions on a simply connected region with boundary is completely independent of the special choice of a region. When investigating a special class of functions we can choose a convenient special region, say the upper halfplane.

Riemann's sketch of a proof in §21 is cryptic, and not only because he used the Dirichlet principle. Efforts to fully justify the idea of his proof failed. Given the importance of the theorem for applications, this failure stimulated attempts to develop new methods of proof. These remarks also apply to the uniformization theorem, which generalizes Riemann's mapping theorem. The geometric formulation promoted the acceptance of the notion of a Riemann surface. Riemann himself spoke (W. 40) of "geometric clothing" ("geometrische Einkleidung") used for "illustration and more convenient wording" (zur "Veranschaulichung und bequemeren Fassung"), formulations hardly ever encountered elsewhere in his writings. The use of complex methods for the computation of definite integrals opened up a new field for the applicability of complex function theory, and that is why complex analysis became a fixed component of the mathematical education of physicists and engineers. As for mathematics itself, the question of admissible boundaries of simply connected regions provided essential impulses for the evolution of point set theory.

For the effects of the dissertation in the first fifty years after Riemann, see J. Gray 1994. For later developments see Remmert 1991 Band 2, 157–163. We recommend Ahlfors 1953 and especially Courant 1950, a book saturated with Riemann's style of thinking. It is safe to say that even had Riemann's dissertation consisted of just the mapping theorem, its influence would have ultimately been considerable.

The effect of (5) was unexpected. Riemann's justification of the existence of a minimal solution is inadequate. This was noted by Weierstrass, whose 1870 criticism was devastating and seemed to destroy the very basis of Riemann's justification of complex analysis. But this had also very positive consequences.

One consequence was that people tried, successfully, to prove the relevant results without using the Dirichlet principle. Actually they would have tried to find such proofs regardless of doubts about this principle. Such attempts reflect the wish to construct complex function theory in a "purely complex" way and to avoid the use of tools from real analysis, functions u and v of two real variables x and y. This too was achieved. Incidentally, this does not signify the rejection of Riemann's development of function theory. In view of its conceptual basis, it is closer to our way of thinking than is, say, the Weierstrass approach.

Another consequence of the criticism directed at Riemann's justification of the Dirichlet principle was even more important than the first one. Since there were no counterexamples and the principle itself was believable, people felt that it must be provable. Hilbert obtained a proof after 1900, and in doing so developed the so-called direct methods of the calculus of variations, which avoid the detour through the partial differential equations associated with the variational problem. One begins instead with a sequence of functions for which the values of the integral, or more generally of the functional, to be minimized approximate the infimum. One must show that the space of admissible functions has a compactness property which justifies the conclusion that a subsequence converges to a function for which the functional takes on its minimum. In this way a method was developed that not only saved the Dirichlet principle but has progressively become more important in the 20th century.

But let us go back briefly to the attempts to avoid the Dirichlet principle. Much was achieved by H. A. Schwarz and C. Neumann. As for the mapping theorem, the conclusive result was obtained independently by Poincaré and by Koebe in 1907. It asserts that every simply connected Riemann surface is holomorphically equivalent to one of following three surfaces: $\mathbb{C} \cup \{\infty\}$ (the number sphere or complex projective straight line), $\mathbb{C}$ (the number plane or complex straight line), or the open disk $|z| < 1$. The key that leads one to this group of problems in the literature is the uniformization theorem. This problem and its easy-to-formulate answer were almost obvious to Riemann, but half a century was needed to obtain it.

We do not know whether Riemann expected a stronger response. After all, he did say

> "However, we now refrain from the realization of this theory ... for we rule out, at present, consideration of an expression of a function."
>
> ("Die Ausführung dieser Theorie ... unterlassen wir indes jetzt, da wir die Betrachtung eines Ausdruckes einer Function gegenwärtig ausschliessen.")

He set aside for a few years the task of investigating concrete functions and classes of functions, and tackled it in connection with lectures devoted to these matters. Of course, this did not happen during his first year as university instructor.

1.3 The elaborations

Riemann's dissertation was to a large extent programmatic. He postponed its realization for a few years, turning first to physics and then also to real analysis and geometry. The two papers of 1857, which he began to prepare in his lectures beginning in the fall of 1855, were the first indication of the immense range of this program. The second paper especially, devoted to Abelian functions, redeemed the promise of the dissertation and at the same time opened vast new areas and posed weighty questions. Below we discuss the complexes of problems from 1857 as well as the relevant deliberations on minimal surfaces first published on the basis of the *Nachlass*. The zeta function will be discussed in a separate section (Section 1.4), partly because of the special interest taken in it and partly because it is somewhat outside Riemann's function theory proper, belonging rather to the area of Cauchy's methods.

In the next few sections we discuss developments in different areas to the extent to which they are directly linked with Riemann, and in 1.3.7 we try to sketch later assessments of his complex analysis.

1.3.1 Ordinary differential equations

It was not until 1857, six years after his dissertation, that Riemann again published work on complex analysis. The shorter of his two papers of 1857 appeared in the *Nachrichten der Göttinger Gesellschaft der Wissenschaften* (W. 67–83). He again went on from the work of Gauss, who in 1812 investigated in great detail the hypergeometric series

$$F(\alpha,\beta,\gamma,z) = 1 + \frac{\alpha\beta}{1\cdot\gamma}z + \frac{\alpha(\alpha+1)\beta(\beta+1)}{1\cdot 2\cdot\gamma(\gamma+1)}z^2 + \frac{\alpha(\alpha+1)(\alpha+2)\beta(\beta+1)(\beta+2)}{1\cdot 2\cdot 3\cdot\gamma(\gamma+1)(\gamma+2)}z^3 + \cdots$$

Riemann had also looked over Gauss' *Nachlass* and concluded that it contained results related to those of Kummer in *Crelle* 15 (1835), (W. 84/85).

The reader can easily verify that many well-known functions can be represented by such series. Three examples are

$$(1+z)^n = F(-n, \beta, \beta, -z)\,,$$

which is a geometric series for $n = -1$ and which is the source of the name "hypergeometric,"

$$\log(1+z) = zF(1, 1, 2, -z)\,,$$

and

$$e^z = \lim_{\beta\to\infty} F(1, \beta, 1, z/\beta)\,.$$

In general, the radius of convergence of the series is 1.

The hypergeometric series was in many respects an ideal subject which Riemann could use to display his views and methodology: they had been investigated by Gauss and are important for applications; specifically, it was discovered in the meantime that many functions of mathematical physics are special cases of this series. Riemann wanted to get rid of the series representations in order to get a better understanding of the functions in question.

In a summary of his paper (W. 85) he points out that he used the method spelled out in Article 20 of his dissertation

> "by means of which all the results found earlier can be obtained virtually without computation."
>
> ("durch die sich sämtliche früher gefundenen Resultate fast ohne Rechnung ergeben.")

It is obvious that analytic continuation by means of chains of circles is impracticable. He writes (W. 67):

> "In the present paper I have investigated this transcendental function by means of a new method that is essentially applicable to every function that satisfies a linear differential equation with algebraic coefficients. Using this method it is possible to obtain almost immediately from the definition results obtained earlier, partly by rather laborious computations, and this is done in the part of this paper that is here at hand, primarily with a view to providing a convenient overview of the possible representations of this function for its manifold applications in physical and astronomical investigations."
>
> ("In der folgenden Abhandlung habe ich diese Transcendente nach einer neuen Methode behandelt, welche im Wesentlichen auf jede

Function, die einer linearen Differentialgleichung mit algebraischen Coefficienten genügt, anwendbar bleibt. Nach derselben lassen sich die früher zum Theil durch ziemlich mühsame Rechnungen gefundenen Resultate fast unmittelbar aus der Definition ableiten, und dies ist in dem hier vorliegenden Theile dieser Abhandlung geschehen, hauptsächlich in der Absicht für die vielfachen Anwendungen dieser Function in physikalischen und astronomischen Untersuchungen eine bequeme Übersicht über ihre möglichen Darstellungen zu geben.")

The differential equation, which Riemann initially does not even write down explicitly, is

$$z(1-z)w'' + [\gamma - (\alpha + \beta - 1)]w' - \alpha\beta w = 0.$$

We recall that Riemann lectured on the hypergeometric series in the winter terms 1856/57 and 1858/59. Parts of the elaborations were included in the supplementary volume to the collected works published in 1902 (N. 665–691). At a congress of mathematicians in Heidelberg in 1904, Wirtinger spoke about these matters as well as about further developments (N. 720–738). We shall use somewhat modernized language to describe briefly the principal ideas, and we shall do this, more generally, for the differential equation

$$a_0(z)w^{(n)} + a_1(z)w^{(n-1)} + \cdots + a_{n-1}(z)w' + a_n(z)w = 0. \qquad (a)$$

The coefficients $a_k(z)$ are rational functions of z, and M denotes the set of poles of the a_k, the zeros of the first coefficient a_0, and the point ∞. We will refer to the elements of M as the singular points of (a). In a neighborhood of every nonsingular point $z_0 \notin M$ there are n linearly independent solutions $w_1, \ldots, w_n$ of (a) that span the n-dimensional complex vector space of all solutions. If C is a curve that begins and ends at z_0 and avoids all singular points, then analytic continuation of w_j along C leads to a solution $\hat{w}_j$ for which we must have

$$\hat{w}_j = \sum_k m_{jk} w_k$$

with constant complex m_{jk}. These coefficients depend on C, but do not change if instead of C we take a curve homotopic to it, i.e., one obtained from C by continuous modifications that avoid the points of M. The endomorphism of the vector space $V(z_0)$ of the local solutions at z_0 defined by the m_{jk} depends only on the homotopy class of C; this assertion is a consequence of the Cauchy integral theorem. One can introduce in the set of paths C a multiplication (by joining paths and taking into consideration their orientations) and one obtains in this way the group of homotopy classes of paths. If we associate with each homotopy class the appropriate endomorphism of $V(z_0)$, then we obtain a homomorphism of this fundamental group into the group of all homogeneous linear transformations of $V(z_0)$. But then the endomorphisms represented by the matrices (m_{jk}) also form a group, known as the monodromy group of the equation (a) at the point z_0. Continuation along a path from z_0 to z_1 (in the complement of M) generates an isomorphism of $V(z_0)$ to $V(z_1)$ that also depends just on the homotopy class of the path. The monodromy groups at z_0 and z_1 are conjugate subgroups of the general linear groups of $V(z_0)$ and $V(z_1)$ respectively.

To return to Riemann's paper of 1857. He did not have at his disposal the algebraic terminology that makes understanding easier for us. He had no need to use an analytic expression for w and could even dispense with the differential equation. He needed the set M and the behavior of w at the singular points in order to show that $V(z)$ and this behavior determine a solution w to within rational transformations.

Riemann defines axiomatically a set of functions ("P-functions")

$$P\begin{Bmatrix} a & b & c & \\ \alpha & \beta & \gamma & z \\ \alpha' & \beta' & \gamma' & \end{Bmatrix}$$

without referring directly either to the hypergeometric function or to the differential equation:

1. Every P-function has just the singularities a, b, c.
2. Any three branches of a P-function are linearly dependent.
3. $P = c_\alpha P_\alpha + c_{\alpha'} P_{\alpha'}$ in some neighborhood of $z = a$

so that $P_\alpha(z-a)^{-\alpha}$ and $P_{\alpha'}(z-a)^{-\alpha'}$ are holomorphic at $z = a$ and do not vanish. Similar conditions hold at b and c, the exponent differences $\alpha - \alpha'$, $\beta - \beta'$, $\gamma - \gamma'$ are not integers, and the sum of all six exponents is 1.

The hypergeometric case corresponds to the values $a = 0$, $b = \infty$, $c = 1$.

Riemann finds a number of "transformations," i.e., modifications of the ten parameters in the symbol P that leave the set P invariant. He shows that the P-functions are precisely the solutions of a differential equation which reduces, in a special case, to the hypergeometric equation. Moreover, he shows how one can arrive at integral representations, and he fits in the Gaussian series representation. Eventually, he says that in this way he can find, almost without computing, hundreds of relations between pairs of hypergeometric functions, including all the known ones.

Kummer was the only one who could fully appreciate this achievement, the replacement of his laborious derivations by sparse and transparent deliberations. But we have not been able to find a relevant comment by him.

26: Ernst Eduard Kummer

This 20-page note is all Riemann wrote about ordinary differential equations. But his P-symbolism has been adopted, and, in addition to his method of integration of a special kind of partial differential equation, it is the most important object applied mathematicians associate with his name.

Riemann pursued this topic in lectures and in manuscripts published later. A fragment dated 20 February 1857 is titled "Zwei allgemeine Sätze über lineare Differentialgleichungen mit algebraischen Coefficienten" ("Two general theorems on linear differential equations with algebraic coefficients") (W. 379–390). His aim was to determine an nth-order differential equation of this kind by prescribing a set M of singularities and a system of n-dimensional complex vector spaces $V(z)$ for $z \notin M$ with prescribed algebraic behavior at the points of M. He obtained differential equations whose singular points are, in Fuchs' later terminology, regular. The Riemann–Hilbert problem is an extension of these investigations. We explain it briefly.

Hilbert formulated Problem 21 as follows (*Werke* III, 322): "Proof of the existence of linear differential equations having a prescribed monodromy group" ("Beweis der Existenz linearer Differentialgleichungen mit vorgeschriebener Monodromiegruppe"). He explained that one had to show "that there always exists a linear differential equation of the Fuchsian class with given singular points and monodromy group. The problem requires the production of n

functions of the variable z, regular throughout the complex z-plane except at the given singular points; at these points the functions may become infinite of only finite order, and when z describes circuits about these points the functions shall undergo the prescribed linear substitutions." ("daß es stets eine lineare Differentialgleichung der Fuchsschen Klasse mit gegebenen singulären Stellen und einer gegebenen Monodromiegruppe gibt. – Die Aufgabe verlangt also die Auffindung von n Funktionen der Variablen z, die sich überall in der komplexen z-Ebene regulär verhalten, außer etwa in den gegebenen singulären Stellen: in diesen dürfen sie nur von endlich hoher Ordnung unendlich werden, und beim Umlauf der Variablen z um dieselben erfahren sie die gegebenen linearen Substitutionen.") Since that time one speaks of the Riemann–Hilbert problem. Hilbert himself immediately contributed to its treatment. An overview of the treatment of this problem, and a solution that includes the case of noncompact Riemann surfaces, is found in H. Röhrl, "Das Riemann–Hilbertsche Problem der Theorie der linearen Differentialgleichungen," *Math. Annalen* 133 (1957), 1–25. Quite some time before that, the problem underwent an inessential modification and the modified version became established. Instead of a single differential equation of higher order people treated first-order systems. The restriction to Fuchsian equations was dropped, and the coefficients could be meromorphic functions on the Riemann surface. Röhrl managed to solve this more general problem by using the relatively recent theory of functions of several complex variables. Further developments have been brought together by Narasimhan (N. 18).

Riemann's idea, which he put before his listeners from the very beginning, was easy to formulate: the solutions are characterized by the position and nature of the singularities (of the differential equation). The road to the realization of this idea was very long; it stretched deep into the 20th century.

In the 19th century another approach, one that could also be linked with Riemann's lectures on the hypergeometric series and the monodromy group, was initially more fruitful. We describe it briefly, adopting somewhat different notations. In particular, our x and y are complex variables. Let y_1 and y_2 be linearly independent solutions of the differential equation

$$a_0(x)y'' + a_1(x)y' + a_2(x)y = 0$$

in a neighborhood of one of its regular points. The continuation of $z = y_1/y_2$ along a closed curve yields $(az + b)/(cz + d)$ with matrix

$$M = \begin{bmatrix} a & b \\ c & d \end{bmatrix}$$

in the monodromy group.

If f is the inverse of the function $x \mapsto z(x)$, then f is invariant under the transformation $z \mapsto (az+b)/(cz+d)$. This is so because, in the course of the analytic continuation, x returns to its initial value and z goes over into $(az+b)/(cz+d)$ (N. 676 ff.). Now we invert the order of things and ask the following question: given a group of linear fractional transformations of the variable z, which functions are invariant under it and what differential equation characterizes them? In this connection Riemann arrived at what came to be known as the Schwarzian derivative (N. 676).

The lectures were published for the first time in 1902, in a supplementary volume to the collected works prepared by Noether and Wirtinger. But as Wirtinger pointed out in his Note 3 (N. 678), some of the ideas in the lectures were effective earlier. It was especially C. Neumann and F. Klein who referred to parts of them. In the lectures, Riemann used explicitly the number sphere and the stereographic projection (N. 1990, 678–679); conformal mappings of circular arc polygons into the upper halfplane; and ideas covered by the key word "automorphic functions," extensively developed by Klein and Poincaré (N. 595 and N. 720–738, Wirtinger's lecture of 1904. An excellent account of the development of automorphic functions, accessible to students, is given in Gray 1986.)

By now it is clear how regrettable it is that Riemann had few listeners who could propagate his methods and ideas. The opposite was true of Weierstrass, and this helps explain the greater initial effectiveness of his ideas. We note that, after Riemann, the first significant researcher in the area of linear differential equations was Lazarus Fuchs (1833-1902) who called himself Weierstrass' student but whose mode of thought paralleled Riemann's.

1.3.2 Analysis as the source of topology

It is an amazing fact that fundamental parts of modern mathematics have their origin in Riemann's analysis. Thus Cantor came to set theory while considering problems in Riemann's paper on trigonometric series, and in Riemann's complex analysis we find for the first time systematic fundamental ideas of topology. The latter include the set-theoretic topology of manifolds with convergence structure and a discussion of the significance of numerical invariants such as the order of connectivity, the genus, and what later came to be known as the first Betti number. Enrico Betti (1823–1892) was a friend of Riemann and was greatly influenced by him.

A now almost obvious idea is the one-point (∞) compactification of the complex plane. In Riemann's time this idea was contrary to the general trend; at that time projective geometry had just successfully adopted the notion of the line at infinity. With hindsight we see that Riemann's number plane, including ∞, is the same as the number sphere, or the real 2-sphere, or the complex projective straight line. And this statement implies the topological assertion that these three manifolds are homeomorphic. The concept of a manifold will be discussed in Chapter 3. Here we limit ourselves to the papers on complex analysis.

The word "topology" was probably first used by J.B. Listing, Gauss' student and later a professor of physics at Göttingen. We know of no mathematical contacts between him and Riemann. After its use by Riemann, the term "analysis situs" lived on in mathematics for a long time.

Already in his dissertation Riemann introduced a topological invariant, namely the connectivity number (see 1.2.2). In his *Theorie der Abel'schen Functionen* (1857) he discusses it in the section titled "Lehrsätze aus der Analysis situs" ("Theorems from analysis situs") in connection with the theory of integrals of two-term perfect differentials. Already at this point he sees topology, or analysis situs, as an independent research area and announces further investigations (in this connection see W. 91–96), but his immediate aim is to sketch what is needed for the paper. This means that he restricts himself to two-dimensional connected manifolds that are locally homeomorphic to open sets in $\mathbb{R}^2$ but have a boundary. If such a surface has no boundary — a case of special interest when dealing with Riemann surfaces — then it must be

> "changed into one with a boundary by removal of an arbitrary point."
>
> ("durch Ausscheidung eines beliebigen Punktes in eine begrenzte verwandelt werden") (W. 94).

This was not stated so clearly in 1851, perhaps because of uncertainty over the point ∞. A crosscut is a curve that connects two (not necessarily distinct) boundary points.

> "An $(n+1)$-tuply connected surface F can be changed into an n-tuply connected surface F' by means of a crosscut, i.e., a cut in the interior leading from one boundary point to another."
>
> ("Eine $(n+1)$-fach zusammenhängende Fläche F kann durch einen Querschnitt — d.h. durch eine von einem Begrenzungspunkte durch das Innere zu einem Begrenzungspunkte geführte Schnittlinie — in eine n-fach zusammenhängende Fläche F' verwandelt werden") (W. 93).

In this way, Riemann obtains in the end a simply connected surface to which he can apply his method of the Dirichlet principle because it has a single closed boundary curve.

> "Now imagine that we are given in the z-plane an everywhere n-tuply extended surface T without boundary, one to be regarded as closed in accordance with what was said earlier, and that this surface has been dissected into a simply connected surface T'. Since the boundary of a simply connected surface consists of one piece, whereas a closed surface ends up with an even number of boundary pieces as a result of an odd number of cuts and with an odd number of boundary pieces as a result of an even number of cuts, this dissection requires an even number of cuts. Let the number of these crosscuts be $2p$ "
>
> ("Man denke sich jetzt in der z-Ebene allenthalben n-fach ausgebreitete unbegrenzte und nach dem Obigen als geschlossen zu betrachtende Fläche T gegeben und diese in eine einfach zusammenhängende Fläche T' zerschnitten. Da die Begrenzung einer einfach zusammenhängenden Fläche aus einem Stücke besteht, eine geschlossene Fläche aber durch eine ungerade Anzahl von Schnitten eine gerade Anzahl von Begrenzungsstücken, durch eine gerade eine ungerade erhält, so ist zu dieser Zerschneidung eine gerade Anzahl von Schnitten erforderlich. Die Anzahl dieser Querschnitte sei $= 2p$ ") (W. 104).

In 1864, Clebsch (1833–1872), Riemann's successor in Göttingen, called the number p the genus of the surface. Riemann surfaces were thought to lack intuitive appeal. Klein used spheres with p handles as models for surfaces of genus p; $p = 0$ gives the 2-sphere and $p = 1$ the torus. As early as 1874 (*Math. Ann.* 7, 549–557) he showed that two orientable closed surfaces are homeomorphic if and only if they have the same genus.

Hindsight enables us to discover all manner of things topological in Riemann's papers of 1851 and 1857. Dissection leads to homology and the closed paths used in integration to homotopy; the latter was important before Riemann, especially in Cauchy's work. (In this connection see also R. van den Eynde, "Historical evolution of the concept of homotopic paths," *Arch. Hist. Ex. Sci.* 45 (1992), 127–188, and M. Bollinger, "Geschichtliche Entwicklung des Homologiebegriffs," ibid., 9 (1972), 94–170.)

That Riemann himself had in mind an "analysis situs," or topology, independent of applications to analysis is also borne out by the fragment XXIX from the *Nachlass* (W. 479–483). But more important for the development of topology than a few hints was the new mode of thought, grounded in understandable analytic facts, and initiated by the concept of a Riemann surface. In addition to Clebsch, it was above all C. Neumann who adopted this mode of thought and propagated it after 1863. In 1868 Clebsch and Neumann founded *Mathematische Annalen*, later published for a long time by Klein. Initially, the journal did not do well, possibly because it propagated novel ideas, such as Riemann's. It pays to look through the first few volumes of this journal to get an idea of the influence of his papers. Of course, this influence was not limited to the new journal.

For a thorough survey of the development of topology in connection with Riemann see E. Scholz 1980. Moreover, this work explores other areas of Riemann's work more thoroughly than we are able to do here.

1.3.3 Abel's theorem

There is a fundamental question in analysis that confronts a student at an early stage of his university studies. To pose it we start with a preliminary remark.

One sees without much difficulty that there are classes of functions that are closed under differentiation, i.e., the derivative of an element of the class is an element of the class. For example, the derivatives of rational functions are also rational functions, and if P is a polynomial and f is a rational function of x and $\sqrt{P(x)}$ then so is f'. The converse assertion is false: the indefinite integral of a rational function can be a transcendental function. Still, the worst that can happen is that a logarithm may turn up; if we use complex numbers then we can even dispense with inverse trigonometric functions, for

$$\frac{1}{1+x^2} = \frac{1}{2i}\left(\frac{1}{x-i} - \frac{1}{x+i}\right).$$

Now the question. Are there classes of functions that are closed under integration? In 1821 Cauchy gave an affirmative answer to this question for real functions and soon after for complex functions: the class of continuous functions has this property! From the modern standpoint this is a satisfying answer. After all, we are used to characterizing classes of functions, or spaces as we often call them today, by means of properties, and continuity is the historically first-studied property of functions, and an eminently reasonable one. It is also the first to be studied in lectures on analysis.

The formula of Cauchy (and Hadamard) implies that the class of everywhere convergent power series is closed under differentiation and integration, and in the complex domain Cauchy established the corresponding result for functions without singularities in the plane. All these are instances of characterization in terms of concepts or properties.

But Cauchy's contemporaries saw these matters in a different light. They attached great importance to definitions of functions by means of explicit expressions. And in spite of Riemann's emphatic pronouncements against such definitions, most of his work in function theory is devoted to the classical problem of integration of functions given by expressions.

Beginning in 1826, without the concept of a Riemann surface, Abel knew how to build a far-reaching result out of partial classical results. He considered integrals of the form $\int R(x, y)\, dx$, R a rational function of x and y, for algebraic functions y of x, i.e., $f(x, y) = 0$ for an irreducible polynomial f. Very simple examples show that such integrals cannot be conveniently expressed as functions of the upper limit. On the other hand, they also show that certain sums of such integrals satisfy certain simple relations.

Let $R = 1/y$ and $f(x, y) = y^2 - P(x)$, P a polynomial. We are to find for the integral

$$u = I(x) = \int_0^x \frac{dt}{\sqrt{P(t)}}$$

an "addition theorem" of the form

$$I(x_1) + I(x_2) = I(x_3),$$

or, equivalently, an addition theorem of the form

$$J(u_1 + u_2) = x_3 = \text{an expression in } x_1\,,\ y_1 = \sqrt{P(x_1)},\ x_2, \text{ and } y_2 = \sqrt{P(x_2)}$$

for the inverse function $J = I^{-1}$. Let $P(x) = 1-x^2$. Then we have the addition theorem for the arcsine with $x_3 = x_1y_2 + x_2y_1$, or for the inverse function

$$\sin(u_1 + u_2) = \sin u_1 \cos u_2 + \sin u_2 \cos u_1\,.$$

For the lemniscatic integral $y^2 = P(x) = 1 - x^4$, and Euler found in 1751

$$x_3 = (x_1y_2 + x_2y_1)/(1 + x_1^2x_2^2)\,.$$

This was initially a curiosity, from which one could nevertheless read off construction possibilities for the lemniscate. The reason for this is that the only operations applied to x_1 and x_2 in the expression for x_3 are rational operations and extraction of square roots. Gauss took this up in the *Disquisitiones* of 1801 and pointed out the analogy between division of the circle and division of the lemniscate. Abel too worked on such problems.

Legendre's comprehensive work on elliptic integrals was available in Abel's time. For $P(x) = (1 - x^2)(1 - k^2x^2)$ it implied the relation

$$x_3 = (x_1y_2 + x_2y_1)/(1 - k^2x_1^2x_2^2)\,.$$

We see that in all three cases there is an algebraic equation $\varphi(x_1, y_1, x_2, y_2, x_3) = 0$ for the addition theorem $I(x_1) + I(x_2) = I(x_3)$, where the integrand $R(x, y)$ (here $1/y$) is rational and x and y are connected by an algebraic relation $f(x, y) = 0$ (here $y^2 - P(x) = 0$ with P a polynomial of degree not exceeding 4).

Abel's astounding achievement was that he not only guessed the general statement underlying these examples but formulated it correctly and proved it. The result was contained in a note submitted three months before his death and published in *Crelle's Journal* 4 (1829).

Let $R(u, z)$ be a rational function and let $f(u, z)$ be a polynomial. By an Abelian integral we mean

$$\int R(u, z)\, dz, \quad \text{where } f(u, z) = 0;$$

thus u is assumed to be an algebraic function of z. Taken literally, Abel seems to think in terms of the reals, but the complex shines through. Roughly speaking, Abel's theorem asserts that a sum of such integrals can be written as a sum of p such integrals plus algebraic and logarithmic terms. Abel noticed that the number p depends only on the function f; it turned out that it is the genus of the Riemann surface of f. We illustrate Abel's theorem for the hyperelliptic case. Here $f(u, z) = u^2 - P(z)$, where P is a polynomial of degree 6, and p is 2. We have

$$\sum_{m=1}^{M} \int_0^{z_m} R(u, z)\, dz = \int_0^{a} R(u, z)\, dz + \int_0^{b} R(u, z)\, dz + R_0 + \sum_k c_k \log R_k\,,$$

where R_0 and R_k are rational functions of $z_m, a, b, u_m, u(a)$, and $u(b)$; in turn, for a given M, a and b are algebraic functions of z_m and u_m.

Abel had submitted his great paper to the Paris Academy as early as 1826, but the referees, Legendre and Cauchy, neglected it, with the result that it appeared as late as 1841. But some of Abel's results *were* published in his lifetime. Jacobi obtained similar results at about the same time.

Just as in the case of elliptic integrals, what interested Abel and Jacobi more than the integrals was the corresponding inverse functions, the analogs of the elliptic functions. Beginning in 1832, Jacobi made progress with a new idea. Let us consider with Jacobi at first just the two hyperelliptic integrals

$$w = \int_0^z \frac{dt}{\sqrt{P(t)}}, \quad w = \int_0^z \frac{t\, dt}{\sqrt{P(t)}}$$

that correspond to the "polynomial degree" 6. Jacobi noted that the trouble with the inverse functions was that they were infinitely many-valued, and thus seemingly intractable; after all, Riemann surfaces had not yet been discovered. Jacobi's idea was to consider ordered pairs of arguments (z_1, z_2) and of

27: Niels Henrik Abel

functional values (w_1, w_2):

$$\int_0^{z_1} \frac{dz}{\sqrt{P(z)}} + \int_0^{z_2} \frac{dz}{\sqrt{P(z)}} = w_1 ,$$
$$\int_0^{z_1} \frac{z\,dz}{\sqrt{P(z)}} + \int_0^{z_2} \frac{z\,dz}{\sqrt{P(z)}} = w_2 .$$

He was then able to show that the symmetric expressions $z_1 + z_2$ and $z_1 z_2$ are always single-valued functions of the two variables w_1 and w_2, and this makes the pair (z_1, z_2) a function of the pair (w_1, w_2). He was also able to obtain for this function an addition theorem. This was an essential step forward in the general "Jacobi inversion problem." Incidentally, Jacobi discovered that in the case of hyperelliptic functions (polynomial degree 6) four periods turn up.

Obviously, in Riemann's case everything takes place on the Riemann surface T of genus p associated with $f(w, z) = 0$, and $\int R(w, z)\,dz$ is an integral of a rational function on this surface. To get the drift of the matter the reader may find it useful to recall the deliberations at the end of Section 1.1.8, which make it possible to see through the relations in the case of the elliptic functions "virtually without computing."

28: Carl Gustav Jacobi

On the Riemann surface T, $R(w, z)$ is a single-valued function. Indeed, one can compute w uniquely as a function of z on T. But the integral can be manyvalued. We already know this from the example of $R = \frac{1}{1+z^2}$ and $T = \mathbb{C} \cup \infty$. Circuits around the poles $\pm i$ add integral multiples of π to the value of a line integral. A crosscut from $-i$ to $+i$ makes the surface simply connected after the removal of i from $\mathbb{C} \cup \infty$. Two integrals from z_1 to z_2, taken along different paths between these points, differ by an integral multiple of the "modulus of periodicity." The value of this factor depends on the number of times a path crosses the crosscut.

With each required crosscut there is associated a complex number h_j as its modulus of periodicity. The function R "lives" on the Riemann surface of f. Incidentally, it is possible to show that every meromorphic function on a closed Riemann surface $f(w, z) = 0$ is a rational function of w and z. This shows that the method is tailored to Abelian integrals.

Riemann (W. 105–107) classified the Abelian integrals. If a (possibly many-valued) integral is finite throughout T then we say it is of the first kind. Riemann showed that there are exactly p such linearly independent integrals. Integrals of the second kind are those with a simple pole at a point z_0; for a fixed z_0 there are $p + 1$ linearly independent functions of this kind. Integrals of the third kind are those with two logarithmic singularities; there cannot be just one logarithmic singularity because the sum of the residues of the integrand must be 0. For a given pair of singular points there are again $p + 1$ linearly independent functions of this kind.

Throughout the discussion what counted was f and the surface T, not the function R. Every Abelian integral turns out to be a finite sum of integrals of the three kinds:

> "Thus all the functions w are algebraic functions of z, ramified like T, or integrals of such functions. Given the surface T, this system of functions is determined and depends only on the position of its branch points."
>
> ("Es sind daher sämtliche Functionen w algebraische wie T verzweigte Functionen von z oder Integrale solcher Functionen. Dieses System von Functionen ist bestimmt, wenn die Fläche T gegeben ist und hängt nur von der Lage ihrer Verzweigungspunkte ab") (W. 109).

We see how Abel's theorem and the role of the number p become plausible. These hints will have to suffice.

Riemann investigated the Jacobi inversion problem for Abelian integrals, i.e., the determination of z as a function of u, where

$$u = \int^{z} R(w, \zeta)\, d\zeta \quad \text{with} \quad f(w, \zeta) = 0.$$

He adopted the method introduced by Jacobi for the case $p = 1$ (we sketched it earlier), and, like Jacobi, used the addition theorems the latter had obtained earlier for his Abelian integrals. He took sums of p such integrals and defined new Abelian functions in p variables with a periodicity property.

We describe briefly one more question. It is associated with the key Riemann-Roch theorem. G. Roch (1839–1866), a student of Riemann, extended the deliberations in the 1857 paper in a short note in *Crelle's Journal* 64 (1865), 372–376. At issue is the number of linearly independent functions on a surface of genus p that are meromorphic and have simple poles at m prescribed points. In general, as Riemann put it, such a function w is determined up to $m-p+q+1$ arbitrary constants. Here q is the number of linearly independent functions that vanish at the m prescribed positions of the poles.

Basically, the counts of constants or the determinations of the dimensions of certain function spaces on a Riemann surface are realizations of the program in Article 20 of the dissertation. Riemann addressed this matter again in the continuation (W. 212), published in 1865, where he spoke of

> "our method, based on the determination of functions through their discontinuities and infinities."
>
> ("Unsere Methode, die sich auf die Bestimmung der Functionen durch ihre Unstetigkeiten und ihr Unendlichwerden stützt.")

His results achieved far more than the justification of Abel's theorem and the solution of the Jacobi inversion problem. In 1.3.4 we return to the discussion of the additional contents of the 1857 paper *Theorie der Abel'schen Functionen*.

We note that this paper was written in competition with Weierstrass, whose contributions Riemann (W. 101) appreciated. But he stressed certain priority claims. Specifically, he pointed out that he dealt with the essential parts of the paper in his lectures in 1855/56 and specified precisely which parts he had obtained in 1851/52 (W. 102);

> "but [I] was then diverted from this investigation by another issue."
>
> ([Ich] "ward aber dann durch einen andern Gegenstand von dieser Untersuchung abgezogen.")

The issues discussed in this section have lost none of their attractiveness for researchers. There are excellent accounts of them by experts. The development up to the time of Abel is described by R. Cooke in "Abel's Theorem," in Rowe-McCleary 1989, Vol. I, 389–421. A report that begins in the spirit of Abel and leads up to recent research is P. A. Griffiths, "Variations on a Theorem of Abel," *Inventiones Mathematicae* 35 (1976), 321–390. Many chapters in Koch 1986 describe the methods and results in modern language but include the historical context. Narasimhan (in N) and Freudenthal 1975 deal more fully with this circle of topics than with other parts of Riemann's work.

Brill and Noether 1892 is written in the spirit of the end of the 19th century. This extremely broad and well-substantiated account includes algebraic geometry and the beginnings of function theory. Its authors played an essential role in developments after Riemann. They show Riemann from another standpoint than did Felix Klein.

1.3.4 Algebraic curves

Riemann's function theory had an invigorating effect on what is known as algebraic geometry. The area initially influenced was the theory of curves $F(x, y) = 0$. Already in the 18th century Euler and (even more) Bézout noted that, in questions such as the number of points of intersection of two such curves, admission of complex values of x and y leads to general statements. In accordance with Article 20 of Riemann's dissertation, it is reasonable to expect the Riemann surface to provide substantial insights into the associated real algebraic curve. Klein claims (1926, 296) that "from the very beginning [Riemann] had a very good appreciation of the significance of his theories for algebraic geometry." (Riemann habe "die Bedeutung seiner Theorien für die algebraische Geometrie von Anfang an sehr wohl erkannt"). But only hindsight interpretation of the publications lends any credence to such assertions.

Klein reports (p. 297) on the stir caused by A. Clebsch's paper "Über die Anwendung der Abel'schen Functionen in der Geometrie" ("On the application of the Abelian functions in geometry") *Crelle* 63 (1863), 189–243. Initially at issue was $F(x, y) = y^2 - a_3x^3 - a_2x^2 - a_1x - a_0$ and its connection with elliptic functions, but Clebsch soon treated more general cases. He realized the importance of Riemann's p and named it the genus. The name has endured.

In this area too Riemann's ideas were not a direct source of enlightenment. Brill and Noether (1892, 320) mention a letter written by Clebsch to Roch in August 1864 in which "Clebsch declares that he himself, after tremendous efforts, understood very little of Riemann's paper, and that he also failed to understand the essence of Roch's dissertation" ("Clebsch erklärt, daß er selbst nach den größten Bemühungen von Riemann's Abhandlung nur sehr wenig verstanden habe, und daß ihm auch Roch's Dissertation der Hauptsache nach unverständlich geblieben sei"). Nevertheless Clebsch recognized the importance of the genus p for the classification of algebraic curves at a time when it was still degree-based.

Recently, people who otherwise avoid algebraic geometry but are attracted by Fermat's so-called last theorem on the solvability in integers of the equation $a^n + b^n = c^n$ have taken an interest in the genus of algebraic curves. Upon division of this equation by c^n we see that Fermat's problem can be restated as the problem of the number of rational points on the curve $x^n + y^n = 1$. Quite generally, a number-theoretic question about integral solutions is often translatable into the geometric question of rational points on a curve $F(x, y) = 0$. In the latter context we mention Faltings' 1983 proof of the so-called Mordell conjecture. Specifically, Faltings showed that there are at most finitely many rational points on a curve of genus $p > 1$. This implies that for $n \geq 5$ the equation $a^n + b^n = c^n$ can have at most finitely many integral solutions. The proof of Fermat's last theorem, announced by A. Wiles in 1993, depends in an essential way on the investigation of the elliptic curve $y^2 = x(x + a^n)(x - b^n)$, an idea due to G. Frei.

Of the many results obtained by Clebsch we mention just one: "in general," p of the points of intersection of the curve $F = 0$ with another curve are determined by the remaining ones. The reason for mentioning this result is that in proving it Clebsch exploited Riemann's formulations of Abel's theorem and of Jacobi's inversion problem.

The kind of algebraic geometry initiated by Clebsch played a significant role in the second half of the 19th century. Klein (1926, Ch. VII) wrote about it with great enthusiasm. Well-founded information is found in the 1892 report of Brill and Noether; both were key figures in this area.

An algebraic curve and a Riemann surface are actually the same thing. Nevertheless the terminological difference-"curve" in the case of geometers and "surface" in the case of analysts - has survived to this day.

Those who for the first time come across this confluence of analysis, geometry, and algebra must marvel at it. Each of these areas has its own kind of objects: functions with their Riemann surfaces, algebraic curves, and extension fields of $\mathbb{C}$. In modern terms, the objects of each kind have their isomorphism classes and there are canonical bijections among the sets of these classes. Riemann's merit is that he brought together these three modes of thought, distinguishable, according to Grattan-Guinness, in the first half of the 19th century, by uncovering their common structure.

Riemann formulated the class construction for algebraic equations (W. 119–120) in terms of birational equivalence. Let (s, z) depend rationally on (s_1, z_1) and conversely. $F(s, z) = 0$ and $F_1(s_1, z_1) = 0$ belong to the same equivalence class if these two equations, or, to put it differently, the sets of zeros of the two polynomials, go over into each other under a birational substitution. The effect of the substitutions outside the curves is irrelevant.

Riemann described the set of birational equivalence classes for a given p by investigating the multisheeted structure of the corresponding Riemann surface. For $p > 1$ this set depends on $3p - 3$ parameters (constants) which he called class moduli. For $p = 1$ there is just one modulus.

His next problem was (W. 122) to determine the lowest possible degree of an equation in each class. He did this in a few lines.

Brill and Noether (1892, 274) acknowledged the significance of these considerations for the algebraic geometry of the subsequent decades in these words: "The concepts of 'class' and 'rational transformation' [Riemann speaks of substitutions, and this is clearer to us!], which Riemann justified herewith, are among the most significant achievements of the whole of the newer mathematics. Like every transformation, so too a rational transformation gives one grounds for setting aside, from the huge and bewildering number of properties of an object, those that are peculiar to it, and for emphasizing those that it has in common with other related objects and that unite it with the latter in a 'field'; in this way [the rational transformation] paved the way for a systematic buildup and for viewpoints for a natural classification." ("Die Begriffe 'Klasse' und 'rationale Transformation' [Riemann spricht, für uns klarer, von Substitution!]), die hiermit Riemann begründet gehören zu den bedeutendsten Errungenschaften der neueren Mathematik überhaupt. Wie jede Transformation, so bietet auch die rationale eine Handhabe, um aus der grossen, verwirrenden

Anzahl von Eigenschaften eines Gebildes solche beiseite zu schieben, die dem Individuum als solchem anhaften, und diejenigen hervorzuheben, die es mit anderen verwandten gemeinschaftlich hat und mit diesen zu einem 'Körper' vereinigt; sie bahnte so den Weg zu einem systematischen Aufbau und zu Gesichtspunkten für eine naturgemässe Classification.")

A clear account of recent viewpoints, written for nonspecialists, is found in D. Mumford's lectures *Curves and their Jacobians*, The University of Michigan Press, Ann Arbor, 1975. Riemann's role and the related development are extremely well described in J. Gray, "Algebraic geometry in the late 19th century," in Rowe-McCleary 1989, Vol. I, 361–385.

1.3.5 Minimal surfaces

Another question, which Riemann dealt with at about the same time as Weierstrass, refers to minimal surfaces, i.e., *surfaces of least area for a given boundary* (W. 301–333). The relatively long elaboration by Hattendorff, published in 1867, is based on Riemann's formulas. Riemann worked on this issue in 1860 and in 1861. The problem involves geometry and physics, and its treatment uses real and complex analysis. In other words, problem and treatment involve almost all of Riemann's work areas. Because of this, and because of the parallelism with the work of Weierstrass, the topic deserves more attention than is usually paid to it.

The earliest book on the calculus of variations, Euler's *Methodus inveniendi* of 1744, includes the surface of rotation of a catenary as an example of a minimal surface. Later Lagrange provided the differential equation of these surfaces. As a second-order quasi-linear equation, it was not readily amenable to the methods of integration developed in the first half of the 19th century. The physical illustration of this surface by means of a (weightless) soap bubble set in the boundary curve must have been known long before it was presented in 1873 by the physicist Plateau. In 1785 Meusnier found that minimal surfaces are characterized by the vanishing of the mean curvature. Using the elements of modern differential geometry one can derive from this, in just under two pages, the Weierstrass representation (published in 1866/67) of a minimal surface by means of a holomorphic function (Laugwitz 1977, 131/132).

Riemann realized that the key to dealing with a minimal surface is the following: the Gaussian spherical mapping of the minimal surface into the unit sphere (centered at the origin), effected by the normals to the surface, is conformal (W. 308). Hattendorff shows explicitly that the stereographic projection of the (p, q)-plane onto the sphere is conformal, which implies that it is possible to parametrize by means of $\eta = p + iq$ first the sphere, and then, using the inverse of the Gaussian spherical mapping, the minimal surface itself (W. 306). It follows that the position vector is the real part of a vector-valued function that depends on $p + iq$ alone and is therefore holomorphic. From this we can go over to the Weierstrass representation. Riemann's equivalent representation (W. 310) gives the position vector in integral form; the integrand involves a holomorphic function $u(\eta)$ in the form $U = \frac{du}{d \log \eta}$.

As always, conformal mapping is an important tool for Riemann. If the boundary of the required surface consists of rectilinear pieces, then their images on the sphere are arcs of great circles. Riemann wants to pay special attention to the case when U is an algebraic function of η, for this brings him to closed Riemann surfaces. Of course, one must first employ a trick: the minimal surface, and thus also its image, have boundaries, but in the case of piecewise rectilinear boundaries the surface can be continued in a "symmetric and congruent" manner (W. 314). The examples amenable to this approach include some that lead to hypergeometric or elliptic problems (W. 322 ff.).

An easily formulated example has as boundary a quadrilateral in 3-space obtained by removing two skew edges from a regular tetrahedron (W. 327). The image on the sphere is a regular quadrangle with angles $2\pi/3$. The diagonals are mutually perpendicular and bisect each other.

> "The points diametrically opposite to the vertices are the vertices of a congruent quadrilateral. Between the two lie four quadrilaterals that are also congruent to the original one. Each of them shares two vertices with the original [quadrilateral] and two with the one opposite to it. These six quadrilaterals fill the sphere simply. Hence $du/d \log \eta$ is an algebraic function of η."
>
> ("Die den Eckpunkten diametral gegenüberliegenden Punkte sind die Ecken eines congruenten Vierecks. Zwischen beiden liegen vier dem ursprünglichen ebenfalls congruente Vierecke, die je zwei Eckpunkte mit dem ursprünglichen, zwei mit dem gegenüberliegenden gemein haben. Diese sechs Vierecke füllen die Kugeloberfläche einfach aus. Es wird also $du/d \log \eta$ eine algebraische Function von η sein.")

We replace Hattendorff's involved formulation with a single sentence: from the center of a sphere project a cube inscribed in it to this sphere.

The required piece of minimal surface is rotated through 180° about a segment of its boundary. This operation can be continued at will. The quadrilaterals on the sphere are subject to a corresponding operation. After two applications of the operation a quadrilateral goes over into the opposite one.

"Therefore $(\frac{du}{d\log\eta})^2$ must remain unchanged when η is replaced by $-1/\eta$."

("Daher muß $(\frac{du}{d\log\eta})^2$ unverändert bleiben wenn η mit $-1/\eta$ vertauscht wird.")

Additional symmetry considerations and clever substitutions lead to the coordinates of the required surface whose existence is thereby established. One such coordinate is

$$x_1 = \frac{C}{\rho_1}\int\frac{d\omega}{\sqrt{\omega(1-\rho_2\omega)(1-\rho_3\omega)}}.$$

Here ρ_1, ρ_2, ρ_3 are the cube roots of unity with, say, $\rho_1 = 1$. The coordinates x_2 and x_3 are obtained from x_1 by cyclic permutation of the indices.

This minimal surface is usually named after H. A. Schwarz. At Weierstrass' suggestion, Schwarz studied such questions from 1865 on, and he published his results in 1871. In W. 334 there is a reference to A. Schondorff's Göttingen prize paper of 1867, which was directly linked with Riemann.

H. Weber, the editor of Riemann's works, attached additional discussions of examples to two of Riemann's short notes (W.445–454). But there is hardly any indication of a direct influence of his notes on the subsequent development of the theory of minimal surfaces. R. Courant's (1950) account comes close to Riemann's conception. The topic has become an extensive research area and many young researchers have been awarded Fields medals for relevant contributions. We recommend J.C.C. Nitsche's *Lectures on Minimal Surfaces* I, C.U.P., Cambridge 1989 with its beautiful illustrations; the ones on p. 238 ff. bear on the Schwarz minimal surface.

1.3.6 Riemann's students and their notes on function theory

It is clear that there were very few students and scientists who had close contacts with Riemann and could immediately continue his work. We discuss Dedekind in some detail below. He may well have been the only person who appreciated the full breadth and depth of Riemann's ideas.

What about Riemann's students? Freudenthal (1975, 451) thought that it must be possible to find in Göttingen a list of them, but according to Neuenschwander (1981a, 237) this is not the case. But Neuenschwander did find out something about the *number* of students and ascertained some names. In the winter of 1854/55, in a course on partial differential equations and mathematical physics, there were eight students. In subsequent years the average number of students in the function-theoretic courses was four. From Abbe's letter, dated August 1861, Neuenschwander (1981a, 238) deduced that in the summer of that year, in an introductory course on function theory, there were initially 13 and at the end about ten students; a few of them came from outside Göttingen expressly because of Riemann's courses. By comparison, in the same term, the number of students in Stern's course was 16 to 18 at the beginning and six at the end. We know the names of quite a number of students because relatively many of their lecture notes have been preserved (Neuenschwander 1987).

From the present standpoint it is clear that the student of Riemann who was predestined to further develop his function theory was *Gustav Roch*. But Roch died just a few months after Riemann. It seems that Hattendorff did not particularly care for function theory; he already had difficulties with minimal surfaces (see Section 1.3.5).

We must mention *Friedrich Prym* (1841–1915) (incidentally, like Dirichlet, a native of Düren). In connection with Prym we see with great clarity the change in the evaluation of Riemann's significance in analysis. A. Krazer's obituary notice, "Friedrich Prym," *Jahresb. d. DMV* 25 (1917), 1–15, is revealing.

Prym studied in Berlin, Heidelberg, and Göttingen, where he took the second half of the course in function theory in the winter of 1861/62 and prepared an elaboration of the first half from Hattendorff's notes. He later distributed lithoprint copies of this material. He worked with Kummer in Berlin and obtained his doctoral degree in February 1863. His dissertation, *Theoria nova functionum ultraellipticarum*, dealt with the case of genus $p = 2$, which he tried to handle strictly along Riemann's lines. At the beginning of 1865 Prym spent a number of weeks in Pisa with Riemann, who gave him suggestions for the treatment of hyperelliptic theta functions.

According to Krazer (p. 3), after Roch's death Prym was initially the only person to whose lot it fell "to carry on Riemann's science" ("die Riemannsche Lehre weiterzuführen"). Krazer repeats Christoffel's statement to the effect that Prym had earned "indescribable merit for the appreciation of Riemann" ("unbeschreibliche Verdienste um das Verständnis Riemanns"). Actually, what was required was to show one's contemporaries that the methods developed by Riemann in his papers of 1857 were effective in other cases. Prym became a full professor at Würzburg in 1869. Together with his student Rost, he published in 1911 a two-volume work of 550 pages titled *Theorie der Prymschen Funktionen erster Ordnung im Anschluß an die Schöpfungen Riemann's* (*The theory of Prymian functions of the first order following the work of Riemann*). Today we can say that here a thing clear in principle to every connoisseur of Riemann's works was done to death. At any rate, the book appeared decades too late "to counteract the break with Riemann's methods deplored [by Prym himself]" (der von Prym selbst "beklagten Abkehr von den Methoden Riemann's entgegenzuwirken") (Krazer p. 4). In the meantime, in 1882, Prym eliminated from the theta formula all function-theoretic elements and based his arguments completely on algorithmic procedures in the sense of Jacobi, a development that must be viewed as regressive.

While Prym's elaboration of the course on elliptic and Abelian functions became effective, at least as a basis for lectures, for an admittedly tiny group of students of the provincial university in Würzburg, the manuscript of the other great course on hypergeometric functions slept a hundred years' sleep. In the winter of 1858/59 *Wilhelm von Bezold* (1837–1907), the future physicist and meteorologist, "took this course, and, in view of Riemann's fame and high standing at the university, recorded it carefully in Gabelsberg shorthand" (hatte "diese Vorlesung besucht und aus Achtung vor dem Rufe und Ansehen Riemann's an der Universität sorgfältig in Gabelsberger Stenographie aufgezeichnet). This is how Wirtinger put it (N. 723) in his presentation to the Third International Congress of Mathematicians in Heidelberg in 1904 (N. 719–738). It is clear that Bezold did not take notes out of interest in the material. He later recalled the episode and, in 1894, left the notes to his Berlin colleague Fuchs. Wirtinger says that in the meantime the results and problems in the notes had been rediscovered from a different direction, but his comparison of Riemann's work with the subsequent development is instructive. In 1902 Noether and Wirtinger published excerpts from the manuscript (N. 667–691).

In this supplementary volume of 1902 the authors also took into consideration Riemann's lectures on integrals of algebraic differentials, given in the winter of 1861/62. Here they relied on the notes of Prym and *Minnigerode*. Stahl (1899) used the notes and elaborations of *Hattendorff* and *Schering* but made many changes.

Around 1900 people were primarily interested in the exploration of Riemann's methods and results; today we are also interested in the didactic aspects of his work. The introductory parts of his lectures, important in this respect, have been edited (in 1987) by Neuenschwander, who relied chiefly on Abbe's notes. We mentioned them in Section 1.1. Neuenschwander's thorough search for manuscripts of Riemann's lectures, not just those devoted to function theory, led him to the results that he presented in N. 859–864. He also found out that there are in Göttingen Dedekind's notes of Riemann's lectures. These are on partial differential equations (winter 1854/55), definite integrals (summer 1855), hypergeometric series (winter 1856/57), and elliptic and Abelian functions (winter 1857/58) (N. 864). We recall that Dedekind habilitated (i.e., qualified as a university lecturer) in the summer of 1854, at about the same time as Riemann, and went to Zürich at the beginning of 1858.

Riemann's teaching was particularly extensive in the winter of 1858/59 (12 hours a week) because he covered for the ailing Dirichlet. There are notes for that term written by H. Naegelsbach: "Partielle Differentialgleichungen nach

Lejeune Dirichlet vorgetragen von Professor Riemann in Göttingen" ("Partial differential equations according to Lejeune Dirichlet presented by Professor Riemann of Göttingen") (N. 865, note 6).

1.3.7 Later evaluations

In lectures given between 1915 and 1919, five decades after the death of Riemann and two after the death of Weierstrass, Felix Klein complained (Klein 1926, 312):

"When I was a student, the Abelian functions were regarded, an aftereffect of the Jacobi tradition, as the undisputed apex of mathematics, and each of us had the obvious ambition to advance in this area. And now? The young generation hardly knows the Abelian functions. -

How has this come about? In mathematics, just as in the other sciences, one can observe over and over again the same course of events. At a certain time, for internal and external reasons, new formulations of questions come into being which tempt young researchers and turn them away from the old questions. And the mastery of the old questions, just because they were worked on many times, calls for truly extensive study. This is inconvenient, and for this reason alone people gladly turn to problems that have been little developed and thus require less previous knowledge, problems such as formal axiomatics, or set theory, or some such!"

("Als ich studierte, galten die Abel'schen Funktionen — in Nachwirkung der Jacobischen Tradition — als der unbestrittene Gipfel der Mathematik, und jeder von uns hatte den selbstverständlichen Ehrgeiz, hier selbst weiterzukommen. Und jetzt? Die junge Generation kennt die Abel'schen Funktionen kaum mehr. -

Wie ist das gekommen? In der Mathematik, wie in anderen Wissenschaften, können immer wieder dieselben Vorgänge beobachtet werden. Einmal treten neue Fragestellungen aus inneren und äußeren Gründen auf, welche die jüngeren Forscher reizen und von den alten Fragen ablenken. Dann aber erfordern die alten Fragen, eben weil sie vielfach bearbeitet worden sind, zu ihrer Beherrschung ein nachgerade immer umfangreiches Studium. Das ist unbequem, und man wendet sich schon darum gerne Problemen zu, welche noch weniger ausgebildet sind und darum weniger Vorkenntnisse erfordern, mag es sich nun um formale Axiomatik oder Mengenlehre oder sonst etwas handeln!")

We refer the reader to Narasimhan (N. 14–16) for an overview of the modern contributions that are certainly far deeper, and of a different kind, than what Klein may have expected. True, there are now a hundred times more active research mathematicians than at the beginning of the 20th century, but it is nevertheless astounding how many of them advance the "old questions." And they do this using sophisticated and advanced versions of what Klein dismissed as formal axiomatics, or set theory, or some such!

On the other hand, we cannot expect the great majority of beginning students to acquire a special taste for the original motives behind the study of Abelian functions. That integration of familiar functions leads to new classes of functions is an easily won insight. But who will want to pursue ever deeper the study of these new classes of functions? Even if the integrals turn out to be important in analysis and in applications, a pure mathematician is not likely to go beyond existence theorems. And for applications there are numerical methods. Things are different on the level of complex function theory; more specifically, in the theory of compact Riemann surfaces. If anything, interest in Riemann's insights has grown. But today the original formulation of the question, which we adopted and treated here at the beginning because of our historical approach, is no longer the main source of this interest. It is now one of many circles of similar questions, linked, of course, by numerous cross-connections.

On the occasion of the 150th anniversary of the birth of Weierstrass, Heinrich Behnke said, besides other things, that "what is left is ... the supporting structure that Weierstrass built in order to secure his theory. But the special functions that he treated are of little interest" ("Geblieben ist ... das Gerüst, das Weierstraß zur Sicherung seiner Theorie konstruierte. Wenig Interesse aber finden die speziellen Funktionen, die er behandelte") (H. Behnke, K. Kopfermann, Hsg.: *Festschrift zur Gedächtnisfeier für Karl Weierstraß 1815–1965*. Köln und Opladen: Westdeutscher Verlag 1966). What Behnke said then applies in general to our present attitude to 19th-century analysis.

In the long run, the supporting structure built by Riemann turned out to be sturdier. After all, it has also proved its worth, after its further intensive development, in the case of the theory of functions of several complex variables. But initially things looked different. While it was possible to do some computing with the power series in several variables introduced by Weierstrass in his earliest papers, generalization of the concept of a Riemann surface seemed unmanageable. Weierstrass recalls (1886, 176–177) that he was the first to introduce concrete functions of several variables, namely in connection with the treatment of the Jacobi inversion problem, and that he successfully used theta series for this purpose. "[The fact] that, once the results had been found, one could, eventually, work one's way through to them also in different, and possibly simpler ways is another matter; but we doubt whether the problem could have ever been brought to conclusion in the latter ways had it not been treated earlier." ("Daß man schließlich, nachdem die Resultate einmal gefunden waren, auch auf anderen, vielleicht einfacheren Wegen zu ihnen hat vordringen können, ist eine Sache für sich; wir möchten aber bezweifeln, ob man auf letzteren Wegen die Aufgabe jemals hätte zum Abschluß bringen können, wenn sie nicht schon vorher behandelt worden wäre.")

Incidentally, it was with this pointed remark against Riemann and his solution of the Jacobi inversion problem that Weierstrass concluded his lectures in the summer of 1886. As noted earlier, Riemann acknowledged Weierstrass' contributions (W.101) and said that the problem

> "had already been solved in many ways by the persistent works of Weierstrass crowned by such splendid success ... "
>
> ("schon auf mehreren Wegen durch die beharrlichen mit so schönem Erfolge gekrönten Arbeiten von Weierstrass gelöst worden ... ")

Riemann's 1857 paper (W. 127 ff.) enables one to see how the coefficients in the exponents of the terms of the theta series materialize. But Weierstrass may well be right when he says that his own papers were a premise for this. The theta functions of several variables appeared suddenly in Riemann's work, and C. Neumann (1884, IV in the introduction) notes that the only flaw of the presentation is that it does not show "the inner necessity that leads from the Abelian integrals to the construction of those theta functions." (die "innere Nothwendigkeit, welche von den Abelschen Integralen zur Bildung jener Thetafunctionen hinleitet.")

A century later we see that, on the one hand, it is precisely the theta functions of several variables that have become a large area with many applications, and that, on the other hand, the general function theory of several variables must be viewed as a further development of Riemann's conception. At this point we must mention, however briefly, the deep investigations of complex spaces. Initially, "genuine" functions were in the foreground in the work of Hartogs, Carathéodory, Behnke, and Osgood. Then everything shifted, very much in Riemann's sense, to manifolds (H. Cartan, K. Stein, J.-P. Serre, and the Münster school). H. Grauert ("The Methods of the Theory of Functions of Several Complex Variables," *Misc. Math.* (1991), 129–143) gave a laconic answer to the question of where the functions are now and what is complex analysis: in the case of several complex variables it is more difficult to construct holomorphic and meromorphic functions than in the case of a single variable. And complex analysis (of several variables) is more a special kind of geometry than an analysis of the properties of functions. And yet one can see in this a consistent resumption of Riemann's ideas! This also applies to the role of topology, as vaguely expressed in the so-called Oka-principle: if we can show that complex-analytic problems have continuous solutions, then they often have holomorphic solutions, and obstacles to solutions are of a topological and not of an analytical nature. For a short sketch see R. Remmert, "Komplexe Analysis in Sturm und Drang (Laudatio auf H. Grauert)" ("Complex analysis in storm and stress (A eulogy of H. Grauert)"), *DMV Mitteilungen* 1 (1993) 5–13. Manifolds, as introduced by Riemann, are the *lebensraum* of complex analysis.

Further to the evaluation of Riemann. Felix Klein did a great deal for the propagation of Riemann's complex analysis. It is the leitmotiv of the large third volume of his collected works, and the main ideas are clearly set out in his lectures (Klein 1926). Since Klein is important as a witness of the effect of Riemann's ideas in the last decades of the 19th century, we quote some of his summarizing statements (Klein 1926, especially Ch. 6). It is with a measure of surprise that one reads his frequent allusions to intuition and perception, which give the impression that these were admissible means of proof for Riemann. And remarks like the following are not likely to show Riemann to have been a rigorous mathematician:

"As for logical rigor, it is out of the question that we should set the same requirements for Riemann's works as for the productions of Weierstrass. Rather, Riemann is effective by virtue of his wealth of ideas and the fullness of his viewpoints, which always hit the essential."

("Wir dürfen an Riemann's Arbeiten überhaupt nicht die Anforderungen stellen, was logische Strenge anbelangt, wie an de Entwicklungen von Weierstraß. Vielmehr wirkt Riemann durch seinen Ideenreichtum und die Fülle seiner Gesichtspunkte, die immer das Wesentliche treffen") (p. 256).

There is a misunderstanding here that is reflected in the imprecise use of the word "rigor." A *proof* can and should be rigorous, but what is meant by rigorous assumptions or definitions? Substantial mathematics is *local ordering*. It begins with assumptions that may certainly be questioned. But if it is to arrive at propositions of some interest, then it must make of these assumptions a basis for subsequent rigorous deductions. The question is: which assumptions are admissible in the light of the general consensus at a particular time? An elementary example is that of the real numbers, and in particular their completeness property. In Riemann's time this was not an issue. Everything could be obtained from the decimal representation of the real numbers and no explicit proof was required. The assertion that a bounded monotonic sequence converges can be viewed as originating in the perception of the number line, but this is a well-founded perception, which can be formally underpinned by anyone who finds it necessary. Weierstrass spent many weeks of lectures in unsatisfactory attempts to provide such an underpinning; and this was misunderstood as rigor.

We see this on a higher level in the case of a Riemann surface. Weyl's book provided the first rigorous justification of this concept, and today the "abstract" Riemann surface is introduced in textbooks. But this adds little to the rigor of the *proofs* in Riemann's papers.

Riemann's use of the Dirichlet principle is a different matter. He stated it first (W. 30) as a theorem but later downgraded it to just a plausibility consideration. All that is needed to ensure the requirement of rigor for the results that he based on this principle is to add the clause: "Subject to the validity of the Dirichlet principle." This is how Weierstrass' spiritual grandson Landau proceeded in the case of the Riemann hypothesis, from which he deduced many consequences.

Carl Neumann proceeded in a similar way in 1884. But we must first mention his earlier publications, in which he presented Riemann's complex analysis in a didactically excellent way. Incidentally, he used "Riemann spheres," and as proof of their authenticity referred to communications on Riemann's lectures. These earlier publications are: *Die Umkehrung der Abel'schen Integrale* (Halle 1863, 16pp.) (*Inversion of Abelian integrals*); *Das Dirichlet'sche Princip in seiner Anwendung auf die Riemannschen Flächen* (Teubner, Leipzig 1865, 80pp.) (*The Dirichlet principle as applied to Riemann surfaces*); *Vorlesungen über Riemann's Theorie der Abel'schen Integrale* (*Lectures on Riemann's theory of Abelian integrals*). After the first edition of 1865 he published in 1884 a much enlarged second edition, where as a kind of appendix he included his own proof of the Dirichlet principle. He put the three essential existence theorems in the main text and relegated their proofs to the appendix. In this way he contrived to concentrate on what was essential in Riemann's analysis proper. The three existence theorems (p. 238 ff.) signify in effect the existence of the three kinds of Abelian integrals.

Even if we took the Dirichlet principle as given, it would be difficult to describe Riemann's deliberations on his mapping theorem in Article 21 of his dissertation as more than a sketch or a set of directions for a proof. As Gauss put it in his report, all this had to be fleshed out.

But what about those of Riemann's deductions that were based on detailed proofs? His contemporaries had an algorithmic orientation and were unused to, and uncomfortable with, conceptual arguments. It is easy to see that someone who did his best to avoid the fundamental concept of a Riemann surface would have no access to Riemann's thought. Weierstrass (1886, 144) rejected the infinite-sheeted Riemann surface at a time when he was willing to acknowledge the n-sheeted one; we, today, are unable to see the conceptual difference between the two. Of course, he was right when he said in this connection that it was difficult to see how this "means of sensualization" ("Versinnlichungsmittel") could be transferred to functions of several variables. Here the superiority of his power series was undeniable. The subsequent function theory of several complex variables had, and continues to have, difficulties with higher-dimensional complex manifolds, but it is all worthwhile.

It is conceivable that the criticism of the Dirichlet principle provided the majority of late-19th-century analysts with a welcome excuse for not dealing with Riemann's concepts — Neumann's work notwithstanding — and, above all, for exploiting his algorithmically accessible results.

A detailed discussion of the evolution of the Dirichlet principle is found in Monna 1975.

1.3.8 Dedekind and the algebraization of complex function theory

It is no exaggeration to say that in contrast to Riemann, in whose hands all mathematics turned into analysis, Dedekind invariably tried to see mathematics as algebra. And here the word "algebra" must be interpreted completely in the sense of the 20th century, and not as algebraic analysis or as the science of algebraic equations. If, following modern usage, we mean by algebra the totality of structures definable by means of relations, then Dedekind's theory of real numbers is such a structure because it uses just the binary order relation on the rationals. As we will see, even when dealing with Riemann's differential geometry Dedekind used (linear) algebra for clarification.

Dedekind familiarized himself with Riemann's function theory in the latter's lectures. When he collaborated with Heinrich Weber on the edition of Riemann's collected works he again immersed himself in this theory. Later the two published the fundamental paper "Theorie der algebraischen Functionen einer Veränderlichen" ("The theory of algebraic functions of a single variable") (*Crelle* 82, 1882, 181–290). At issue was the elimination of transcendental (today we would say: topological) concepts and methods from Riemann's theory of algebraic functions; it was felt that algebraic subject matter should be treated in as algebraic a manner as possible. In addition to the algebraic approach of Dedekind and Weber there were the analytic approach (C. Neumann, H.A Schwarz) and the geometric one (Clebsch, Brill, Noether). At the time one spoke of the arithmetic conception of Dedekind and Weber because of the essential role of the many analogies with Dedekind's theory of algebraic numbers. Before the shift in the meaning of the word "algebra" people also regarded the Weierstrass method of power series, viewed as a more precise continuation of the old algebraic analysis, as an algebraic conception.

If one restricts oneself from the outset to algebraic functions with $F(z, w) = 0$, F a polynomial, then one can obtain the partial derivatives F_z and F_w and the differential equation $F_z\,dz + F_w\,dw = 0$ without the use of limits. But

29: Richard Dedekind (1868)

Dedekind and Weber went further: the Riemann surface as the habitat of algebraic functions was also to be accessible in a purely algebraic way, without the use of (set-theoretic-) topological, "transcendental" considerations. The required ideas came from Dedekind's number theory.

We now introduce Heinrich Weber (1842–1913), a many-sided mathematician who studied at Heidelberg and then taught, in addition to other places, at Königsberg, Göttingen, and Strassburg. Weber was a solid and diligent worker who was able to acquire a thorough and profound understanding of new ideas, and could write masterly and readable expositions. He held in trust and developed Riemann's heritage in, besides other works, the two-volume *Die partiellen Differentialgleichungen der mathematischen Physik* (*The Partial Differential Equations of Mathematical Physics*). For decades this was a unique standard work known as "Riemann-Weber." Dedekind could leave to Weber not only most of the work connected with their edition of Riemann's collected works but also the final editing of their great joint paper. Dedekind appreciated this very much and expressed his gratitude in a warm letter of 30 December 1880 (Dedekind, *Werke* III, 488). Also relevant for us is Weber's profound knowledge of algebra, reflected in his three-volume textbook that went through many editions and was simultaneously a handbook of "modern" algebra of around 1900 and a prelude to "Modern Algebra" à la van der

30: Heinrich Weber

Waerden. As a mathematician, Weber was completely independent and rich in ideas. Beginning in his early papers, he aimed to reconstruct Riemann's theory of Abelian integrals in a purely algebraic manner. His "Neuer Beweis des Abel'schen Theorems" ("A new proof of Abel's theorem") appeared in *Mathematische Annalen* 8 of 1874, and his investigations of the inversion problem appeared as early as 1869/70, that is, before Dedekind's first great publication on ideal theory in the second edition of Dirichlet's number theory of 1871. An appreciation of Weber's contributions, and a biography, are found in G. Frei's "Heinrich Weber and the emergence of class field theory," Rowe-McCleary 1989 Vol. I, 425–450.

Dedekind liked to use his Braunschweig official duties as an excuse for avoiding "slave labor," although by modern standards the lot of even a director of a Technical University was not then such a hard one. But there is no doubt that he was the spiritual initiator of the joint work, and that the following introduction reflects his fundamental ideas and motives.

"The purpose of the investigations communicated below is to justify the theory of algebraic functions of a single variable, which is one of the main achievements of Riemann's creative work, from a simple as well as rigorous and completely general viewpoint. In previous investigations of this subject one made, as a rule, certain restrictive assumptions about the singularities of the investigated functions, and either mentioned the so-called exceptional cases in passing as limiting cases or set them completely aside. Similarly, certain fundamental theorems on continuity and expandability are admitted whose obviousness is based on geometric intuition of one kind or another. It is possible to obtain a secure basis for the fundamental notions as well as for a general and exception-free treatment of the theory if one starts from a generalization of the theory of rational functions of a single variable, in particular of the theorem that every integral rational function of a single variable can be decomposed into linear factors. This generalization is simple and well known in the case in which the number denoted by Riemann by p (called genus by Clebsch) has the value zero. In the general case, which relates to the case just mentioned in a way similar to the relation of the most general algebraic numbers to the rational numbers, the right approach was suggested by the methods, used with optimal results in number theory, that are connected with Kummer's creation of ideal numbers and can be applied to the theory of functions.

If, as in number theory, one means by a field of algebraic functions a system of such functions constituted so that the application of the first four rules of arithmetic to the functions of the system always leads to functions of that same system, then this concept coincides completely with that of the Riemann class of algebraic functions. An arbitrary function of such a field can be regarded as an independent variable and the others as dependent on it. For each of these 'modes of representation' ('Darstellungsweisen') there is a system of functions of the field, to be designated as integral functions, whose quotients exhaust the whole field. Among these integral functions it is again possible to single out groups of functions with the characteristic mark of integral rational functions with a common divisor. It is true that such a divisor does not exist in the general case, but if one connects the relevant theorems on rational functions not with the divisor itself but with the functions divisible by the latter, then they admit a perfect transfer to the general algebraic functions. In this way we get to the concept of an ideal, a term derived from Kummer's number-theoretic works, where the nonexistent divisors are introduced into the theory as 'ideal divisors' ('ideale Teiler')...

After an appropriate definition of multiplication one can compute with these ideals according to the same rules as those for rational functions. In particular, one can prove the theorem that every ideal is uniquely decomposable into factors which themselves cannot be further decomposed and are therefore called prime ideals. These prime ideals correspond to the linear factors in the theory of integral rational functions. On the basis of the latter [i.e., the prime ideals] one arrives at a completely precise and general definition of 'a point of the Riemann surface,' i.e., of a perfectly determined system of number values that can be assumed by the functions of the field in a consistent manner.

Based on this, a formal definition of the differential quotient leads to the genus number and to a completely general and elegant representation of differentials of the first kind. To this is added the proof of the Riemann–Roch theorem on the number of arbitrary constants in a function determined by the points where it becomes infinite, as well as the theory of differentials of the second and third kinds. Up to this point, neither the continuity nor the expandability of the investigated functions enter in any way. For example, no gap would arise anywhere if one wished to restrict the domain of employed numbers to the system of algebraic numbers. In this way, a well-delimited and relatively comprehensive part of the theory of algebraic functions is treated solely by means belonging to its own sphere.

Of course, all these results can be obtained from Riemann's theory with a far more modest expenditure of means and as special cases of a very comprehensive general scheme; but it is well known that a rigorous justification of this theory still presents difficulties, and until one succeeds in completely overcoming these difficulties, our approach, or at least a related one, may well be the only one that leads to the goal, as far as the theory of algebraic functions is concerned, with satisfactory rigor and generality. Ideal theory itself would be greatly simplified if one were to assume the concept of a Riemann surface, and in particular that of a point of the latter together with the conceptions based on the continuity of the algebraic functions. In our work the opposite is done: a long detour is used to provide an algebraic justification of ideal theory from which one obtains a perfectly precise and rigorous definition of 'a point of the Riemann surface' which can serve as a basis for the investigation of continuity and of questions connected with it. For the time being, these questions, including questions on Abelian integrals and on moduli of periodicity, are excluded from our investigation."

("Die im nachstehenden mitgeteilten Untersuchungen verfolgen den Zweck, die Theorie der algebraischen Funktionen einer Veränderlichen, welche eines der Hauptergebnisse der Riemannschen Schöpfung ist, von einem einfachen und zugleich strengen und völlig allgemeinen Gesichtspunkt aus zu begründen. Bei den bisherigen Untersuchungen über diesen Gegenstand werden in der Regel gewisse beschränkende Voraussetzungen über die Singularitäten der betrachteten Funktionen gemacht, und die sogenannten Ausnahmefälle entweder als Grenzfälle beiläufig erwähnt oder auch ganz beiseite gesetzt. Ebenso werden gewisse Grundsätze über die Stetigkeit und Entwickelbarkeit zugelassen, deren Evidenz sich auf geometrische Anschauung verschiedener Art stützt. Eine sichere Basis für die Grundvorstellungen sowie für eine allgemeine und Ausnahmslose Behandlung der Theorie läßt sich gewinnen, wenn man von einer Verallgemeinerung der Theorie der rationalen Funktionen einer

Veränderlichen, insbesondere des Satzes, daß jede ganze rationale Funktion einer Veränderlichen sich in lineare Faktoren zerlegen läßt, ausgeht. Diese Verallgemeinerung ist einfach und bekannt in dem ersten Falle, in welchem die von Riemann mit p bezeichnete Zahl (das Geschlecht nach Clebsch) den Wert Null hat. Für den allgemeinen Fall, welcher sich zu dem eben genannten ähnlich verhält, wie der Fall der allgemeinsten algebraischen Zahlen zu demjenigen der rationalen Zahlen, wiesen die mit bestem Erfolge in der Zahlentheorie angewandten Methoden, die sich an Kummer's Schöpfung der idealen Zahlen anschließen, und der Übertragung auf die Theorie der Funktionen fähig sind, auf den richtigen Weg.

Versteht man, analog der Zahlentheorie, unter einem Körper algebraischer Funktionen ein System solcher Funktionen von der Beschaffenheit, daß die Anwendung der vier Spezies auf Funktionen des Systems immer zu Funktionen desselben Systems führt, so deckt sich dieser Begriff vollständig mit dem der Riemannschen Klasse algebraischer Funktionen. Unter den Funktionen eines solchen Körpers kann eine beliebige als unabhängige Veränderliche und die übrigen als von ihr abhängig betrachtet werden. Für jede dieser 'Darstellungsweisen' ergibt sich ein System von Funktionen des Körpers, die als ganze Funktionen zu bezeichnen sind, deren Quotienten den ganzen Körper erschöpfen. Unter diesen ganzen Funktionen lassen sich nun wieder Gruppen von Funktionen aussondern, welchen die charakteristischen Merkmale solcher ganzen rationalen Funktionen zukommen, die einen gemeinschaftlichen Teiler haben. Ein solcher Teiler existiert zwar im allgemeinen Falle nicht, wenn man aber die bezüglichen Sätze über rationale Funktionen nicht an den Teiler selbst, sondern an das System der durch denselben teilbaren Funktionen knüpft, so gestatten sie eine vollkommene Übertragung auf die allgemeinen algebraischen Funktionen. Auf diese Weise gelangt man zu dem Begriff des Ideals, ein Name der aus Kummers zahlentheoretischen Arbeiten stammt, wo die nicht existierenden Teiler als 'ideale Teiler' in die Rechnung eingeführt werden ...

Mit diesen Idealen läßt sich nach gehöriger Erklärung der Multiplikation ganz nach denselben Regeln rechnen, wie mit rationalen Funktionen. Insbesondere ergibt sich der Satz, daß jedes Ideal auf eine einzige Weise in Faktoren zerlegbar ist, welche selbst nicht weiter zerlegt werden können und daher Primideale genannt weden. Diese Primideale entsprechen den linearen Faktoren in der Theorie der ganzen rationalen Funktionen. Auf Grund derselben gelangt man zu einer völlig präzisen und allgemeinen Definition des 'Punktes der Riemannschen Fläche', d.h. eines vollkommen bestimmten Systems von Zahlwerten, welche man den Funktionen des Körpers widerspruchslos beilegen kann.

Eine darauf gegründete formale Definition des Differentialquotienten führt sodann zu der Geschlechtszahl und zu einer ganz allgemeinen, eleganten Darstellung der Differentiale erster Gattung. Hieran schließt sich der Beweis des Riemann-Rochschen Satzes über die Anzahl der willkürlichen Konstanten in einer durch ihre Unendlichkeitspunkte bestimmten Funktion, und die Theorie der Differentiale zweiter und dritter Gattung. Bis zu diesem Punkte kommt die Stetigkeit und Entwickelbarkeit der untersuchten Funktionen in keiner Weise in Betracht. Es würde z.B. nirgends eine Lücke bleiben, wenn man das Gebiet der benutzten Zahlen auf das System der algebraischen Zahlen beschränken wollte. Dadurch wird ein wohl abgegrenzter und ziemlich umfassender Teil der Theorie der algebraischen Funktionen lediglich durch die seiner eigenen Sphäre angehörigen Mittel behandelt.

Freilich ergeben sich alle diese Resultate durch einen weit geringeren Aufwand von Mitteln und als Spezialfälle einer vielumfassenden Allgemeinheit aus Riemann's Theorie; allein es ist bekannt, daß diese Theorie bezüglich einer strengen Begründung noch gewisse Schwierigkeiten bietet, und bis gelungen ist, diese Schwierigkeiten vollständig zu überwinden, dürfte der von uns betretene Weg oder wenigstens ein verwandter, wohl der einzige sein, der für die Theorie der algebraischen Funktionen mit befriedigender Strenge und Allgemeinheit zum Ziele führt. So würde sich die Theorie der Ideale selbst außerordentlich vereinfachen, wenn man den Begriff der Riemannschen Fläche und insbesondere den eines Punktes derselben samt den auf die Stetigkeit der algebraischen Funktionen gegründeten Anschauungen voraussetzen wollte. In unserer Arbeit ist umgekehrt auf einem langen Umwege die Theorie der Ideale algebraisch begründet und aus dieser eine vollkommen präzise und strenge Definition des 'Punktes der Riemannschen Fläche' gewonnen, welche auch als Basis für die Untersuchung der Stetigkeit und der damit zusammenhängenden Fragen dienen kann. Diese Fragen, wozu auch die auf die Abelschen Integrale und die Periodizitätsmoduln bezüglichen gehören, bleiben von unserer Untersuchung einstweilen ausgeschlossen.")

Dedekind and Weber did not fulfill the hope expressed at the end of their introduction, a hope that refers to Riemann's principal concern. But they made it abundantly clear that this called for a topologization ("investigation of continuity") of the point-set Riemann surface obtained by them. In fact, later on it took a great deal of work to arrive at the periods of Abelian integrals while pursuing the Dedekind-Weber approach. In this connection we refer the reader to the last chapter in C. Chevalley's *Introduction to the Theory of Algebraic Functions of One Variable*, New York 1951. But the method adopted by Dedekind and Weber turned out not to be a blind alley, and they were able, as much as possible, to bypass topological (i.e., continuity-related) considerations. At any rate, they obtained the Riemann–Roch theorem.

Dedekind's essential first step also marked a revolutionary turning point. For Riemann, a closed Riemann surface was inhabited by a class of algebraic functions. He was aware of the field property of this class. Dedekind *started* with a field that is to be viewed as the field of quotients of a ring (of "integral" elements). The prime ideals of this ring are the points of the associated Riemann surface. In the simplest case, that of genus $p = 0$, things look as follows. Consider all rational expressions in a symbol ζ with coefficients in $\mathbb{C}$. In this field the ring of integral elements is the set of expressions of the form $c_0 + c_1\zeta + c_2\zeta^2 + \cdots + c_n\zeta^n$, $n \in \mathbb{N}_0$, $c_k \in \mathbb{C}$. The fundamental theorem of algebra implies that the prime ideals are precisely those generated by each of the linear expressions $\zeta - z_0$, $z_0 \in \mathbb{C}$. Thus the prime ideals yield all $z_0 \in \mathbb{C}$. Instead of ζ we write z and interpret the field as that of the rational functions of z. Instead of ζ we could have started with any rational expression ω in ζ (but $\omega \notin \mathbb{C}$). This corresponds to a rational variable substitution, and Riemann saw it as such in the general case as well (W.119). True, we would then have another subring of integral elements, but the final outcome would be the same. The point ∞ gives rise to a minor complication. Dedekind and Weber immediately got around this complication as well as the dependence of the prime ideals on the choice of the element ω. We will explain this briefly in modern terms, with reference to a general algebraic function field.

Once we know that we are dealing with complex-valued functions f on a set M, $f \mapsto f(P_0)$, for $P_0 \in M$, defines a "homomorphism" of the field onto $\mathbb{C}\cup\{\infty\}$ that leaves $\mathbb{C}$ elementwise fixed and behaves "reasonably" with respect to ∞, i.e., in accordance with $c + \infty = \infty$, $c/\infty = 0$, and so on. Conversely, given an extension field of $\mathbb{C}$ we can assign the role of z to any element not in $\mathbb{C}$ and name every "homomorphism" $z \mapsto zP$ that acts like the identity on $\mathbb{C}$ a "place." (There are also finitely many places that correspond to the prime ideals containing z^{-1}.)

The best we could do here was indicate the fundamental idea that made it possible to largely eliminate the topological element, i.e., the element of "continuity." To go further the reader should best consult W.-D. Geyer, "Die Theorie der algebraischen Funktionen einer Veränderlichen nach Dedekind und Weber" ("The theory of algebraic functions of a single variable according to Dedekind and Weber") in Scharlau 1981, 109–133, and chapter 23 in Koch 1986. A book that goes on from Dedekind and Weber is K. Hensel and G. Landsberg, *Theorie der algebraischen Funktionen einer Variablen und ihre Anwendung auf algebraischen Kurven und abelsche Integrale* (*The theory of algebraic functions of a single variable and its application to algebraic curves and Abelian integrals*), Leipzig 1902. As the title indicates, the book also takes up the part of Riemann's paper that calls for analytic treatment (Abelian integrals) as well as algebraic geometry. Contemporaries found the book easier to read than the work of Dedekind and Weber, but from a foundational point of view it represents a watering down of Dedekind's idea in that it largely replaces the ideal-theoretic by the function-theoretic. This is one more instance of a common phenomenon, which is that many people are likely to reject an ever-so-beautiful algebraic theory in favor of a (possibly more complicated) analytic treatment. A relevant and relatively recent example is provided by the treatment of differential equations, where the analytic method of the Laplace transform stood its ground successfully against the algebraic operator calculus of Mikusiński.

We turn once more to Klein (1926, 323 ff.) as a contemporary witness. He had this to say about Dedekind's ideal-theoretic papers on number and function theory: "At this point Dedekind adopted a turn to the abstract, which in principle simplifies the issue greatly and has therefore become, on many occasions, a model for the mode of thought and mode of presentation for members of the younger generation, whereas the older researchers, for example Kronecker, have been unable to reconcile themselves to it ... I have always found Dedekind's terminology unpleasant. He calls these totalities ideals ... He should have talked about 'reals'." ("Dedekind nimmt dabei eine Wendung zum Abstrakten welche die Sache im Prinzip sehr vereinfacht und daher für die Denkweise und Darstellungsweise der Jüngeren vielfach vorbildlich geworden ist, während die älteren Forscher, z.B. Kronecker, sich damit nicht befreunden konnten ... Unangenehm ist mir nur immer Dedekinds Terminologie gewesen. Er nennt diese Gesamtheiten Ideale ... Er hätte von 'Realen' sprechen sollen.") The reason for Klein's objection is that the sets in question actually exist in the domain under discussion. Even in the second decade of the 20th century he could not bring himself to utter the word "set." Nor was he able to internalize the fundamental thought that in dealing with ideals one must not think about a set, a collection. Rather, one should realize that a new object is involved. Here his sympathy may well have been with Kronecker, and it may also be relevant to recall Siegel's letter to Weil which we quoted at the beginning of the preface.

In conclusion we note that Dedekind and Weber also made an attempt to justify Riemann's results. In fact, in the context of the special circle of questions about algebraic functions that they dealt with, they actually managed to avoid using the Dirichlet principle. At the same time, the aspect that they allude to in the introduction is of interest: they need no "restrictive assumptions pertaining to the singularities," and recall that people used to "mention the so-called exceptional cases in passing as limiting cases or set them completely aside." They actually eliminate the considerations that Riemann indicates briefly as follows (W.103):

> "The case when a function becomes infinitely small or infinitely large of νth order at a point of the surface T can be thought of as the case in which the function becomes infinitely small or infinitely large of first order at ν points that coincide at that point (or are infinitely close to it), as will occasionally occur in the sequel."
>
> ("Der Fall, wo eine Function in einem Punkte der Fläche T unendlich klein oder unendlich gross von der νten Ordnung wird, kann so betrachtet werden, als wenn die Function in ν dort zusammenfallenden (oder unendlich nahen) Punkten unendlich klein oder unendlich gross von der ersten Ordnung wird, wie in der Folge bisweilen geschehen soll.")

Here one notes in Riemann's case remnants of an infinitesimal mode of thought which he not only used heuristically for himself but also put in a scientific paper as an abbreviation of an argument presumably well known to the reader.

Hermann Weyl, who was one of the great mathematicians of the 20th century, and who pursued research in many fields, including Riemann's, objected to "the tendency occasionally expressed in the literature to build up the theory of Riemann surfaces 'without topology' ": "I cannot see in this either a difficulty or a special merit" (die "in der Literatur gelegentlich zum Ausdruck gekommene Tendenz, die Theorie der Riemannschen Flächen 'ohne Topologie' aufzubauen": "Ich kann darin weder eine Schwierigkeit noch ein besonderes Verdienst erblicken") (Weyl 1955, 107 footnote). Of course, this remark was aimed at Bieberbach's attempt of 1918 to connect uniformization with Weierstrass' function theory. Weyl summarized his view (1955, 135) in these words:

"In the theory of algebraic function fields, two approaches clearly stand out that lead to a deeper understanding of their dominant laws. One is that of *abstract algebra* ... and the fields of coefficients are not restricted to those of characteristic 0 ... The other approach is the *topological* one followed by Riemann ... Weierstraß' standpoint can be described as the algebraic-function-theoretic one: the explicit construction predominates, but on the other hand one operates completely in the continuum of complex numbers."

("Für die Theorie der algebraischen Funktionenkörper zeichnen sich deutlich zwei Wege ab, auf denen man zu einem tieferen Verständnis der in ihnen obwaltenden Gesetze kommen kann. Der eine ist der der *abstrakten Algebra* ... und man wird als Koeffizientenkörper nicht nur solche der Charakteristik

0 zulassen ... Der andere Weg ist der von Riemann beschrittene *topologische*. Weierstraß' Standpunkt kann man als einen zwischen diesen beiden Extremen liegenden algebraisch-funktionentheoretischen bezeichnen: die explizite Konstruktion waltet vor, andererseits aber wird durchweg im Kontinuum der komplexen Zahlen operiert.")

1.4 The zeta function and the distribution of primes

1.4.1 Preliminary remarks

The zeta function is known to a great many people largely because of the as yet unresolved Riemann hypothesis, but what really matters is that after the gamma function it is one of the most important nonelementary functions. This judgment is independent of its importance for number theory. The zeta function is meromorphic with just one simple pole, and so the notion of a Riemann surface is not necessary for its treatment. As we will see, it is an integral part of the realization of the program intimated by Riemann in Article 20 of his dissertation.

Riemann's paper "Über die Anzahl der Primzahlen unter einer gegebenen Größe" ("On the number of primes that are less than a given quantity") (W. 145–153) of 1859 is his only publication on number theory. It was his way of thanking the Berlin Academy for appointing him a corresponding member. Here he used his function-theoretic abilities in an area that one would not, at first glance, expect to have anything to do with analysis. After all, the natural numbers represent the discrete, the disconnected, in mathematics, whereas analysis has to do with the representation and treatment of the continuous.

In addition to illustrating the surprising connection between numbers and complex analysis, the paper also tells us a great deal about Riemann's view of function theory and, together with the *Nachlass* examined by Siegel and others, gives us a particularly clear picture of his skill and perseverance in dealing with problems of analysis.

The paper is more sketchy than anything Riemann himself submitted for publication. His letter to Weierstrass (N. 823–825) shows that he was aware of its provisional nature. Given its methods and subject matter, we should regard it as a piece of great good luck that he brought himself to publish the paper in the first place.

Gauss provided a basis for his fame with his *Disquisitiones arithmeticae*. All his life he collected empirical data on the distribution of primes. Dirichlet's results on primes in arithmetic progressions are among his most significant mathematical achievements. Riemann's publication falls in the year in which he succeeded both of his teachers at Göttingen. We could think of it as an act of homage to his two models.

He begins by assembling the tools he needs for his number-theoretic objective. A detailed discussion of the paper in the order chosen by Riemann himself is presented in Edwards 1974. A new summary, short but readable, and an overview of subsequent developments is given by R. Narasimhan (N. 2–6).

We present first a plausible historical approach. We know little about Riemann's own way.

1.4.2 An approach

Following Riemann, we mean by the zeta function the function defined by the generalized harmonic series

$$\zeta(s) = \sum_{n=1}^{\infty} \frac{1}{n^s}, \tag{1}$$

where $s = \sigma + it$ and $\sigma > 1$ (σ and t are real; this is by now standard notation for the complex variable s). Already long ago mathematicians regarded it as formally related to the geometric progression

$$g(s) = \sum_{n=1}^{\infty} \frac{1}{s^n}, \tag{2}$$

which converges to $\frac{1}{s-1}$ for $|s| > 1$. In the middle of the 14th century, in a widely read book, N. Oresme proved that $\zeta(1)$, the harmonic series proper, diverges. Since the middle of the 17th century one could use, say, the integral test to conclude that the zeta function converges for $s > 1$. But then $|n^{-s}| = n^{-\sigma}|n^{-it}| = n^{-\sigma}$ implies its convergence for $\sigma > 1$. But computation of its values stumped the greatest analysts for a long time, so Euler's 1734 discovery that $\zeta(2) = \pi^2/6$ elicited universal admiration.

Presently Euler discovered remarkable connections between $\zeta(s)$ and the primes. This marked the beginning of analytic number theory. Euler published some of his results in his widely known *Introductio*, and it was this work that Riemann went on from (W. 145):

> "Euler's observation that the product
>
> $$\Pi \frac{1}{1 - \frac{1}{p^s}} = \sum \frac{1}{n^s}, \tag{3}$$
>
> for p ranging over all primes and n ranging over all natural numbers, served me as a starting point in this investigation."
>
> ("Bei dieser Untersuchung diente mir als Ausgangspunkt die von Euler gemachte Bemerkung, daß das Produkt
>
> $$\Pi \frac{1}{1 - \frac{1}{p^s}} = \sum \frac{1}{n^s} \tag{3}$$

31: Leonhard Euler

wenn für p alle Primzahlen, für n alle ganzen Zahlen gesetzt werden.")

Riemann states in a matter-of-fact way what the discoverer Euler found admirable (*admirabile*): the connection between the sequence of natural numbers n and the primes, whose succession is abstruse ("quorum progressio est abstrusa"; Euler Op. I, 14, 227). In the more than 100 years between Euler's discovery and Riemann's continuation, most mathematicians regarded these things as mere curiosities, although Euler considered them worthy of inclusion in his *Introductio*!

But it is next to impossible to see how this product representation can help us give an analytic statement for the number $\pi(x)$ of primes less than x. The distribution of primes is really abstruse. There are prime-free gaps of arbitrary length — all the integers between $n! + 2$ and $n! + n$ are composite — and we know from experience that twin primes p and $p + 2$ crop up every now and then.

Riemann constructed his paper in a way that makes it difficult to see how he arrived at his solution. When it comes to covering his tracks he is here a disciple of Gauss! This being so, we will not explain his paper section by section, all the more because this is done in an exemplary way by Edwards 1974, 1–38. But we will follow Euler a while longer and try heuristically to find asymptotic representations for $\pi(x)$. We recall that $f(x)$ is said to be asymptotically equal to $g(x)$, $f(x) \sim g(x)$, if $f(x)/g(x) \to 1$ for $x \to \infty$.

How do we verify (3)? For $\sigma > 1$ everything is absolutely convergent. We expand the factors on the left in geometric series and use the fact that the natural number n is uniquely representable as a product of prime powers. For $\sigma > 1$, (3) implies

$$\log \zeta(s) = -\sum \log\left(1 - \frac{1}{p^s}\right). \tag{4}$$

Expanding $\log\left(1 - \frac{1}{p^s}\right)$ and summing we obtain

$$\log \zeta(s) = \sum_p \frac{1}{p^s} + \frac{1}{2}\sum_p \frac{1}{p^{2s}} + \frac{1}{3}\sum_p \frac{1}{p^{3s}} + \cdots . \tag{5}$$

From now on we regard Euler's deliberations as heuristic. He puts $s = 1$ and accepts divergent series. For him

$$\zeta(1) = \sum_{k=1}^{\infty} \frac{1}{n} = \log \infty + .577\ldots .$$

Since, again for $s = 1$, the sums on the right in (5), other than the first, add up to a relatively small finite number, Euler writes (*Introductio* §279)

$$\log \log \infty = \sum \frac{1}{p}, \tag{6}$$

which we interpret as

$$\log \log x \sim \sum_{p \le x} \frac{1}{p}. \tag{7}$$

For the sake of convenience we continue to use Euler's way of writing and try to find out something about $\pi(x)$ from (6); this in spite of the fact that it looks hopeless and was not tried by Euler!

A natural logarithm brings to mind its integral definition, so we write

$$\log \log \infty = \int_1^{\log \infty} \frac{dt}{t}.$$

The upper limit suggests the substitution $t = \log x$, which gives

$$\log \log \infty = \int_e^{\infty} \frac{dx}{x \log x}. \tag{8}$$

Now we also want to write the right side of (6) as an integral. To this end we use Riemann's observation that

$$\frac{1}{p} = \int_p^{\infty} \frac{dx}{x^2}$$

and obtain

$$\sum \frac{1}{p} = \int_2^\infty \pi(x) \frac{dx}{x^2}. \tag{9}$$

This is so because $\pi(x)$ tells us in how many of the individual integrals the value of x lies between the limits p and ∞. Eureka: we have linked $\pi(x)$ with Euler's $\sum \frac{1}{p} = \log \log \infty$!

If we get even bolder and compare, in view of (6), the integrands in (8) and (9), then it is not unreasonable to expect that

$$\pi(x) \sim \frac{x}{\log x}. \tag{10}$$

Surprisingly enough, this is the so-called prime number theorem. It was proved in 1896, independently, by Hadamard and de la Vallée-Poussin, both of whom used extensions of Riemann's methods. Legendre, whose number theory Riemann read as a gymnasium student (N. 852), had tried from 1798 on to underpin a relation of the kind appearing in (10).

One can try, at first heuristically, to determine the density of the primes, i.e.,

$$\frac{\pi(x+h) - \pi(x-h)}{2h}$$

for large x and $h \ll x$. If we replace here $\pi(x)$ by $\frac{x}{\log x}$ then, after a brief computation, we find that this density is asymptotically equal to $\frac{1}{\log x}$. The 72-year-old Gauss wrote in 1849 to his friend Encke that he had guessed this in 1792 or 1793 and had tried repeatedly to shore up this guess by counting primes; that he took no note of Legendre until 1849; and that empirical evidence supported his formula

$$\pi(x) \sim \int^x \frac{dt}{\log t} \tag{11}$$

rather than Legendre's

$$\pi(x) \sim \frac{x}{\log x - A}, \qquad A = 1.08366.$$

We shall see that Riemann used the integral in Gauss' formula (11) and referred to the counts of Gauss and Goldschmidt, mentioned in the letter to Encke.

Riemann does not mention Legendre either. And we miss the name of Chebyshev, who, a decade before Riemann's paper, came very close to (10). Specifically, he used real analysis to show that for large x

$$.92 \le \frac{\pi(x)}{x/\log x} \le 1.11.$$

Chebyshev's name does appear in one of Riemann's extant notes (see Edwards 1974, 5).

Incidentally, by applying integration by parts to the integral logarithm it is easy to show that

$$\operatorname{Li} x = \int_2^x \frac{dt}{\log t} = \frac{x}{\log x} - \frac{2}{\log 2} + \int_2^x \frac{dt}{\log^2 t} \sim \frac{x}{\log x}, \tag{12}$$

so that $\pi(x) \sim \operatorname{Li} x$ and (10) are equivalent.

By now we are at least heuristically convinced that there is an intimate connection between the zeta function and $\pi(x)$. But equating the integrands of two divergent definite integrals is suspect. Hence the need to get rid of (1) and to look at $\zeta(s)$ globally. This is Riemann's basic idea.

The replacement of $1/p$ by an integral for the derivation of (9) proved its worth. We can apply the same procedure to $1/p^s$ in (5). In so doing, not much effort is required to take into consideration the extra sums on the right ignored by Euler. Riemann uses

$$F(x) = \frac{\pi(x+0) + \pi(x-0)}{2}$$

and states (W. 148):

If one replaces in (5)

"p^{-s} by $s\int_p^\infty x^{-s-1}\,dx$, p^{-2s} by $s\int_{p^2}^\infty x^{-s-1}\,dx, \ldots$, then one obtains

$$\frac{\log \zeta(s)}{s} = \int_1^\infty f(x) x^{-s-1}\,dx \ldots\text{"} \tag{13}$$

In so doing, Riemann puts

$$f(x) = F(x) + \frac{1}{2}F(x^{1/2}) + \frac{1}{3}F(x^{1/3}) + \cdots. \tag{14}$$

If we think of $\zeta(s)$ as given and of $f(x)$ as required, then (13) represents an integral equation to be solved for $f(x)$. At that time the only known method for the solution of integral equations was the Fourier inversion formula, and Riemann used it here for the first time in connection with a non-physical problem.

Before carrying this out we note that we are primarily interested in $\pi(x)$ rather than in $f(x)$, and that $F(x)$ is equal to $\pi(x)$ apart from the points at which it has jumps. But it is not much trouble to solve (14) for $F(x)$ by an easily verified inversion formula (W. 152):

$$\text{"}F(x) = \sum (-1)^\mu \frac{1}{m} f(x^{1/m}), \tag{15}$$

where for m we put in succession the squarefree numbers and μ denotes the number of primes in m."

(worin für m der Reihe nach die durch kein Quadrat außer 1 teilbaren Zahlen zu setzen sind und μ die Anzahl der Primfaktoren von m bezeichnet.")

Euler could have come this far. The next, obviously essential, step is the solution of the integral eqation (13). First we must see what information about the zeta function we need in order to obtain something usable for the number-theoretic problem.

According to Fourier, the integral equation

$$\psi(\mu) = \int_{-\infty}^{\infty} \phi(\lambda) e^{-i\lambda\mu} \, d\lambda \tag{16}$$

has the unique solution

$$\phi(\lambda) = \frac{1}{2\pi} \int_{-\infty}^{\infty} \psi(\mu) e^{i\mu\lambda} \, d\lambda \,. \tag{17}$$

It is easy to check that, in our case, the sufficient conditions for the applicability of this result are satisfied; Riemann does not go into this matter. Keeping in mind that $f(x) = 0$ for $x < 1$, we put in (13), for a fixed $\sigma > 1$, $x = e^{\lambda}$, $s = \sigma + i\mu$, and obtain, with

$$f(x)x^{-s} = f(e^{\lambda})e^{-\lambda\sigma}e^{-i\lambda\mu} = \phi(\lambda)e^{-i\lambda\mu}$$

for $\lambda \geq 0$, and with $\phi(\lambda) = 0$ for $\lambda < 0$,

$$\psi(\mu) = \frac{\log \zeta(\sigma + i\mu)}{\sigma + i\mu} = \int_{-\infty}^{+\infty} \phi(\lambda) e^{-i\lambda\mu} \, d\lambda \,, \tag{18}$$

i.e., by (17),

$$\phi(\lambda) = f(e^{\lambda})e^{-\lambda\sigma} = \frac{1}{2\pi} \int_{-\infty}^{\infty} \psi(\mu) e^{i\mu\lambda} \, d\mu \,,$$

which comes down to

$$f(x) = \frac{1}{2\pi} \int_{-\infty}^{\infty} \frac{\log \zeta(\sigma + i\mu)}{\sigma + i\mu} e^{(\sigma + i\mu)\log x} \, d\mu \,, \tag{19}$$

or, written differently, with a path of integration parallel to the imaginary axis,

$$f(x) = \frac{1}{2\pi i} \int_{\sigma - i\infty}^{\sigma + i\infty} \frac{\log \zeta(s)}{s} x^s \, ds. \tag{20}$$

Actually, the complex domain does not really enter here. Indeed, there is a real version of the Fourier formulas, and, since $f(x)$ is real, we could take the real part in (19). But the complex notation is more transparent. For all that, we have managed to express $f(x)$, and because of (15) also $F(x)$, in terms of the zeta function. The remarkable connection between the distribution of primes and the series $\sum n^{-s}$ rooted in the natural numbers, a connection intimated by Euler, has been realized!

The actual evaluation of the integral (19) looks hopeless. But one of the essential impulses for the development of complex analysis from Cauchy's time on has been precisely the evaluation of definite integrals, and his integral and residue theorems are the tools available for this purpose. We need to know the singularities of $\log \zeta(s)$, i.e., the zeros of $\zeta(s)$, the residues, and its behavior on plausible paths of integration.

For $\sigma > 1$, $\zeta(s)$ has no zeros. This can be directly inferred from its product representation; a convergent infinite product can vanish only if one of its factors does. Hence to compute the integral we must consider analytic continuation beyond the halfplane $\sigma > 1$. This brings us finally to the beginning of Riemann's paper.

For $g(s)$ in (2), we have $g(s) = 1/(s-1)$, which gives us explicitly the analytic continuation for $s \neq 1$; the latter was also used in the the 18th century for the evaluation of divergent geometric series. If one were foolish enough not to use this sum, then one could make do with the differential equation $g' = -g^2$, or one could derive from the series, in some way, the functional equation

$$g(1+t) = -g(1-t) \quad \text{for } |t| > 2,$$

without leaving the domain of convergence. This functional equation could be used for the purpose of analytic continuation, and this would be similar to one of the methods discussed in the following section.

Euler's conception that the value of a (divergent) series is determined by the value of the "expression" ("Ausdruck") that gives rise to it, refers to *functions* (given by expressions). It does not extend to *numerical equations* such as $1-1+1-\cdots = 1/2$, but it does cover the *functional relation* $g(s) = 1/(s-1)$.

1.4.3 The functional equation

First we must continue $\zeta(s)$ for all complex s. The general method that we learn today following the Weierstrass school consists in power series expansions and their transformation for chains of circles. In our case this would be laborious and contrary to Riemann's conception that a function must be viewed globally rather than as a collection of local representations. The expansion (1) is valid only for $\sigma > 1$,

> "whereas one can easily find an expression for the function that is always valid".

> ("es lässt sich indess leicht ein immer gültiger Ausdruck für die Function finden".)

For the Euler gamma function, for which Riemann still used the symbol Π, we have for $\sigma > 0$

$$\Gamma(s+1) = \Pi(s) = \int_0^\infty e^{-t} t^s \, dt; \quad \Gamma(n+1) = \Pi(n) = n! \tag{21}$$

and the functional equations

$$\Gamma(s+1) = s\Gamma(s) \text{ and } \Gamma(s)\Gamma(1-s) = \frac{\pi}{\sin \pi s}. \tag{22}$$

Substitution of $nx = t$ in the integral for $\Gamma(s)$ yields

$$\frac{\Gamma(s)}{n^s} = \int_0^\infty e^{-nx} x^{s-1} \, dt,$$

and summation over n and the geometric series yield

$$\Gamma(s)\zeta(s) = \int_0^\infty \frac{x^{s-1} \, dx}{e^x - 1}. \tag{23}$$

Together, (23) and (21) show how the zeta function fits into the program in Article 20 of the dissertation, namely, it is obtained from the exponential function by "elementary operations." One could have obtained this for $s > 0$ already in the 18th century, and for $s = 2m$ the formula is found at the latest in Abel's collected works in 1839, with which Riemann was familiar. But he knew "easily" how to define the integral in (23) for all complex s. And it then follows that the expression

$$\Gamma\left(\frac{s}{2}\right) \pi^{-s/2} \zeta(s) \tag{24}$$

> "remains unchanged when s is replaced by $1 - s$."
>
> ("bleibt ungeändert, wenn s in $1 - s$ verwandelt wird.")

In his derivation in W.146 Riemann uses complex integrals and the Cauchy residue theorem. In the language of Article 20 of the dissertation, here too transition to complex arguments allowed "harmonies and regularities" ("Harmonien und Regelmäßigkeiten") to manifest themselves.

From this functional equation one immediately obtains $\zeta(s)$ in the halfplane $\sigma < 1$, and therefore also the fact that for $\sigma < 0$, $\zeta(s)$ has no zeros other than the "trivial" zeros at the negative even integers, whose presence follows from the gamma function.

Riemann seems not to have known that more than a century earlier Euler had struggled with the issue of the functional equation. Following his youthful triumph, the computation of $\zeta(2k)$, Euler tried to compute the values $\zeta(2k+1)$.

This problem remains wide open to this day. It was only a few years ago that $\zeta(3)$ was shown to be irrational.

Euler had a time-tested method: he used series expansions for the representation of properties of a function even if the series diverged, and used summation procedures to obtain functional values. As early as 1740 he obtained results for the alternating series

$$\phi(s) = \sum \frac{(-1)^{n+1}}{n^s}.$$

For these series we have

$$\phi(s) = (1 - 2^{1-s})\zeta(s). \tag{25}$$

Moreover, they are amenable to summation procedures. To this end Euler applied repeatedly the operator $x\frac{d}{dx}$ to the function

$$f(x) = 1 + x + x^2 + \cdots = \frac{1}{1-x},$$

found

$$f_1(x) = x\frac{d\,f(x)}{dx} = x + 2x^2 + 3x^3 + \cdots = \frac{x}{(1-x)^2},$$

$$f_2(x) = x\frac{d\,f_1(x)}{dx} = x + 2^2x^2 + 3^2x^3 + \cdots = \frac{x(1+x)}{(1-x)^3},$$

and obtained for $x = -1$ the values of $\phi(-n)$, and thereby also of $\zeta(-n)$; in particular $\zeta(-2k) = 0$. In 1749, on the basis of a comparison with a few values for positive s that he knew, he ventured the bold guess

$$\frac{\phi(1-s)}{\phi(s)} = \frac{-\Gamma(s)(2^s - 1)\cos \pi s/2}{(2^{s-2} - 1)\pi^s}.$$

In view of (24), the latter expression yields the functional equation

$$\zeta(1-s) = \pi^{-s}2^{s-2}\Gamma(s)\cos\frac{\pi s}{2}\zeta(s) \tag{26}$$

for the gamma function, valid, according to Riemann, for all complex s.

Although this result did not get Euler closer to the evaluation of $\zeta(2k+1)$, he rejoiced in it anyway. This is obvious from the title of the paper: *Remarques sur un beau rapport entre les séries des puissances tant directes que réciproques*. For more detailed accounts involving Euler we refer the reader to R. Ayoub, "Euler and the zeta function," *Am. Math. Monthly* 81 (1974), 1067–1086; A. Weil, "L'oeuvre arithmétique d'Euler," in Euler 1983, 111–134; and *Number Theory. An approach through history*. Birkhäuser 1983, 261–276. Weil provides persuasive corrections of earlier misjudgments of Euler's deliberations.

Euler found (26) by means of clever tricks and verified it for special cases, while Riemann had a key which opened a door to general methods that lead to formulas and make the connections understandable. The key was complex analysis, and especially complex integration.

This seems to have been the method with which Riemann found the functional equation, and it is this approach that he wished to impart to the reader. The proof was later rendered superfluous by a second proof of the functional equation. The latter relies on another substitution in the gamma integral that leads to a Jacobi theta function and uses its known functional equation.

The symmetry about $s = 1/2$, expressed in the functional equation, suggests the introduction of the function

$$\xi(t) = \Pi\left(\frac{s}{2}\right)(s-1)\pi^{-s/2}\zeta(s) \tag{27}$$

with $s = \frac{1}{2} + it$. This is an entire function, and its everywhere convergent expansion in powers of t contains only even powers. Its coefficients are determined by the theta function just mentioned and they tend very quickly to zero.

Riemann stated that there is a product representation

$$\xi(t) = \xi(0)\,\Pi_\alpha\left(1 - \frac{t^2}{\alpha^2}\right)$$

with zeros α of $\xi(t)$ that correspond to the "critical" zeros ρ of the zeta function. He merely indicated the decisive steps and they were later clarified. The order of the factors in the product matters, the ρ must be ordered according to increasing values of $|\mathrm{Im}\,\rho|$; the convergence of the product is equivalent to the (conditional) convergence of $\sum \frac{1}{\rho}$ used later.

While Riemann was unaware of Euler's attempts to obtain the functional equation, there were direct influences on him. We know about them through André Weil 1989.

We recall first of all Dirichlet's investigations, dating to about 1837 and bearing on the infinity of primes in arithmetic progressions, in which he used his famous L-series $L_\chi(s) = \sum \chi(s)n^{-s}$ in the real domain for $s \to 1+$. These investigations do not involve analytic continuation and functional equations, but there is a special case in which the latter are immediately investigated, namely in connection with

$$L(s) = 1 - \frac{1}{3^s} + \frac{1}{5^s} - \frac{1}{7^s} + - \cdots$$

on the interval $0 < s < 1$, ruled by conditional convergence. In Vol. 38 (1849) of *Crelle's Journal*, available to Riemann, Malmstén investigated the functional equation for $L(s)$ and mentioned that a similar proof could be given for $\phi(s)$; he also recalled that both results had been announced by Euler. Obviously unaware of the papers of Euler and Malmstén, O. Schlömilch published the same result in the same year. His paper was titled "Übungsaufgaben für Schüler. Lehrsatz von dem Herrn Professor Dr. Schlömilch" ("Exercises for schoolboys. Theorem of Herr Professor Dr. Schlömilch") and it appeared in *Archiv der Mathematik und Physik*, Vol. 12, p. 415. Another proof, by T. Clausen, appeared in the same *Archiv* 30 (1858) 166–169. Schlömilch published his own proof in *Zeitschrift für Mathematik und Physik* 3 (1858) 130–132.

Riemann may have been prompted by all these publications. But A. Weil has found an additional source, namely Eisenstein's copy of a French edition of Gauss' *Disquisitiones*. We recall the close contact Riemann had with Eisenstein in Berlin. On the last (blank) page of his copy of the book Eisenstein wrote down a proof of the functional equation for $L(s)$ and set down the date 7 April 1849. He used the Poisson summation method for the series and the Fourier transform for the function, which is equal to x^{q-1} for $x > 0$ and vanishes for $x < 0$. He referred to an item in Dirichlet that is now accessible in *Werke* Vol.1, p. 401. Thus the special version of the Fourier transform, later known as the Mellin transform, used by Riemann, appeared earlier. Weil thinks it not only possible but likely that Eisenstein discussed this proof with Riemann in 1849, before the latter left Berlin. "If so, this might have been the origin of Riemann's paper of 1859." Be that as it may, we see that Riemann's number-theoretic paper did not drop out of the sky. Promptings came from the circle around Dirichlet when Riemann was still a student, and they were renewed in 1856 when Dirichlet came to Göttingen. And Riemann was able to go far beyond these promptings.

1.4.4 Riemann's explicit formula for the prime number function

Riemann's evaluation of the decisive formula (19) is sketched on less than two pages and was later exhaustively verified. His result, with a minor correction in the constant, is

$$f(x) = \operatorname{Li} x - \sum_{\rho} \operatorname{Li}(x^{\rho}) - \log 2 + \int_x^{\infty} \frac{dt}{(t^2-1)t\log t}. \tag{28}$$

Here ρ ranges over the nontrivial zeros in the order mentioned earlier. Moreover, for $x > 1$

$$\operatorname{Li} x = \lim_{\epsilon \to 0} \left(\int_0^{1-\epsilon} \frac{dt}{\log t} + \int_{1+\epsilon}^{x} \frac{dt}{\log t} \right) = \int_2^x \frac{dt}{\log t} + 1.04\ldots.$$

This key part of $f(x)$ in (28) appears because $\zeta(s) = \frac{1}{s-1} +$ a power series in $(s-1)$, so that $\log \zeta(s)$ contributes the principal part $-\log(s-1)$ to (19). Riemann shows that for $x > 1$ it yields the value of $\operatorname{Li} x$.

If we assume that $f(x) \sim F(x)(= \pi(x))$ and ignore in (28) the other terms, then (28) looks almost like the prime number theorem, $f(x) \sim \operatorname{Li} x$.

In 1895 H. von Mangoldt gave a careful proof of (28) as well as the following equivalent, but more transparent, statement of (28) for $x > 1$:

$$\psi(x) = \sum_{p^n < x} \log p = x - \sum_{\rho} \frac{x^{\rho}}{\rho} + \sum_{n} \frac{x^{-2n}}{2n} + \text{const.} \tag{29}$$

Equation (28) shows clearly the decisive role of the zeros ρ : $\frac{1}{x}\sum x^{\rho}/\rho \to 0$ is necessary and sufficient for $\psi(x)/x \sim 1$, which, as proved earlier by Chebyshev, is equivalent to the prime number theorem.

We give a brief argument which makes it plausible that (29) follows from Riemann's formula (13). (13) can also be written as

$$\log \zeta(s) = \int_0^{\infty} f(x) s x^{-s-1}\, dx = -\int_0^{\infty} f(x) \frac{dx^{-s}}{dx} dx\,, \tag{30}$$

which simply begs to be integrated by parts. This yields the simpler formula

$$\log \zeta(s) = \int_0^{\infty} x^{-s}\, df(x)\,.$$

Here we used the Stieltjes integral. Differentiation gives

$$-\frac{\zeta'(s)}{\zeta(s)} = \int_0^{\infty} x^{-s} \log x\, df(x) = \int_0^{\infty} x^{-s}\, d\psi(x)\,. \tag{31}$$

The differential admits the elegant interpretation

$$d\psi = \log x\, df(x)\,. \tag{32}$$

The function

$$\psi(x) = \sum_{p^n < x} \log p$$

satisfies condition (32). This function is piecewise continuous. We define it at the jump discontinuities by its mean value. The size of the jump at $x = p^n$ is

$$\log p = \frac{1}{n} \log p^n = \frac{1}{n} \log x$$

and zero otherwise. The same is true of $\log x\, df(x)$. Indeed, according to its definition (14), $f(x)$ has jump discontinuities only at $x = p^n$, and their size is $1/n$. This verifies (32).

If we write the Riemann formula (28) as

$$df(x) = \left(\frac{1}{\log x} - \sum \frac{x^{\rho-1}}{\log x} - \frac{1}{(x^2-1)x \log x} \right) dx \quad (\text{for } x > 1),$$

multiply by $\log x$, and substitute the geometric series for

$$\frac{1}{x^2-1} = \frac{1}{x^2}\frac{1}{1-x^{-2}},$$

then we obtain

$$\log x \, df(x) = \left(1 - \sum x^{\rho-1} - \sum x^{-2n-1}\right) dx\,, \tag{34}$$

which comes down to (29). Incidentally, the constant in (29) is $-\log 2\pi$. Clearly, the last sum in (29) is $\frac{1}{2}\log(1-x^{-2})$, and everything hinges on $\sum x^\rho/\rho$.

It is surprising that Riemann did not hit on the notion of the Stieltjes integral, now taught to all beginning students. The sum definition of an integral in his habilitation paper can be easily extended to $\int g(x)\,dh(x)$ if, as in this case, we are dealing with monotonic functions $h(x)$.

1.4.5 The zeros and the Riemann hypothesis

By now we know how important the zeros ρ are. Riemann discusses them between the functional equation and the derivation of the prime number formula. His remarks come down to the following.

The nontrivial zeros have real parts between 0 and 1. (That these limits are not taken on was first proved by Hadamard.) According to Riemann, the number of zeros ρ with imaginary part between 0 and T is

$$\text{“about (etwa)} = \frac{T}{2\pi}\log\frac{T}{2\pi} - \frac{T}{2\pi}\,.\text{”} \tag{35}$$

This was proved by Mangoldt in 1905. In this connection Riemann states that the number of roots between these limits with real part 1/2 is about the same [as the number given in (35)]; this has not yet been resolved. Next comes the Riemann hypothesis:

> “and it is likely that all roots” have real part 1/2: “Of course, it would be desirable to have a rigorous proof of this; in the meantime, after a few perfunctory vain attempts, I temporarily put aside looking for one, for it seemed unnecessary for the next objective of my investigation.”
>
> (“und es ist wahrscheinlich, daß alle Wurzeln” den Realteil 1/2 haben: “Hiervon wäre allerdings ein strenger Beweis zu wünschen; ich habe indess die Aufsuchung desselben nach einigen flüchtigen vergeblichen Versuchen vorläufig bei Seite gelassen, da er für den nächsten Zweck meiner Untersuchung entbehrlich schien.”)

(In fact, no further attempts are found in the *Nachlass*.)

Riemann himself downplayed his hypothesis. But many accomplished mathematicians, beginning with Stieltjes, have tried to prove it. Like the hypotheses of Fermat and Goldbach, which can be explained to every layman, the Riemann hypothesis may well have contributed to the development of mathematics, but the great advances of analytic number theory are due to other motivations as well. The frequently heard defense of work on hypotheses, namely its contribution to the evolution of methods, is certainly questionable. It reminds one very much of the defense of expenditures on space travel and on the search for elementary particles, based on the suspect promise of technological spinoff benefits. One argument in favor of mathematical pursuits is that they cost society little and everybody can take part in them.

What if the Riemann hypothesis were true? It would imply that $|\frac{x^\rho}{\rho}| = x^{1/2}|\frac{1}{\rho}|$, and then, in view of the known density of the roots, (29) would yield

$$\psi(x) = x + O(x^{\frac{1}{2}+\epsilon}) \tag{36}$$

for every $\epsilon > 0$; as usual, $O(x^k)$ is a function such that the quotient $O(x^k)/x^k$ is bounded. The Riemann hypothesis implies the sharp formulation (36) of the prime number theorem $\psi(x)/x \sim 1$ (and conversely). On the other hand, if the Riemann hypothesis is false and there is a zero $\rho_0 = \sigma_0 + it_0$ with $\frac{1}{2} < \sigma_0 (< 1)$, then $\psi(x) = x + O(x^{\rho_0})$ is false.

Without the Riemann hypothesis we know today that, for example,

$$\psi(x) = x + O(x \cdot \exp(-c \cdot \log^{1/10} x))$$

with an explicitly computable constant c. With the Riemann hypothesis we have the following improved version of (36):

$$\psi(x) = x + O(x^{\frac{1}{2}} \log^2 x).$$

1.4.6 The Nachlass

Obviously, only the printed paper (and not the *Nachlass*) could exert a lasting influence. Riemann notes in the paper that he would like to make further attempts to understand the distribution of the zeros. Already in H. Weber's 1876 edition of Riemann's works one finds on p. 155 a few lines from his letter to Weierstrass in which he explains what is obvious to any careful reader of his paper, namely, that he had not gone into sufficient detail about some things. He adds that he has a new expansion of the function from which certain things follow, an expansion which he

> "has not yet sufficiently simplified to announce it"

("noch nicht genug vereinfacht hatte, um sie mitteilen zu können")
(N. 822–825).

One would have expected that this hint would impel prospective editors and mathematicians with historical interests to look at Riemann's *Nachlass*, the more so since the Göttingen librarian Distel, Klein's contemporary, pointed out that the *Nachlass* in the university library contains relevant notes. In 1932, after attempts by Bessel-Hagen, Carl Ludwig Siegel carefully reviewed the papers and presented them in a systematic and essentially improved way. Siegel's work is a towering mathematical achievement. It would be highly desirable that a proficient historian of mathematics should clarify what in this work is Riemann's own.

The very examination and ordering of the *Nachlass*, not just of the part related to the zeta function, was a daunting task, modestly described by Siegel in these words (Siegel 1932, 46): "No part of Riemann's writings related to the zeta function is ready for publication; occasionally one finds disconnected formulas on the same page; frequently just one side of an equation has been written down; remainder estimates and investigations of convergence are invariably missing, even at essential points." ("In Riemann's Aufzeichnungen zur Theorie der Zetafunktion finden sich nirgendwo druckfertige Stellen; mitunter stehen zusammenhanglose Formeln auf demselben Blatt; häufig ist von Gleichungen nur eine Seite hingeschrieben; stets fehlen Restabschätzungen und Konvergenzuntersuchungen, auch an wesentlichen Punkten.")

Summing up, Siegel objects strongly to the impression that one could share with Felix Klein and Edmund Landau as long as one was unaware of the notes on the zeta function: "The fairytale that Riemann found the results of his mathematical paper by means of 'great general' ideas, without using the formal tools of analysis, is probably not as widespread now as it was in Klein's lifetime." ("Die Legende, Riemann habe die Resulate seiner mathematischen Arbeit durch 'große allgemeine' Ideen gefunden, ohne die formalen Hilfsmittel der Analysis zu benötigen, ist wohl jetzt nicht mehr so verbreitet wie zu Klein's Lebzeiten.")

It seems that Riemann made no further attempts to prove his hypotheses about the location of the zeros and about the infinitude of zeros with $\sigma = \frac{1}{2}$. He derived semiconvergent series expansions that could be used to investigate the behavior of the zeros for $t \to \pm\infty$ on $\sigma = \frac{1}{2}$, and, more generally, on strips parallel to it. In this way he was able to obtain quite accurately a few zeros $\frac{1}{2} + it_1$. In particular, he found the least value of the imaginary part to be $t_1 = 14.1347$. This is just $4 \cdot 10^{-3}$ too much. Siegel regarded Riemann's integral representation of the zeta function of even greater theoretical interest. It is based on a complex integral that was subsequently used on many occasions.

In 1920 Hardy and Littlewood rediscovered and used the principal term of Riemann's semiconvergent expansion.

Landau had easy access in Göttingen to the *Nachlass* but may not have thought Riemann capable of very much. True, he speaks (1909, 29/30) of the "most brilliant and fruitful paper" (von der "genialsten und fruchtbarsten Abhandlung") but immediately adds the qualification: "Riemann's formula is far from the most important thing in prime number theory ... Thus one should not think that Riemann's formula, even if he had proved it, would have solved the Chebyshev problems ... He just [!] created the tool which when refined made it possible later ... to prove many other things." ("Die Riemannsche Formel ist bei weitem nicht das Wichtigste an der Primzahltheorie ... Man darf also nicht glauben, daß Riemann's Formel, selbst wenn er sie bewiesen hätte, die Tschebyschefschen Probleme gelöst hätte ... Er hat nur [!] das Werkzeug geschaffen, durch dessen Verfeinerung man später ... vieles andere hat beweisen können."). Siegel, who was familiar with the *Nachlass*, provides the correction: "The power of Riemann's analytic technique emerges with special clarity from his derivation and transformation of the semiconvergent series for $\zeta(s)$." ("Wie stark Riemann's analytische Technik war, geht besonders deutlich aus seiner Ableitung und Umformung der semikonvergenten Reihe für $\zeta(s)$ hervor").

1.4.7 The evaluations

One should keep in mind that in Klein's time the Mathematical Seminar in Göttingen was located close to the university library building on the Prinzenstrasse, which had custody of Riemann's *Nachlass*. Hilbert was active in Göttingen from 1895 until his death in 1943, Felix Klein from 1886 until his death in 1925, and Landau from 1909 until his forced departure in 1933. They expressed opinions about Riemann's paper without acquiring knowledge of the *Nachlass*.

In 1900 Hilbert included the resolution of the Riemann hypothesis as the 8th of his famous problems and referred to it as an "extremely important assertion" (eine "äußerst wichtige Behauptung"). He also recommended the further investigation of Riemann's explicit formula.

Klein had an odd view of Riemann. Siegel's cautious criticism of Klein applies as well to other mathematicians who were alive in 1932. Apropos of Riemann's prime number paper Klein said in 1894: "Riemann must have very often relied on his intuition" ("Riemann muß vielfach mit der Intuition gearbeitet haben"). This is certainly not meant in a negative sense, for Klein prized intuition. But in the case of Riemann's paper, what was obviously required in addition to mathematical imagination was particularly stubborn effort.

As for Landau, one can readily imagine that Riemann's very style did not appeal to him. In addition to rhetorical praise one finds a pedantic kind of criticism (Landau 1909, 29 ff.) that certainly detracts unfairly from the significance of Riemann as a developer of new methods and as the first discoverer of an exact analytic function for $\pi(x)$. According to Landau, all Riemann proved was the functional equation and the fact that $(s-1)\zeta(s)$ and $\zeta(s)-\frac{1}{s-1}$ are entire functions. As for the rest of the paper, he provides the following summary: "Riemann also conjectured, without being able to provide reasons for correctness other than heuristic ones, that the zeta function has the following six properties ... " ("Riemann vermutete ferner, ohne mehr als heuristische Gründe für die Richtigkeit anführen zu können, daß die Zetafunktion folgende sechs Eigenschaften besitzt ... "). He repeatedly stressed that even if these properties were proved, they would not imply the prime number theorem.

Even after the appearance of Siegel's paper, which was published in a historical series and not in a "regular professional journal," Landau's pronouncements, probably intended to modify Hilbert's positive formulations, were repeatedly featured in professional works. Actually, one gets the impression that the only mathematical accomplishments he was willing to acknowledge were the ones that displayed a precise execution of details, and that he did not recognize the trend-setting ideas of Euler and Riemann as mathematics. Thus in Landau's 1909 *Historische Übersicht über die Entwicklung des Primzahlproblems* (*Historical survey of the evolution of the prime number problem*) Euclid is followed by Legendre while Euler is not even mentioned. And when he states on p. 102 of his main work of 1909 the result

$$\sum_{p\leq x}\frac{1}{p}=\log\log x+B+O(1/\log x),$$

he fails to mention that this is a precise formulation of a statement due to Euler. Reading the section devoted to Euler's series one is bound to get the impression that Euler was not a mathematician. But one must add that when he wrote in 1906 in a *historical* journal, Landau fully acknowledged Euler's relevant investigations.

An outsider contemplating the expert literature on analytic number theory in general and on the prime number theorem in particular must conclude that Euler's amazement, his "admirabile," has all but vanished among all the functions and formulas, the catalogued constants and estimates, the o's and O's. And even if neither Riemann's hypothesis nor his explicit formula are really needed for what has been achieved today, the fact remains that both contain traces of the amazing origins.

Finally, one must not limit oneself to asymptotic formulas. It is of principal interest that by obtaining the relation (28), or even the relations (20) and (15), Riemann found a precise characterization of the primes. An $x > 0$ is a prime if and only if $F(x+) - F(x-) = 1$. Moreover, in the sense of §20 of the dissertation, F can be expressed in terms of elementary operations. This too is an astounding discovery.

2. Real Analysis

Preliminary remarks

Riemann's habilitation paper *Über die Darstellbarkeit einer Funktion durch eine trigonometrische Reihe* (W. 227-265) (*On the representability of a function by means of a trigonometric series*) was completed in 1853, but was published by Dedekind only in 1868. If we consider not just its topic but also its origin, contents, and influence, then we must assign the paper to the foundations of real analysis.

From the beginning of the 19th century it became ever clearer that real analysis presented foundational problems that were very different from those of the theory of complex analysis. One could see this with special clarity in the case of trigonometric series, whose behavior differs radically from that of power series, which are of key importance in the complex domain. Now Dirichlet, who had addressed this issue in his paper of 1829, prevailed on Riemann, who had dealt with complex analysis in his dissertation, to concern himself in his second academic qualifying paper with the justification of real analysis using the set of concrete questions related to trigonometric series.

Riemann may have felt that his results on these series were not conclusive and only partly satisfactory, and he did not publish the paper himself. But the paper contains many methods and examples bearing on functions and series, and, last but not least, his reflections on the "Riemann" integral. Immediately after its publication, the paper provided strong impulses for the initial discussion of the foundations of real analysis: numbers, sets, functions, and integrals.

Riemann included a historical sketch that influenced subsequent historical writing in a strong but very onesided manner. We will say more about this in 2.2 from the viewpoint of the present.

2.1 Foundations of real analysis

2.1.1 The concept of an integral

Practically everybody has encountered Riemann's name for the first time in connection with the definition of an integral. But is it not true that Newton and Leibniz invented the integral in the 17th century? What is new in Riemann's contribution?

For Leibniz, to whom we owe the notations used to this day, the definite integral was a sum of infinitely many infinitely small summands, $\int y\,dx = \sum y_i \Delta x_i$. The indefinite integral with variable upper limit turned out to be "the inverse tangent problem" in the sense that its differential quotient led in all known cases to the integrand. In the 18th century this fundamental property was adopted as a convenient definition of the integral. The sum definition was used infrequently. Euler used it to compute approximately a solution of the differential equation $y' = f(x, y)$.

If one knows a primitive function F with $F' = f$, then one can express $\int_a^b f(x)\,dx$ as the difference $F(b) - F(a)$. But at the beginning of the 19th century, in connection with the solution of differential equations by Fourier's method, people obtained many integrals for whose integrands no primitive functions were known. Fourier himself made do with arguments involving areas.

Cauchy's conceptual justification of real analysis included a proof of the integrability of a whole class of functions: the continuity of $f(x)$ in the interval $a \leq x \leq b$ implies the existence of $\int_a^b f(x)\,dx$. Cauchy used the Leibniz sum definition. His continuity concept was: $f(x+i) - f(x)$ is infinitely small for an infinitely small i. Conversely, Cauchy could prove the fundamental theorem $F'(x) = f(x)$ for $F(x) = \int_a^x f(t)\,dt$, and this became a derived rather than a defining property of the integral (Cauchy 1823).

Following Cauchy's approach, one could easily establish the integrability of piecewise continuous functions. This was used by Dirichlet in 1829. The general view, shared also by Riemann, was that this took care of all functions "occurring in nature." The mathematical requirement of proving the integrability of a larger class of functions was indicated by Dirichlet in 1829 and left by him as a task for Riemann.

In addition to the desire to clarify the principles of the infinitesimal calculus, Riemann (W. 238) mentioned as an additional motive the fact that in number theory there turn up functions more general than those arising "in nature." In his famous paper on the zeta function (1859) Riemann gave a very general formulation of the Fourier integral formula (W. 149).

Riemann defined $\int_a^b f(x)\,dx$ as follows: let $a = x_0 < x_1 < x_2 < \ldots < x_{n-1} < x_n = b$, $\delta_k = x_k - x_{k-1}$, $0 < \epsilon_k < 1$.

> "Then the value of the sum
>
> $$S = \sum_{k=0}^{n-1} f(x_k + \epsilon_{k+1}\delta_{k+1})\delta_{k+1}$$
>
> will depend on the choice of the intervals δ and of the magnitudes ϵ. If it has the property that, regardless of the choice of δ and ϵ, it

approaches indefinitely a fixed limit A as all δ's become infinitely small, then this value is called $\int_a^b f(x)\,dx$."

("Es wird alsdann der Werth der Summe

$$S = \sum_{k=0}^{n-1} f(x_k + \epsilon_{k+1}\delta_{k+1})\delta_{k+1}$$

von der Wahl der Intervalle δ und der Grössen ϵ abhängen. Hat sie nun die Eigenschaft, wie auch δ und ϵ gewählt werden, sich einer festen Grenze A unendlich zu nähern, sobald sämmtliche δ unendlich klein werden, so heißt dieser Werth $\int_a^b f(x)\,dx$.")

Having given the definition of the integral (which we took over from him), Riemann wanted to give necessary and sufficient conditions for its existence. He denoted by D_k the oscillation of the function between x_{k-1} and x_k, i.e.,

"the difference between its largest and smallest values in this interval."

("Unterschied ihres grössten und kleinsten Werthes in diesem Intervalle.")

Today we would speak of the supremum of the $|f(x) - f(x')|$ for $x_{k-1} \leq x, x' \leq x_k$. The sum $\sum \delta_k D_k$ must

"become infinitely small together with the magnitudes δ."

("mit den Grössen δ unendlich klein werden.")

Now suppose that $f(x)$ is "always finite" in the interval and all δ's are less than d. The necessary and sufficient condition is that

"it must be possible to make the sum of the lengths of the intervals in which the oscillations are $> \sigma$, regardless of the value of σ, arbitrarily small by an appropriate choice of d."

("daß die Gesammtgrösse der Intervalle, in welchen die Schwankungen $> \sigma$ sind, was auch σ sei, durch geignete Wahl von d beliebig klein gemacht werden kann.") (W. 241).

Here too Riemann's statements are not easy to follow. The accepted formulations are those of Darboux 1875 (*Ann. de l'Ecole Norm. Sup.* (2) 4, 57–112), who set down detailed proofs.

Riemann showed immediately that his condition is usable and that his concept of an integral is more comprehensive than Cauchy's. He defined $r(x) = x$ for $-\frac{1}{2} < x < \frac{1}{2}$, $r(\pm\frac{1}{2}) = 0$ and extended it to a periodic function with period

1 for all x. The function

$$f(x) = \sum_{n=1}^{\infty} \frac{r(nx)}{n^2}$$

converges for all x because $\sum \frac{1}{n^2}$ is convergent and $|r(nx)| < \frac{1}{2}$. But for $x = \frac{p}{2m}$, $\frac{p}{2m}$ reduced, we have

$$f(x+0) = f(x) - \frac{\pi^2}{16m^2} \text{ and } f(x-0) = f(x) + \frac{\pi^2}{16m^2}.$$

This is so because $1 + \frac{1}{3^2} + \frac{1}{5^2} + \cdots = \frac{\pi^2}{8}$. The discontinuities form a dense set but there are only finitely many jumps $\geq \sigma$ in every finite interval. The criterion tells us that $f(x)$ is integrable.

Riemann noted the integrability of piecewise continuous functions. He seems to have taken for granted the integrability of continuous functions. Following the example of the Weierstrass school, we tend to establish this by proving uniform continuity in a separate lemma. Cauchy's concept of continuity can be interpreted so that it directly includes uniform continuity. Riemann also seems to have thought not of pointwise continuity but of uniform continuity on intervals. This is shown by a note added to his dissertation (W. 46, Remark (1)).

The function just given is neither piecewise continuous nor piecewise monotonic, so that Riemann could use it to demonstrate the range of his concept of an integral.

It was later realized that a function of bounded variation can be characterized as the sum or difference of two monotonic functions and thus is integrable. It is easy to see that Riemann's function does not belong to this class either, for the sum of the absolute values of its jumps in any finite interval is infinite. In fact, if m is an odd prime, then there are $\frac{m-1}{2}$ jumps of size $\frac{\pi^2}{8m^2}$ in the unit interval. The divergence of the sum of the reciprocals of the primes implies that already the primes m lead to unbounded variation.

Riemann also discussed, in a predictable manner, the extension to improper integrals whose integrands became infinite at isolated points.

> But he rejected "other assertions of Cauchy because of their great arbitrariness," even if they were occasionally expedient.
>
> ("Andere Festsetzungen von Cauchy" lehnt er aber "wegen ihrer grossen Willkürlichkeit" ab, wenn sie auch gelegentlich zweckmässig seien) (W. 240).

Riemann's discussion of the concept of integral is less than six pages long. While one gets the impression that he regarded his definition as the only reasonable one, it stimulated the development of other concepts of integrals (Stieltjes, Lebesgue), including the so-called generalized Riemann integral.

Riemann did not develop a theory of his integral. Apart from characterizing it he proved no theorems about it.

In Cauchy's concept of an integral the Fundamental Theorem has a symmetric formulation: If $f(x)$ is continuous then so is $F(x) = \int_{x_0}^{x} f(\xi)\,d\xi + C$. Conversely, if $F(x)$ has a continuous derivative and $F'(x) = f(x)$, then $F(x) = \int_{x_0}^{x} f(\xi)\,d\xi + C$. The equalities denote equalities of values. It is clear that in case of discontinuous integrands this formulation is, in general, not valid. Hankel showed in 1870 that the integral of the Riemann function is not differentiable. In his important book of 1878 Ulisse Dini gave a weak version of the Fundamental Theorem for the Riemann integral. Later, the integral introduced in 1902 by Lebesgue gained universal acceptance.

Riemann's brief remarks, intended merely as an answer to Dirichlet's question about the meaning of the Fourier coefficient formulas, provided the impulse for the development of extensive theories of measure and integration. Rather in passing, Lebesgue provided a conclusive answer to Riemann's own question: a function bounded on an interval $[a, b]$ is Riemann integrable if and only if its set of discontinuities has measure 0. Riemann would have probably put it this way: a function finite in $a \leq x \leq b$ is integrable if and only if the total length of the intervals outside of which it is continuous is infinitely small. The influential lectures of Weierstrass on these matters are of interest for their immediate effect in Germany. Already in 1886 Weierstrass found Riemann's integrability condition too restrictive and claimed that "a function with arbitrary discontinuities is integrable provided that it remains finite..." (Weierstrass 1886, 110) ("daß jede mit beliebigen Unstetigkeiten behaftete Funktion, falls sie nur endlich bleibt, integriert werden kann ... ").

We highly recommend the account of the development of integration theory, in and beyond the work of Riemann, presented in Knobloch 1983.

2.1.2 *Rigor in analysis*

Some of Riemann's formulations are bound to strike a reader familiar with a modern basic course in analysis as nonrigorous. They should certainly not be used in an examination!

We have already pointed out that Riemann spoke of the difference between the largest and smallest values of a function in the interval between x_1 and x_2, assuming only that it is "always finite." In modern parlance, the latter condition holds for the function $f(0) = 0$ and $f(x) = x^{-1/2}$ for $0 < x \leq 1$, which even has an improper Riemann integral $\int_0^1 f(x)\,dx = 2$. But in the work of Cauchy,

Riemann, and their contemporaries, "always finite" meant "bounded." Obviously, boundedness does not justify speaking of "largest and smallest values" of a function. In the analysis of Weierstrass the relevant existence theorem for continuous functions was an important lemma. Riemann's deliberations show that the appropriate terms are our "least upper bound" and "greatest lower bound." Until Riemann's time the completeness of the real numbers, required in this context, was taken for granted. Cauchy used it in his proof of the intermediate value property that involved a nested sequence of intervals (Cauchy 1821, Note III) and Bolzano did the same in 1817 in the same context. And yet both of these mathematicians were by intention and practice masters of rigor before Weierstrass.

The need for the incorporation of such obvious items in a deductive theory became a necessity with increased numbers of students of mathematics, as was the case, say, with Weierstrass at Berlin, beginning around 1861. Before that time we find hardly any false propositions in the work of experienced mathematicians; but we do find misleading formulations and contracted proofs that were later regarded as having gaps.

Undoubtedly, an instance of the latter is what Riemann called the Dirichlet Principle, where the taking on of the minimum was based on the existence of an infimum. We have already discussed the importance of this principle for Riemann's complex analysis. In this connection Weierstrass observed that one can use this argument to justify the following conclusion: Consider the totality of continuously differentiable functions $y = f(x)$ on $-1 \leq x \leq 1$ with $f(-1) = -1$ and $f(1) = 1$. The greatest lower bound of the values of the integrals $I[y] = \int_{-1}^{1} (xy')^2 dx$ is 0. And yet, obviously, there is no such function for which $I[y] = 0$.

Far from being artificial, this counterexample has a rather obvious physical significance if we start from a multidimensional formulation. The "Dirichlet" Principle turned up in Gauss' work in connection with the minimization of an energy integral. The integrand is $(\text{grad } u)^2$. If we take instead $(\vec{r} \cdot \text{grad } u)^2$, where $\vec{r}$ is the position vector, then, on restriction to a single dimension, we obtain Weierstrass' example.

Weierstrass was also the first to systematically thematize uniformity arguments. He was introduced to these issues by his teacher Gudermann while he was a student at Münster. From the outset, Riemann, like Cauchy before him, used continuity on intervals, that is, uniform continuity. In this connection it is interesting that in his Berlin lectures of 1854 Dirichlet explicitly proved uniform continuity on closed intervals (under the name "fundamental property") and used it to prove the existence of the definite integral. We have already discussed this in Section 4.6 of the Introduction.

Readers may be even more surprised at obviously false formulations, such as the convergence criterion for a sequence (S_n):

> "The general condition for convergence is ... that $S_{n+m} - S_n$ should decrease indefinitely with increasing n for *every arbitrary m* ... "
>
> ("Die allgemeine Bedingung der Convergenz ist..., daß $S_{n+m} - S_n$ für *jedes beliebige m* mit wachsendem n ohne Ende abnimmt ... ")

(think of the harmonic series!) According to Neuenschwander 1987, 35 (see Section 1.1.5) this is what we find in Riemann's lectures of 1861. Virtually the same misleading turn of phrase is found on a number of occasions in Cauchy's work, beginning with his *Cours d'analyse* of 1821, and, surprisingly enough, in G. Cantor's paper in *Math. Annalen* 21 (1883), 569, and even in the lectures of Weierstrass (1886, 55). Correct formulations of this Cauchy convergence criterion were given by Bolzano in 1817 and by Euler as early as 1734. Obviously, Euler used his own terminology: $S_j - S_i$ is infinitely small for all infinitely large i and j. Of course, all authors used the right criterion in proofs and the example of the harmonic series was ubiquitous!

We have quoted this curious item not out of pedantry but as a clear example of what must be kept in mind when assessing rigor in analysis, especially 19th-century analysis. One must pay attention to the actual handling of proofs and derivations and not just look at statements of theorems and formulas! It is especially in the literature on Cauchy that this simple rule has been disregarded, with the result that people have presumed to discover in his work mistakes in fundamental theorems. At that time rigor did not reside in the as yet poorly differentiated language but in the mathematics of the proofs. In Riemann's case the proofs sometimes present difficulties because they were often just quick sketches. He must be read with care.

2.1.3 The new status of special cases: Examples and counterexamples

In 2.1.1 we investigated the series with which Riemann demonstrated the range of his concept of an integral. This "Riemann function," as it is often called in the literature, showed that more functions are integrable in the Riemann sense than in the Cauchy sense. Moreover, it showed that a function given by an expression need not be piecewise continuous, and, while bounded, can have points of discontinuity in every arbitrarily small interval. As Weierstrass mentioned occasionally, Riemann also gave in his lectures examples of functions, given by uniformly convergent trigonometric series and thus continuous, that had no derivatives "anywhere" ("nirgends" eine Ableitung besitzen). These counterexamples disproved the so-called "Theorem of Ampère," at that time still included in the textbooks, which claimed that every continuous function (given by an expression) is differentiable everywhere with the possible exception of isolated points.

Special examples serve to investigate the scope of concepts: the concept of continuity is more comprehensive than that of piecewise differentiability, and the concept of integrability is more comprehensive than that of piecewise continuity. Today this mode of proceeding is known to every beginner, but in Riemann's time it was new in the history of mathematics. By and large, this too goes back to Cauchy. He gave $y = e^{-1/x^2}$ as an example of a function that is not equal to its Taylor series: not even infinite differentiability ensures the representability of a function by means of a power series.

This state of affairs is worth mentioning because it points in a twofold sense to the new view of the role of concepts in analysis. For one thing, it shows that general concepts began to be taken seriously, and for another, it prepared the transition from the *content* of concepts to their *scope*, i.e., to the characterization of a concept in terms of the "set" of objects that fall under it.

At this point we will try to clarify the first of these two aspects. Euler's *Introductio* of 1748 can serve once more as a typical instance of the older approach. Euler used particular cases to illustrate general situations. When he used concepts, they were essentially descriptions that fit analogous cases. When he formulated in 1734 the convergence criterion for series (with positive terms), later named after Cauchy, he did so not in order to bring out the concept of convergence of series and to characterize its scope but in order to investigate the behavior of harmonic series $\sum n^{-r}$ for fixed positive r and to treat a number of analogous cases. The extent to which he viewed the concept as a side issue is shown by the fact that he did not include it in his textbooks. From the viewpoint of the philosophical controversy involving universals, Euler the mathematician was a nominalist: general concepts are words that integrate concretely available analogous special cases. One considers only "positive" examples, i.e., those for which there are possibly many analogues to which one can apply in similar ways a procedure developed for a particular case. In his philosophy of mathematics Weierstrass, even in old age, shared Euler's position. He maintained that nothing relevant can be obtained from concepts such as continuity, brushed aside his theorems on continuous functions (to which we attach so much importance), and persisted in the view that the proper aim of analysis is the representation of functions by means of expressions.

For a nominalist, concepts have no independent existence. It is usual but incorrect to speak of Euler's function concepts, for what is invariably involved is verbal descriptions of classes of concrete cases. Writings in the history of mathematics devoted to the history of the "function concept" ("Funktionsbegriff") ignore this state of affairs.

In this phase of analysis, examples that did not fit into classes of concrete cases perceived as normal were classified as curiosities without practical significance. Because of their presumed irrelevance they hardly entered the mathematical field of vision. If one stated "theorems" that did not apply to these curious examples then one said at best that the theorems "suffered exceptions" (die Sätze "erleiden Ausnahmen"). This is what is said in the famous footnote, in Abel's paper on the binomial series published in *Crelle's Journal*, containing the example of a Fourier series of a discontinuous function that is (ostensibly) an exception to Cauchy's theorem to the effect that the sum of a series of continuous functions is continuous.

Then came a conceptual turning point. We noted that Cauchy himself had used examples to clarify the scope of concepts. But he ignored Abel's remark: he had proved his theorem using his concepts of continuity and convergence; the example could have no significance; a theorem proved without conceptual flaws admits no exceptions. Actually, Abel's series does not converge everywhere in the sense of Cauchy's definition and the example could have been used to clarify Cauchy's concepts; but this historic chance was missed.

If we used philosophical terminology, then we would have to describe Cauchy here as a promoter of the reality of universals: the concepts of continuity, convergence, and differentiability enjoyed a reality different from that of the things given concretely in terms of formulas, and theorems could be deduced from concepts. It was virtually only in the fundamental theorems of his *Cours d'analyse* of 1821 that Cauchy consistently demonstrated this, and his demonstration produced practically no echo. It was not until the work of Dirichlet and Riemann that general concepts appeared as independent realities, as concepts abstracted from special cases, and gained full recognition in analysis.

Henceforth examples with unexpected properties were regarded not as monstrous curiosities but as indicators of the required clarifications of concepts. Thus Fourier series of discontinuous functions led to the clarification of the uniformity of convergence, and Riemann's function demonstrated the scope of the new integral concept. Real analysis depends for directions on such indicator examples to a far greater extent than does either complex analysis or algebra. It was unavoidable that these examples should themselves become objects of study, as is true of the Riemann function to this very day, but this is without significance for the conceptual change in analysis. It was Hermann Hankel who in 1870 provided a justification for the investigation of unusual, or so-called pathological, functions, apart from their original purpose of clarification of concepts. Hankel's elaboration appeared shortly after the publication of Riemann's habilitation paper and made explicit references to it. After its publication, mathematicians concerned with such questions had no need to feel apologetic.

Even before the appearance of Hankel's deliberations there was some response in Italy to Riemann's examples, specifically on the part of Betti and Casorati. They were prompted either by Riemann himself during his sojourns in Italy or, more likely, by Prym during his visit to Pisa. The interested reader should consult Bottazzini 1986, 250 and the bibliography provided there.

Of greater interest for us are the additional examples investigated by Riemann for the delimitation of concepts. He devoted three full pages (W. 260–262) to a class of integrable functions whose Fourier series he proved to be divergent. He gave as a reason the fact that these functions have infinitely many maxima and minima near $x = 0$; in our terminology, they are not of bounded variation. He compared the behavior of the integral of f with that of the nth term of its Fourier series,

$$\int_0^{2\pi} f(x) \cos n(x - a)\, dx,$$

and noted (W. 262):

> "We see that, as x decreases indefinitely, the increments of the integral $\int_x f(x)\, dx$ cancel each other, in spite of the fact that their ratio to the changes of x grows very rapidly because of the rapid changes of sign of the function $f(x)$; the effect of the appearance of the factor $\cos n(x - a)$ is that these increments add up."
>
> ("Wie man sieht, heben sich in dem Integrale $\int_x f(x)\, dx$ bei unendlichem Abnehmen von x die Zuwachse des Integrals, obwohl ihr Verhältniss zu den Änderungen von x sehr rasch wächst, wegen des raschen Zeichenwechsels der Function $f(x)$ einander auf; durch das Hinzutreten des Factors $\cos n(x - a)$ aber wird hier bewirkt, dass diese Zuwachse sich summiren.")

Riemann did not think it necessary to investigate the possibility of expanding a continuous function in a Fourier series. His construction of examples could have led him to counterexamples in much the same way as it was to lead du Bois-Reymond.

We see that Riemann proceeded like a good physicist. Since he could not decide by reasoning alone between the two theoretically admissible alternatives, the possibility or impossibility of expansion of integrable functions, he thought of a clever experiment and tells us how he hit on it. In this he differed from many who, wrongly citing him, merely experiment with monster functions for their own sake, without pursuing a conceptual aim.

If we pursue the analogy to physics a step further, then we can safely say that special functions can sometimes play the role of good observational material that can serve as a basis for further developments of the theory. Incidentally, Bottazzini (1986, 248) thinks that, after all, Riemann considered a number of important examples that raised more problems than he could apparently solve. This is presumably why he did not publish his habilitation paper himself.

2.2 Trigonometric series before Riemann

2.2.1 Preliminary remarks

It is convenient to use the terminology that has become standard since Riemann. Disregarding convergence, we call the formal expressions

$$\frac{a_0}{2} + \sum_{k=1}^{\infty} a_k \cos kx + b_k \sin kx \tag{1}$$

or

$$\sum_{k=-\infty}^{\infty} c_k e^{ikx} \tag{1'}$$

trigonometric series. If $f(x)$ is a periodic function with period 2π and

$$a_k = \frac{1}{\pi}\int f(x)\cos kx\,dx, \quad b_k = \frac{1}{\pi}\int f(x)\sin kx\,dx\,, \tag{2}$$

or

$$c_k = \frac{1}{2\pi}\int f(x)e^{-ikx}\,dx, \tag{2'}$$

then we say that (1) or (1′) is the *Fourier series* of the function $f(x)$. All integrals are taken over an interval of length 2π.

It is, so to say, a fact of experience, based on many physical applications, that in the case of "reasonable" functions the two kinds of series may be regarded as synonymous. Nevertheless, it is important to be aware from the outset of two circumstances. While they postdate Riemann, they clarify for us the difficulties of the subject.

A *trigonometric* series may well converge for all real x and not be a *Fourier* series of a function $f(x)$. The simplest example is probably $\sum_{k=2}^{\infty} \sin kx/\log k$. On the other hand, we know since du Bois-Reymond 1876 that there are continuous functions whose Fourier series do not converge for all x and thus are certainly not equal to $f(x)$.

These facts give rise to two problems distinguished by Riemann, namely, the problem (A) of representability of a given function $f(x)$ by means of a trigonometric series (W. 238) and the converse problem (B):

> "When a function is representable by means of a trigonometric series, what does this tell us about its behavior, about the change of its value due to a continuous change of the argument?"
>
> ("Wenn eine Funktion durch eine trigonometrische Reihe darstellbar ist, was folgt daraus über ihren Gang, über die Änderung ihres Werthes bei stetiger Änderung des Arguments?") (W. 244).

As for (A), Riemann had at his disposal the celebrated results of Dirichlet 1829: it is sufficient that the function

> "is integrable throughout, does not have infinitely many maxima and minima, and, where its value changes jumpwise, it takes on the mean of the limiting values on either side."
>
> ("durchgehend eine Integration zulässt, nicht unendlich viele Maxima und Minima hat und, wo ihr Werth sich sprungweise ändert, den Mittelwerth zwischen den beiderseitigen Grenzwerthen annimmt") (W. 237).

These are sufficient conditions. In (B) Riemann asked for necessary conditions that must be satisfied by $f(x)$ for it to equal a trigonometric series (1).

A conceivable objective, unachieved to this day, is to find necessary and sufficient conditions for a 2π-periodic function $f(x)$ to be equal for all real x to its Fourier series, or, more generally, to a trigonometric series. The analogy with complex power series obtrudes on the mind: a necessary and sufficient condition for $f(z)$ to be equal to a power series for $|z| < r$ is the existence and continuity of $f'(z)$ for all these z, and then $f(z) = \sum_{k=0}^{\infty} \alpha_k z^k$. Here $\alpha_k = f^{(k)}(0)/k!$ can be determined directly from $f(z)$ by means of the Cauchy integral formulas without recourse to derivatives. In the case of real periodic functions the situation is incomparably more difficult than in the case of complex differentiable functions. It may well be that this state of affairs became part of the general awareness after the appearance of Riemann's account.

Riemann's paper begins with a section titled "Geschichte der Frage über die Darstellbarkeit einer Funktion durch eine trigonometrische Reihe" ("History of the question of representability of a function as a trigonometric series"), which is divided into three subsections. We will adopt Riemann's subdivision. We see certain things differently from Riemann, and so, although his view has clearly influenced the writing of history to our time, we supplement it.

2.2.2 From Euler to Fourier

The development began almost exactly a hundred years before Riemann's paper, with a heated debate conducted primarily in the Memoirs of the Berlin Academy. At issue was the mathematical treatment of the physical problem of a vibrating string. The deviation u of the string from the rest position at a point x at time t is controlled by the partial differential equation

$$u_{tt} = c^2 u_{xx} \tag{3}$$

(we are using later notations).

In 1747 d'Alembert put $r = x + ct,\ s = x - ct$. This changed the equation to $u_{rs} = 0$ with solutions $u = f(r) + g(s)$, or, in the case of (3),

$$u(x, t) = f(x + ct) + g(x - ct)\,. \tag{4}$$

Here f and g are "arbitrary" functions, and Euler regarded (4) as a solution of the physical problem even if f and g are not twice differentiable, as required by the differential equation. D'Alembert objected to this generalization in spite of Euler's undeniable successes.

As an illustration we use Euler's problem of the determination of the vibrations of a string fastened at the points $x = 0$ and $x = L$, with prescribed initial position and velocity at time $t = 0$. With $L = \pi$ and $c = 1$ we have

$$\begin{aligned} &u(x, 0) = a(x), \quad u_t(x, 0) = b(x) \ \text{ for } \ 0 < x < \pi \ \text{ and} \\ &u(0, t) = u(\pi, t) = 0 \ \text{ for } \ t > 0 \ \text{ and } \ u_{tt} = u_{xx}. \end{aligned} \tag{5}$$

While values of x outside the interval from 0 to π have no physical significance, this does not apply to their mathematical role. Specifically, putting $b(x) = 0$ and making use of the boundary and initial conditions, we have:

$$u(x, 0) = f(x) + g(x) = a(x), \quad u_t(x, 0) = f'(x) - g'(x) = 0.$$

Hence

$$g(x) = f(x) = \frac{a(x)}{2}, \quad u(x, t) = \frac{1}{2}[a(x + t) + a(x - t)]\,.$$

Now

$$0 = u(0, t) = \frac{1}{2}[a(t) + a(-t)]$$

implies that a is an odd function, and

$$0 = u(\pi, t) = \frac{1}{2}[a(\pi + t) + a(\pi - t)]$$

shows that a must be 2π-periodic. Thus prescribing a on the interval from 0 to π determines it for all values of the argument.

For a string plucked in the middle we have

$$a(x) = x \quad \text{for} \quad 0 \le x \le \frac{\pi}{2}, \qquad a(x) = \pi - x \quad \text{for} \quad \frac{\pi}{2} \le x \le \pi .$$

Following Euler we can determine the solution $u(x,t)$ geometrically by adding $\frac{1}{2}a(x+t)$ and $\frac{1}{2}a(x-t)$. After time $t = 2\pi$ things revert to the original condition.

At this point Daniel Bernoulli joined the debate and noted, as did his father Johann Bernoulli and Taylor before him, that

$$u(x,t) = \sin nx \cos nt$$

is a solution of the differential equation for every natural number n. Moreover, this solution satisfies the boundary conditions at $x = 0$ and $x = \pi$, and $u_t(x,0) = 0$. In view of the homogeneity of the problem, these solutions can be superposed to yield

$$u(x,t) = \sum_n b_n \sin nx \cos nt.$$

Since he was free to choose the b_n, Bernoulli conjectured that he could satisfy arbitrary initial conditions $u(x,0) = a(x)$. He was right but could not prove his conjecture.

Euler challenged this claim and argued that the Bernoulli solutions were analytical expressions and as such could not encompass all "arbitrary" functions given physically or by means of graphs.

We mention that in 1759 the 23-year-old Lagrange tried to replace Euler's geometric solution by an analytic process. For further details we refer the reader to the literature, which has dealt repeatedly with the whole debate (Bottazzini 1986, 21–33; Kline 1972; Grattan-Guinness 1970).

Riemann was under the impression that before Fourier no one knew the coefficient formulas (2), which would have immediately dispelled Euler's mistaken idea of the limited scope of the Bernoulli solution. It is one of the many ironies of the history of trigonometric series that these formulas were proved, probably for the first time, in 1754 by Clairaut, and published in 1759 (see Kline 1972, 458) not in an obscure journal but by the Paris Academy. It is true that Clairaut made his discovery in the context of perturbation theory, but he did so at a time when the vibrating-string controversy was still raging. His approach deserves attention. For a given (obviously even and 2π-periodic) function $f(x)$ with undetermined coefficients A_n Clairaut put

$$f(x) = A_0 + 2\sum_{n\ge 1} A_n \cos nx .$$

For $x_m = \frac{2\pi m}{k}$ he obtained linear equations for the A_n, which he cleverly transformed and solved. His solutions were

$$A_0 = \frac{1}{k} \sum_{m=1}^{k} f(x_m), \quad A_n = \frac{1}{k} \sum_{m=1}^{k} f(x_m) \cos nx_m ,$$

and for infinitely large k these are the integrals for the Fourier coefficients. At the time Clairaut was a corresponding member of the Berlin Academy and corresponded also with Euler, but no one noticed that his interpolation scheme could have resolved the vibrating-string debate. Besides, the formulas are found in Euler's own papers.

In the deliberation cited by Riemann (W. 233), Lagrange, like Clairaut, dealt with the interpolation problem by dividing the interval from 0 to π into $n+1$ parts. Riemann wrote:

> "Had Lagrange allowed n in this formula to get infinitely large, he would have certainly arrived at Fourier's result."
>
> ("Hätte Lagrange in dieser Formel n unendlich groß werden lassen, so wäre er allerdings zu dem Fourierschen Resultat gekommen.")

Without good reason, and probably to diminish Fourier's achievement, Poisson attributed this step, actually carried out by Clairaut, to Lagrange. Riemann said that Lagrange

> "may well have undertaken the whole work because he believed that these arbitrary functions could not be expressed by means of a formula...."
>
> ("hatte vielmehr die ganze Arbeit gerade unternommen, weil er glaubte, diese willkürlichen Funktionen liessen sich nicht durch eine Formel ausdrücken. ...") (Lagrange 1759, *Œuvres* 1, 39–148).

Lagrange was more concerned about the foundations of analysis than were all of his contemporaries. As already mentioned, at his suggestion the Berlin Academy proposed as a prize problem the clarification of the role of the infinite (and of the infinitely small) in mathematics. He found all of the nearly two dozen submissions wanting and made an attempt to eliminate from analysis not only infinitesimals but also limits and convergence considerations. His 1797 theory of analytic functions was an attempt to establish analytical expressions in the form of power series as a sufficiently rigorous foundation for analysis, which was thus to be based, in Euler's earlier sense, on algebra as metaphysically above suspicion. Euler's vague "analytical expression" was now made precise in a form that was to succeed in complex but not in real analysis, the subject that was Lagrange's primary concern.

In the case of Lagrange one can see with special clarity what was imminent in the vibrating-string debate: a split of mathematics into rigorous and usable. We find this split in the consciousness of the researchers who made the most significant contemporary contributions to both physics and mathematics. It is from this background that we must view the achievements of Fourier, Cauchy, and Riemann.

The three positions adopted in the vibrating-string controversy are worthy of the interest of a psychologist versed in mathematics and in physics. But they also point to essentials in the relation of analysis and physics, a relation of the greatest importance to Riemann. We can most easily understand Daniel Bernoulli's position. His series representation is a description of the physical phenomenon: the fundamental vibration appears with amplitude b_1 and the overtones with amplitudes b_n. The rest is of no interest to physics. D'Alembert's insistence on the analytical expression as the only admissible tool of analysis distances him from a physically reasonable mathematical model. In this prehistory of Riemann, what interests us most is Euler's attitude towards the function concept and its subsequent evolution.

We are concerned, and so was probably Riemann, with these fundamental questions and not with the chatter, especially abundant in the 19th century, that engulfed the controversy and bypassed the essential questions. One can find the facts in Truesdell 1960 (in Euler, *Opera omnia* (II) 10, 237–300). Truesdell begins with the statement of his 'first principle': "What follows confirms the principle that ever the greatest quantity of paper is smeared over with the dullest matter," and talks of the need to "descend to the celebrated and deplorable quarrel which watered the efforts of the principal savants of the middle of the [18th] century." He is "tempted to leave out the whole matter," but "the dilettante essays of the last [i.e., the 19th] century have spread such misconception that it is best to go over the old ground once more, if only to illustrate that second principle that in the history of science nothing is less welcome and less read than an account of the facts." While Truesdell deals primarily with the aspect of mechanics, we concentrate on the mathematics that has merit as a model of the physical. Here another relevant reference is Lützen 1982, 15–24.

2.2.3 *On the development of function concepts*

The implicit assumption in Riemann's question on the representability of a function $f(x)$ by means of a trigonometric series was that two functions are said to be equal if and only if their values $f(x_0)$ and $g(x_0)$ are equal for every numerical input x_0. This notion was obvious to Riemann, but it became generally accepted only at the beginning of the 19th century; it was used consistently by Cauchy in his textbooks after 1821. A companion notion that persisted with great stubbornness was that of a function as an "expression" ("Ausdruck"). Riemann discussed this in his dissertation (W. 3; see Section 1.2.3).

Following his teacher Johann Bernoulli, Euler said in his *Introductio* of 1748 that "A *function* of a variable numerical magnitude is an *analytical expression* composed in some way of the variable numerical magnitude and of actual numbers or of constant numerical magnitudes." ("Eine *Funktion* einer veränderlichen Zahlgröße ist ein *analytischer Ausdruck*, der auf irgendeine Weise aus der veränderlichen Zahlgröße und aus eigentlichen Zahlen oder aus konstanten Zahlgrößen zusammengesetzt ist.") The vibrating-string problem brought with it the notion of a function given by an *arbitrary curve*, or, as we would put it today, by its *graph*. Euler's differential calculus of 1755 deals with such functions. In the case of functions of the first kind Euler occasionally used the expression *functiones continuae*, i.e., functions for which the rule, the formal analytical expression, remains the same throughout its domain of definition. In this sense, the spike function given by $y = x$ for $0 \leq x \leq \pi$, by $y = -x$ for $-\pi \leq x \leq 0$, and extended to a 2π-periodic function was for Euler a *functio discontinua* while its Fourier series

$$\frac{\pi}{2} - \frac{4}{\pi}\left(\cos x + \frac{\cos 3x}{3^2} + \frac{\cos 5x}{5^2} + \cdots\right)$$

was a *functio continua*.

The following note by Gauss (1847; *Werke* X. 1, 398–399) dates from Riemann's student days: "The discontinuous function of t that coincides with $\sin t$ for t between 0 and 180° and vanishes between 180° and 360° is =..." (our dots stand for the Fourier series). ("Die discontinuierliche Function von t welche für t von 0 bis 180° mit $\sin t$ übereinstimmt, von 180° bis 360° aber verschwindet, ist =..."). The law changes, the function is discontinuous! Every teacher is aware of a certain discomfort on the part of the students when they face a function "glued together" out of pieces, even if the function arises out of a physical example. In Section 2.4.2 we will return to the dualism between formulas and graphs; subsequent developments seemed to favor the latter.

If, as was done here in connection with Riemann, we begin the history of trigonometric series with the vibrating-string debate, then we are completely fading out the aspect of analytical expressions. It is this aspect that explains Euler's "error," namely, his objection to D. Bernoulli's position, for graph and expression are different by *nature*. The criterion of equality of values, viewed in the 19th century as a progressive formulation, brought with it also a loss of flexibility. This being so, we must deal with the period before the vibrating-string debate in order to discover the bounds within which Riemann found himself.

In his *Geometry* of 1637 Descartes distinguished between geometric (we would say algebraic) and mechanical curves; Leibniz was proud of the fact that his calculus could also be applied to transcendental curves. For the young Euler, analytical expressions took on a life of their own that was repeatedly justified by the computation of values. Using his formula $e^{it} = \cos t + i \sin t$ he obtained around 1755 many trigonometric series. For example, substituting $z = ae^{it}$ in the geometric series

$$\frac{1}{1-z} = 1 + z + z^2 + z^3 + \cdots$$

he obtained for the real part the expansion

$$\frac{a \cos t - a^2}{1 - 2a\cos t + a^2} = a\cos t + a^2\cos 2t + a^3\cos 3t + \cdots . \tag{7}$$

In accordance with his fundamental proposition that the value of a divergent expansion is equal to the value of the expression that gave rise to it, Euler put $a = 1$ and obtained (Euler, *Opera omnia* (I) 14, 542–584; 15, 435–497)

$$0 = \frac{1}{2} + \cos t + \cos 2t + \cos 3t + \cdots . \tag{8}$$

Integrating (8) from $t = \pi$ he obtained

$$\frac{\pi - t}{2} = \sin t + \frac{\sin 2t}{2} + \frac{\sin 3t}{3} + \cdots , \tag{9}$$

which we must accept as valid for $0 < t < 2\pi$. If Euler had extended (9) to a 2π-periodic function, he would have obtained a representation of a discontinuous function by means of a trigonometric series. Differentiating (8) he obtained

$$0 = \sin t + 2\sin 2t + 3\sin 3t + \cdots . \tag{10}$$

This is nonsense from the viewpoint of the equality of values. But we will see (cf. end of Section 2.4.2) that the subsequent adoption of conceptual criteria other than equality of values made such equalities reasonable and useful.

Historical accounts of divergent series, especially those written in the 19th century, were also full of dilettante chatter. Riemann got rid of this by restricting himself completely to convergent series. We can no longer do this today because this would prevent us from understanding Fourier and his contemporaries.

For Euler, series provided a seamless connection between real and complex analysis: consideration of power series for $z = e^{it}$ yielded trigonometric series. Riemann did not mention this connection in his habilitation paper but referred to it occasionally in other papers (W.110, 82). In his words, the validity of the expansion in terms of power series could be

> "easily established on the basis of Cauchy's work or by means of the Fourier series."
>
> ("nach Cauchy oder durch die Fourier'sche Reihe leicht bewiesen werden.")

If one rewrites the differential equation $u_{xx} + u_{yy} = 0$ in terms of polar coordinates, prescribes the values of u on the circle $x^2 + y^2 = R^2$, and assumes that they can be represented by a Fourier series, then one obtains the only regular solution of the differential equation valid in the interior of the circle by multiplying the nth term of the series by $(\frac{r}{R})^n$. One can proceed in the same way in the case of the imaginary part v of $w = u + iv$. Then one obtains a power series in $z = x + iy$ with $|z| = r$. Today this detour through the domain of the real is no longer common practice, and the likely reason for Riemann's reference to it is that, in his time, Cauchy's approach was not yet well known.

In his paper, especially in its historical part, Riemann frequently speaks of *arbitrarily given* functions. Later it became customary to link this concept with Dirichlet's name. Actually, the essential step was taken by Cauchy in 1821: we can speak of a function $y = f(x)$ if to every x_0 in the domain of variability there corresponds exactly one $y_0 = f(x_0)$. It was this univalence that led Riemann, in the complex domain, to the concept of a Riemann surface.

2.2.4 From Fourier to Dirichlet

J. B. J. Fourier's (1768–1830) starting point was the physical problem of heat conduction. The methods that he first published in 1822 had been known in Paris for more than ten years before this date, and had been further developed and applied to many differential equations of mathematical physics, primarily by S. D. Poisson (1781–1840) and A. L. Cauchy. In our account we pay no attention to priority questions or to detours described in great detail by Fourier. We use the complex notation that was gradually adopted at the time.

Just as in Daniel Bernoulli's case, the first step was provided by a product formula for solutions of linear differential equations, by means of which one obtained superposable special solutions $c(k, t)e^{ikx}$ for every k. Summation or integration over k yielded additional solutions, and $c(k, t)$ was obtained from an ordinary differential equation in the variable t as, say, $c(k)e^{-|k|t}$ for $t > 0$. The solutions

$$\int_{-\infty}^{\infty} c(k)e^{-|k|t}e^{ikx}\,dk \quad \text{or} \quad \sum_{k=-\infty}^{\infty} c_k e^{-|k|t}e^{ikx} \tag{11}$$

had to be made to fit the initial conditions at $t = 0$. It was primarily Poisson who took instead an infinitely small $t = \alpha > 0$. The $c(k)$ or c_k had to be obtained from an integral equation or from a system of linear equations in infinitely many unknowns. The trick is to first determine the initial conditions for which $c(k) = \frac{1}{2\pi} = c_k$. Using known integrals, or summing geometric progressions, one obtains in the case of integrals

$$\delta(x) = \frac{1}{2\pi}\int_{-\infty}^{\infty} c(k)e^{-|k|\alpha}e^{ikx}\,dk = \frac{\alpha/\pi}{\alpha^2 + x^2}, \tag{12}$$

and in the case of series the so-called Poisson kernel $\delta_{2\pi}(x)$, whose form is of no interest to us. The delta function notation is due to Dirac 1927.

It is easy to see that $\int_{-\infty}^{\infty}\delta(x)\,dx = 1 = \int_{-\pi}^{+\pi}\delta_{2\pi}(x)\,dx$, and that for α infinitely small the functions are infinitely small outside an infinitely small neighborhood of $x = 0$ (and of $x = 2\pi g$ with integral g in the case of series). Since the delta functions are invariably positive, we can use the intermediate value theorem for integrals (which Cauchy used as early as 1818) to show that

$$f(x) = \int_{-\infty}^{\infty}\delta(x - x')f(x')\,dx' \quad \text{or} \quad f(x) = \int_{-\pi}^{\pi}\delta_{2\pi}(x - x')f(x')\,dx' \tag{13}$$

to within an infinitely small error; this is so provided that $f(x)$ is continuous, and that it is 2π-periodic in the case of series and sufficiently small for large $|x|$ in the case of integrals. Here we again substitute (11) for the delta function, with $c(k) = c_k = 1/2\pi$, and obtain the Fourier integral formula

$$f(x) = \int_{-\infty}^{\infty} e^{-|k|\alpha}e^{ikx}g(k)\,dk \quad \text{with} \quad g(k) = \frac{1}{2\pi}\int_{-\infty}^{\infty} e^{-ikx'}f(x')\,dx', \tag{14}$$

and the series formula

$$f(x) = \sum_{k=-\infty}^{\infty} e^{-|k|\alpha}g_k e^{ikx} \quad \text{with} \quad g_k = \frac{1}{2\pi}\int_{-\pi}^{\pi} e^{-ikx'}f(x')\,dx'. \tag{15}$$

The equalities in the equations hold to within physically irrelevant infinitely small errors. All this is correct, including the interchanges of summation and integration, which are justified by the presence of the factor $e^{-|k|\alpha}$. With these $g(k)$ or g_k one obtains solutions $u(x, t)$ that are infinitely close to the given $f(x)$ for $t = \alpha$, provided that we put in (11) $c(k) = g(k)$ and $c_k = g_k$.

One was very much tempted to put afterwards $\alpha = 0$ in (14) and (15) because the infinitely small α is of no account. Cauchy was quite right when he observed that this is permissible if the series continues to converge for $\alpha = 0$; for then the transition from (15) to $f(x) = \sum_{-\infty}^{\infty} g_k e^{ikx}$ involves just an infinitely small error. Once α has been eliminated we actually have an equality. Since we know that the Fourier series of a continuous function need not converge, this cautionary remark is justified, and, in general, one cannot dispense with the "convergence-inducing" factors $e^{-|k|\alpha}$.

We know from Gauss' *Nachlass* (*Werke* VII, 470–472; in this connection see Laugwitz 1989b) that even he succumbed to the temptation. His derivation largely parallels Poisson's idea. Instead of Poisson's $e^{-\alpha}$, Gauss used $\frac{1-\epsilon}{1+\epsilon}$ with a positive infinitesimal ϵ, which, in the end, he put equal to 0 without much ado. He probably wrote this in 1816, in connection with a prize paper (in French; something very unusual for Gauss) on planetary perturbations that he was to submit to the Paris Academy but eventually did not. Gauss never disclosed his interest in the representation of periodic functions by their Fourier series, not even when Dirichlet had close contacts in this matter with him and with Riemann on the occasion of his stay at Göttingen in the fall of 1852. From Dirichlet's letter to Gauss, dated 20 February 1853 (Dirichlet, *Werke* 2, 386), we can infer that both were convinced of the representability of continuous functions by Fourier series "provided that one ignores certain very unusual cases" ("wenn man anders von gewissen ganz singulären Fällen absehen will").

This discussion belongs here because it shows that the common interests of Gauss and Riemann extended to this area as well.

More on the methods of Fourier and Poisson is found in Laugwitz 1989a. Riemann ignored completely the summation procedures in his historical overview and mentioned Cauchy (W. 234) in connection with an error that has nothing to do with the summation method. Nor did he mention Fourier's somewhat differently formed method of proof. The likely reason is that he regarded modes of inference involving infinitesimal quantities as unreliable. We discuss them because they clearly influenced Dirichlet.

The summation procedures were rediscovered, in the fullest sense of the word, at the end of the 19th century. The relevant earlier results had been forgotten. We pointed out in Section 1.1.5 that even results obtained in analysis around 1820 by universally recognized methods disappeared later. So it was not the use of infinitesimals that can be blamed for the fact that the results of Poisson and his contemporaries, to the effect that every continuous periodic function can be uniformly approximated by trigonometric polynomials (we are using later terminology), were forgotten. Weierstrass proved this result anew in 1885 (incidentally, before proving the corresponding theorem for algebraic polynomials). He mentioned Fourier, but failed to say that he was proving an old theorem stated in a new language. Even the celebrated theorem of Fejér, proved two decades later, was just old wine in new bottles. The pertinent articles in the *Encyklopädie* hardly ever restate the old results correctly. True, authors such as Poisson regarded these results as virtually obvious and stated them without special emphasis.

We note that Dirichlet used the Fourier *integral* formula in an 1830 paper on mathematical physics, that Riemann used it in his 1859 paper on the distribution of primes, and that neither voiced any reservations about it. On the other hand, both were reticent when it came to the representation of functions by means of Fourier *series*.

Unlike Fourier, we use once more complex notation. Using this notation we can express his main idea in the form of the relation (the delta notation is ours)

$$\delta(x-x') = \frac{\sin p(x-x')}{\pi(x-x')} = \frac{1}{2\pi}\int_{-p}^{p} e^{ik(x-x')}\,dk, \tag{16}$$

which Fourier used (1822 §415, §423) with an infinitely large p. The rest of his argument was geometric, with integrals interpreted as surface areas. The analysts could not accept this as rigorous. Fourier claimed that for a continuous $h(x')$ one had

$$\int \sin p(x-x')h(x')\,dx' = 0,$$

for the contributions between two zeros of the sine function are cancelled by the contributions between the two succeeding zeros; this is so because the distances between the zeros are infinitely small and because the continuity of the function $h(x')$ allows us to regard it as constant. In particular, the relevant function for (16) is

$$h(x') = \frac{1}{\pi(x-x')},$$

which is continuous for $|x-x'| \geq \omega$, where ω is a definite infinitely small quantity. Knowing that $\int_0^\infty \frac{\sin t}{t}\,dt = \frac{\pi}{2}$, Fourier drew from this the conclusion that

$$\int_{x-\omega}^{x+\omega} \delta(x-x')\,dx' = \int_{-\infty}^{\infty} \delta(x-x')\,dx' = 1.$$

In modern terms, he thereby showed that (16) is a delta function. The geometric argument runs the same way when we multiply the integrand by a continuous function. The result, correct to within an infinitesimal error, is

$$\int_{-\infty}^{\infty} \delta(x-x')f(x')\,dx' = f(x).$$

If we replace the delta function by the integral in (16), change the order of integration, and write afterwards ∞ for p, then we obtain the Fourier integral theorem; in (14) we must put $\alpha = 0$. In the case of series, Fourier proceeded in a similar way using the delta function

$$\delta_{2\pi}(x - x') = \frac{1}{2\pi} \sum_{k=-p}^{p} e^{ik(x-x')} .$$

At the end of §417 Fourier (1822) commented on the method as a whole: "The present derivation uses the concept of infinite magnitudes, which was always accepted by mathematicians. It would be easy to give the same derivation in another form by investigating the changes obtained as a result of the continual increase of the factor p in $\sin p(x - x')$. These considerations are too well known to require mention." Darboux noted in his edition of Fourier's works that this was carried out by Dirichlet in 1829.

Fourier's idea, which he found after years of searching, is simple and beautiful, but it must be conceded that his own realization of this idea involved the summation of infinitely many infinitely small errors, and that the behavior of functions such as $h(x')$ and $f(x')$ requires careful reflection. But the key idea should not be suppressed by its epsilontic implementation. The geometric formulation points to the source of counterexamples. Specifically, Fourier's delta function oscillates infinitely fast. If the factor $f(x')$ also oscillates rapidly, then mutual cancellation can cause trouble. Later it turned out that the unbounded variation of $f(x')$ leads to counterexamples.

Riemann gave very short shrift to these Parisian contributions. All he said was that

> "Fourier perceived " the integral representation of the coefficients,
>
> ("Fourier bemerkte" die Integraldarstellung der Koeffizienten (W. 232))

and finally (W. 234):

> "It was owing to Fourier that the true nature of trigonometric series was recognized in a perfectly correct way; since that time they have been frequently used in mathematical physics for the representation of arbitrary functions, and in each particular case one saw readily that the Fourier series actually converged to the value of the function; but it took a long time for this important theorem to be proved in all generality."
>
> ("Durch Fourier war nun zwar die Natur der trigonometrischen Reihen vollkommen richtig erkannt; sie wurden seitdem in der mathematischen Physik zur Darstellung willkürlicher Functionen vielfach angewandt, und in jedem einzelnen Falle überzeugte man sich leicht, daß die Fouriersche Reihe wirklich gegen den Werth der Function

convergire; aber es dauerte lange, ehe dieser wichtige Satz allgemein bewiesen wurde.")

Riemann rejected the delta functions; they appear in Cauchy's work as part of the topic of "singular" integrals over infinitely short intervals in which the functional values can become infinitely large. This is what Riemann meant by "Cauchy's other stipulations" (die "anderen Festsetzungen von Cauchy") bearing on the concept of the integral, stipulations he dismissed (W. 240). Parts of this material were resumed by Weierstrass in 1885/86 with reference to Fourier. In particular, Weierstrass proved the uniform approximability of a continuous function by trigonometric sums by a method that was a translation of the Paris procedure into his own language (Laugwitz 1992).

On the other hand, Riemann devoted to his teacher Dirichlet pages W. 235–239 and reproduced his proof of 1829, a proof presented in lectures for beginners to this day. Dirichlet had spent more than four years (1822–1827) in Paris. During that time he was in close contact with Fourier. This influence, which he acknowledged, is easily recognized in his papers of 1829/30. The 1829 formulation could be called the first influential paper in the epsilontic style. While this method turned up time and again in Cauchy's work, it may not have been clearly recognized because it alternated with infinitesimal arguments. We explained this in Section 4 of the Introduction.

Every historical text on analysis pays special attention to Dirichlet's formulation. One such is Bottazzini 1986, which includes a detailed discussion of Cauchy's contribution. This book gives a good summary, in English, of Riemann's paper and lists its results in the same order as does the original. This being so, we can proceed somewhat differently and try to reconstruct the order of Riemann's thought.

2.3 Riemann's results

Riemann wanted to answer question (B) stated in Section 2.2.1. The relevant part of his paper is §§ 8–13, W. 245–264. Here § 10 occupies a special position. We discuss it first.

2.3.1 Application of the concept of the integral to the Fourier coefficients

Section 10, W. 253–255, shows how Riemann may have arrived at his notion of an integral. The question is whether the Fourier coefficients a_n, b_n of a given function tend to 0. In order to answer this question we consider the integral for b_n,

$$\int_0^{2\pi} f(x) \sin nx \, dx \, .$$

Using the geometric idea of Fourier, to whom he admittedly does not refer, Riemann decomposes the integral "into integrals of extent $\frac{2\pi}{n}$" ("in Integrale vom Umfang $\frac{2\pi}{n}$"). For such a partial integral "The sine is positive in the first half and negative in the second." ("Der Sinus wird in der ersten Hälfte positiv, in der zweiten negativ"). If in the kth interval

$$m_k \leq f(x) \leq M_k,$$

then, because $-M_k \leq -f(x) \leq -m_k$, the sum of the integrals over the two halves is less than $\frac{2}{n}(M_k - m_k)$, or, using the notation of Section 2.2.2, less than $\delta_k D_k/\pi$. Thus if the Riemann integrability condition $\sum \delta_k D_k \to 0$ for $n \to \infty$ is satisfied, then $b_n \to 0$, and, similarly, $a_n \to 0$. We see that the Riemann condition enters here in an almost obvious way!

The result is: the Fourier coefficients of a bounded 2π-periodic integrable function converge to 0.

Given the state of knowledge at the time — recall Dirichlet's letter to Gauss — it may have been plausible to think that such functions, satisfying the additional condition $f(x) = \frac{f(x+0)+f(x-0)}{2}$, were representable by means of Fourier series. Riemann may have concerned himself with this issue. As a matter of fact, he returned to the issue of convergence and representability in §13, W. 260 ff. Now we know that this twin question has an affirmative answer under the additional assumption of bounded variation.

Riemann showed in §13 that if one requires the integrability of the function but not its boundedness then its Fourier series need not converge; in fact, it is possible that "ultimately its terms become infinitely large" ("ihre Glieder zuletzt unendlich groß werden"). He devoted to this issue three printed pages, which shows that he thought it important.

From the viewpoint of the history of mathematics it is interesting that Riemann may be said to have introduced his concept of an integral in an *ad hoc* manner in connection with the question of convergence to 0 of the Fourier coefficients. This was the earliest question that led beyond Cauchy's integral concept for piecewise continuous functions. We owe it to this historical accident that even today lectures on integration begin (almost everywhere) with Riemann's integral. History would have been different if one had asked the question: what kind of integral implies the equality $\lim \int_a^b f_n = \int_a^b f$, where f_n is a monotonically increasing sequence of integrable functions that converge pointwise to the limit f? One could have also required the preservation of the validity of the theorem on integration as the inverse of differentiation. That the Riemann integral does not satisfy this requirement was shown by Volterra in 1881: there are functions that are not Riemann-integrable which nevertheless have bounded derivatives on closed intervals.

Riemann's notion of an integral is not the most felicitous even in the area of trigonometric series. In 1906 Lebesgue showed in his *Leçons sur les séries trigonométriques* that a trigonometric series is a Fourier series if it converges pointwise to a bounded function. This is true if we use for the coefficients the Lebesgue integral and false if we use the Riemann integral.

These examples show that pointwise convergence is a rather unfortunate concept, and, to some extent, Riemann must be blamed for its temporary dominance.

2.3.2 *Riemann's associated function* $F(x)$

From now on a_n and b_n will be any number sequences and not the Fourier coefficients of a function. We will investigate the function

$$\frac{a_0}{2} + \sum_{n=1}^{\infty} a_n \cos nx + b_n \sin nx = \sum_{n=0}^{\infty} A_n(x)$$

defined for all x for which the series on the right-hand side converges. This means that our function need not be defined for all real x. Assume at first that the a_n and b_n are null sequences. According to the previous section, the Fourier series of integrable bounded functions are subsumed into our series provided they converge.

Under the assumption $a_n \to 0, \ b_n \to 0$, the $A_n(x)$ converge uniformly to 0. This implies that the series

$$F(x) = C + C'x + \frac{a_0 x^2}{4} - \sum_{n=1}^{\infty} \frac{A_n(x)}{n^2},$$

obtained by twice integrating the previous series termwise, is uniformly convergent. But then the *associated function* $F(x)$ is continuous. Riemann proved this, for he did not have at his disposal the concept of uniform convergence and obviously did not wish to refer to Cauchy or to others. Today we can say that $F'' = f$ in the sense of the theory of distributions. Riemann considered the second difference quotient

$$D(x) = \frac{F(x+\alpha+\beta) - F(x+\alpha-\beta) - F(x-\alpha+\beta) + F(x-\alpha-\beta)}{4\alpha\beta} \tag{17}$$

and proved that if the series $f(x)$ converges then $D(x)$ converges to $f(x)$:

> "if α and β become infinitely small in such a way that their ratio remains finite, [then $D(x)$ converges] to the same value $f(x)$ as the series."

> ("wenn α und β so unendlich klein werden, daß ihr Verhältnis endlich bleibt, [konvergiert $D(x)$] gegen denselben Werth $f(x)$ wie die Reihe") (W. 246).

In his proof he dealt first with the case $\alpha = \beta$ and found that

$$D(x) = A_0 + \sum_{n=1}^{\infty} A_n \left(\frac{\sin n\alpha}{n\alpha}\right)^2 .$$

The procedure of summation of series for infinitely small α is of a kind much used by Cauchy and Poisson precisely in connection with trigonometric series; see Laugwitz 1989a. Riemann mentioned no precursors, not even Abel, whose method of partial summation he was about to use. A page from the *Nachlass* (W. 460–461) published by Dedekind shows that on 14 September 1852 Dirichlet had called Riemann's attention to Abel's procedure. The page contains Riemann's sketch of a proof of Abel's limit theorem. It seems that at the time neither the method nor the theorem enjoyed universal acceptance, and, as Liouville recalled (Dirichlet, *Werke* 2, 306), Dirichlet helped to propagate them; Riemann's sketch is much clearer than Liouville's account of Dirichlet's proof. Like other notes of that time, this one was written in Latin, which Riemann tried to practice. Dedekind guessed (W. 466) that these deliberations were to lead to examples of pathological functions in the habilitation paper.

Putting $A_n = e_{n+1} - e_n$ Riemann obtained

$$D(x) = f(x) + \sum_{n=1}^{\infty} e_n \left[\left(\frac{\sin (n-1)\alpha}{(n-1)\alpha}\right)^2 - \left(\frac{\sin n\alpha}{n\alpha}\right)^2\right]$$

and showed that the sum on the right converges to 0 "with infinitely decreasing α" ("mit in's Unendliche abnehmendem α"). He reduced the case of more general α and β to that of equal α and β.

As for summation methods, the new element contributed by Riemann was his pointwise approach, whereas Cauchy and Poisson always treated the representability of $f(x)$ on intervals because they did not have Abel's method at their disposal and needed integrations instead.

Riemann obtained two necessary conditions for a 2π-periodic $f(x)$ "whose terms ultimately become infinitely small for every value of x" ("deren Glieder für jeden Werth von x zuletzt unendlich klein werden") to be equal to a trigonometric series. The first necessary condition is:

> There must be a continuous function $F(x)$ such that $D(x)$ in (17) converges to $f(x)$ "when α and β become infinitely small and their ratio remains finite."

("wenn α und β unendlich klein werden und dabei ihr Verhältnis endlich bleibt") (W. 251).

The second necessary condition is:

"Further, with increasing n,

$$n^2 \int_b^c F(x) \cos n(x-a)\lambda(x)dx$$

must ultimately become infinitely small, provided that $\alpha(x)$ and $\lambda'(x)$ are = 0 at the limits of the integral and are always continuous between them, and $\lambda''(x)$ does not have infinitely many maxima and minima."

("Es muß ferner

$$n^2 \int_b^c F(x) \cos n(x-a)\lambda(x)dx\,,$$

wenn $\alpha(x)$ und $\lambda'(x)$ an den Grenzen des Integrals =0 und zwischen denselben immer stetig sind, und $\lambda''(x)$ nicht unendlich viele Maxima und Minima hat, mit wachsendem n zuletzt unendlich klein werden.")

("Become infinitely small" means that we obtain a null sequence.)

All this is hardly elegant; the requirements on λ are of a technical nature because one integrates by parts, and convolutions with the second derivative of the Fourier kernel $\frac{1}{2\pi}\frac{\sin(2n+1)x/2}{\sin(x/2)}$ turn up.

Nevertheless, taken together, the two necessary conditions turn out to be sufficient:

"Conversely, if these two conditions are satisfied, then there is a trigonometric series whose coefficients ultimately become infinitely small and which represents the function wherever it converges."

("Wenn umgekehrt diese beiden Bedingungen erfüllt sind, so giebt es eine trigonometrische Reihe, in welcher die Coefficienten zuletzt unendlich klein werden, und welche überall, wo sie convergiert, die Function darstellt") (W. 251).

Moreover, since the preceding interval from b to c can be arbitrarily small, there follows a localization theorem (W. 253), familiar for more specialized classes of functions in the context of delta functions:

"Thus this investigation has shown that if the coefficients of the series [for $f(x)$] ultimately become infinitely small, then the convergence of the series for a particular value of x depends only on the behavior of the function $f(x)$ in the immediate vicinity of this value."

> ("Aus dieser Untersuchung hat sich also ergeben, daß wenn die Coefficienten der Reihe [für $f(x)$] zuletzt unendlich klein werden, dann die Convergenz der Reihe für einen bestimmten Werth von x nur abhängt von dem Verhalten der Function $f(x)$ in unmittelbarer Nähe dieses Werthes.")

In §§11 and 12 there are attempts to generalize to cases in which the $A_n(x)$ do not converge to 0 for all x. Here Riemann used functions that are continuously differentiable sufficiently many times but vanish outside an interval. This corresponds to the test functions of the modern theory of distributions, and Narasimhan (N. 6–7) stresses this parallel. But it should be pointed out that around 1850 Cauchy made extensive use of just such factors and even introduced for them a special name (*limitateurs*).

2.4 Trigonometric series after Riemann

2.4.1 From trigonometric series to set theory

There is no completely satisfactory answer to Riemann's question (B) concerning the characterization of functions representable by trigonometric series.

It is an irony of the history of mathematics that, immediately after its publication, Riemann's habilitation paper, his least successful work, inspired a development regarded by many as the profoundest change in modern mathematics, namely the rise of set theory. We discuss this also with a view to Chapter 4.

As a student of Kronecker at Berlin, Georg Cantor (1845–1918) worked first in number theory. From 1869 on he taught at Halle, where Eduard Heine (1812–1881) called his attention to problems connected with trigonometric series.

As early as 1870 there appeared Cantor's paper "Beweis, daß für jeden reellen Werth von x durch eine trigonometrische Reihe gegebene Function $f(x)$ sich nur auf eine einzige Weise in dieser Form darstellen läßt" (*Crelle* 72, 139–142) ("A proof that a function $f(x)$, given by a trigonometric series for every value of x, can be represented in this form in just one way"). Using ideas of H.A. Schwarz, Cantor showed that if a trigonometric series converges for all x and its sum is 0, then the associated Riemann function $F(x)$ must be linear, and this implies that all the coefficients of $f(x)$ must vanish. Cantor soon realized that the requirement of convergence with sum 0 can be violated at a finite number of points of the period interval and gradually arrived at the question: what must be the structure of exceptional sets P such that convergence with sum 0 at all points not in P suffices for the vanishing of the coefficients?

Again in collaboration with Heine, Cantor created his theory of irrational numbers based on Cauchy sequences of rational numbers. The theory is contained in the paper "Über die Ausdehnung eines Satzes aus der Theorie der trigonometrischen Reihen" (*Math. Annalen* 5 (1872), 123–132) ("On the extension of a theorem in the theory of trigonometric series"). Here Cantor introduced the derived set P' of a set P as the set of limit points of P and showed that the admissible sets P are those for which one of the derived sets P', P'', P''', ... consists of at most finitely many points. In addition to being one of the early sources of pointset theory, this paper is also the origin of transfinite ordinals. Specifically, $P' \supseteq P'' \supseteq P''' \supseteq \cdots \supseteq P^{(n)} \supseteq \cdots$, and it can happen that the intersection of the derived sets $P^{(n)}$ is not empty (Cantor did not use this terminology). Cantor denoted this intersection by $P^{(\infty)}$ and later by $P^{(\omega)}$. Beginning with this set we can again form derived sets; we can, so to say, go on to count at infinity! Cantor's discovery that one can speak of transfinite cardinals, and that in particular the cardinality of $\mathbb{R}$ is greater than that of $\mathbb{N}$, came somewhat later. What interests us here is that a concrete question related to the work of Riemann provided the motive for transfinite numbers.

If one ignores the deliberations on trigonometric series and the proof of the existence of transcendental numbers based on the countability of the algebraic numbers, then Cantor gave virtually no applications of his ideas about manifolds or sets. In retrospect, one might say that the early idea of transfinite numbers seems to suggest the extension of the principle of proof by complete induction. It was not until 1904 that Zermelo produced a "Proof that every set can be well-ordered" ("Beweis, daß jede Menge wohlgeordnet werden kann") (*Math. Annalen* 59, 514–516), i.e., that it can be mapped in a one-to-one way onto an initial segment of the (finite and transfinite) ordinal numbers.

The principle of proof by transfinite induction turned out to be fruitful in (besides other areas) analysis, the original domain of transfinite numbers. Hamel proved the existence of discontinuous solutions of the functional equation $f(x+y) = f(x) + f(y)$ by showing that $\mathbb{R}$, viewed as a vector space over $\mathbb{Q}$, has a basis: think of $\mathbb{R}$ as well-ordered and include an element of it in the basis if and only if it is linearly independent of its predecessors. Steinitz gave important applications in field theory, and as late as 1935 Rellich used transfinite induction to prove the existence of orthonormal bases in nonseparable Hilbert spaces (one such is the space of almost-periodic functions). In this way the use of transfinite numbers returned repeatedly to the vicinity of its origin.

Max Zorn reports that back in 1933, at Hamburg, he used the lemma, named in the meantime after him, as a working principle in algebra (see P.J. Campbell in *Historia Mathematica* 5 (1978), 77–89). Zermelo realized that the well-ordering theorem is a consequence of the axiom of choice, which asserts that given a system of sets $\{M_\nu\}$ there is a function f such that $f(M_\nu) \in M_\nu$. This axiom, and Zorn's lemma which is equivalent to it, have been used since the 1950s instead of transfinite induction, which is virtually forgotten today. Transfinite numbers have been eliminated from "concrete" mathematics, and people are no longer aware of the origin of set theory in Riemann's questions about trigonometric series.

Recommended books on the history of set theory are Dauben 1979 and Purkert-Ilgauds 1987.

2.4.2 *On the further development of trigonometric series: The arithmetization of functions and their emancipation in functional analysis*

One could view it as an accident that Riemann's investigations of a group of problems derived largely from physics provided the impulse for the development of set theory, which distanced mathematics from physics. But hindsight tells us that this development was implicit in Riemann's position — though of course we are not suggesting that all of its consequences were predictable. We recall Riemann's insistence that Dirichlet's results suffice for all functions "occurring in nature" (für alle "in der Natur vorkommenden" Funktionen). Moreover, as pointed out earlier, Riemann investigated pointwise convergence, whereas earlier mathematicians were interested in the convergence of series of functions on intervals, and thus at least implicitly in uniform convergence. Riemann explicitly included the investigation of series of *numbers* as against the earlier concern with series of *functions*. In the older analysis, oriented towards mathematical physics, rearrangements of series of numbers were devoid of interest. Each term in a series of functions had its number, its fixed position. Riemann's rearrangement theorem (W. 235), which asserts that a conditionally convergent series of numbers can be rearranged so that its sum has a preassigned value, including $+\infty$ and $-\infty$, is important only as an indicator of change. From a purely mathematical viewpoint the theorem is a banality that anyone who poses the question is bound to hit on, and presenting it as "Riemann's rearrangement theorem" gives students a false sense of its intrinsic importance.

The change we mentioned is the replacement of *function* as a fundamental concept by *number*. Dedekind's cuts and Cantor's construction of the reals out of Cauchy sequences of rationals were a continuation of this development. The motto of arithmetization was coined and the concept of a function was reduced to that of a collection of number pairs.

It was different in mathematical physics. Here pointwise convergence turned out to be of no interest. A useful concept was convergence in the mean.

One defines a scalar product $(f, g) = \frac{1}{\pi} \int_{-\pi}^{\pi} f(t)g(t)\,dt$ and a norm $||f|| = \sqrt{(f, f)}$, so that $f_n \to f$ is equivalent to $||f_n - f|| \to 0$. The orthogonality of the trigonometric functions is appropriately fitted in. This structure can be extended to that of a Hilbert space. For this extension Cauchy's notion of an integral suffices, for it yields all that is needed for finite trigonometric sums, and these approximate all the required functions in the sense of the norm. This development could have gone from Fourier and Cauchy straight into the 20th century without Riemann, Cantor, and Lebesgue. In lecturing to physicists and engineers one can proceed in just this way.

If we try to resume the historical thread of Section 2.2 after acknowledging Riemann's paper, then we note that Klein's lectures (Klein 1926/27) contain nothing that bears on this subject beyond skimpy references to Fourier and Dirichlet. The textbooks, exemplified by the French *Cours d'analyse* after Cauchy 1821, avoided the topic of trigonometric series in spite of the fact that their authors, almost all of whom were researchers in mathematical physics, had dealt with these as well as with more general series expansions. Weierstrass, who was very influential because of his Berlin school, approached this topic in his lectures on the approximation theorem as late as 1886. He admitted that he got his main idea from Fourier's theory of heat and switched immediately from trigonometric to ordinary polynomials. In 1857/58, at the beginning of his teaching activity at the polytechnical school, he was obliged to make an exception and wrote about it to H.A. Schwarz on 14 March 1885 while working on the approximation theorem: "Once, in my third semester, I gave here a course on the application of Fourier series and integrals to problems in mathematical physics. But the lack of rigor which I encountered in all pertinent works I then had access to, and the fruitlessness of my efforts at the time to remedy this flaw, annoyed me to such an extent that I decided never again to give this course." ("Ich habe hier einmal, in meinem dritten Semester eine Vorlesung über die Anwendung der Fourierschen Reihen und Integrale auf mathematisch-physikalische Probleme gehalten. Der Mangel an Strenge aber, den ich in allen mir zugänglichen einschlagenden Arbeiten wahrnahm, und die Fruchtlosigkeit meiner damaligen Bemühungen, diesem Mangel abzuhelfen, haben mich so verdrossen, daß ich niemals das Colleg noch einmal zu lesen mich entschließen konnte") (Biermann, 1988, 90).

The topic was too difficult. One tended to withdraw to the seemingly more general complex analysis. Here everything could be done with power series, which behave within their circles of convergence as nicely as do polynomials. Dirichlet had so much confidence in Riemann that he urged him to clarify the subject. But even Riemann returned quickly to functions of a complex variable.

After Riemann we can discern two tendencies. One was to pile up individual facts; such collections are found in Burkhardt 1908 and in Zygmund 1935. The other research direction was guided by the motto "Functions of a real variable" ("Reelle Funktionen") and focussed on the general and fundamental: measure theory and integration theory. Beginning in the middle of the 20th century, the theory of generalized functions (distributions) returned to the historical starting problem, the application of Fourier analysis to differential equations. This is easy to explain in the case of trigonometric series.

A trigonometric series $f(x)$ represents a 2π-periodic distribution of order m if there is a constant M such that the inequality $|a_n|, |b_n| \leq M \cdot n^{m-2}$ holds for the coefficients a_n and b_n. Then the series is the mth termwise derivative of a series that converges absolutely and uniformly and therefore represents a continuous function $F(x)$. For $m = 2$ we have the examples considered by Riemann, and $F(x)$ is his associated function. Distributions of order 0 are the continuous functions, and distributions of order m can be thought of as generalized mth derivatives of continuous functions. The 2π-periodic delta function

$$\delta_{2\pi}(x) = \frac{1}{2\pi} + \frac{1}{\pi} \sum_{n=1}^{\infty} \cos nx ,$$

which we encountered in the work of Euler and Fourier, is also of order $m = 2$.

Today distributions can be regarded as elements of a topological vector space in which the elements to be viewed as actual functions form a dense subspace. In this analysis, formal series acquire legitimacy and the basic objects are the functions. The functional *values* are not very important in the *calculus*, but obviously retain their importance in applications, where we expect the final outcomes to be proper functions. This tendency of functional analysis, running against arithmetization, could already be discerned in Klein's lifetime, as well as at the Hilbert school in Göttingen. Those who developed the new theories of generalized functions (Dirac in the 1920s, Sobolev in the 1930s, and Schwartz in the 1940s) were unaware of the formulations that dated back to the first decades of the 19th century. Just as in the time of Fourier, Poisson, and Cauchy, the most important applications became the partial differential equations of mathematical physics. They must be regarded as a kind of generalization of the ancient vibrating string.

2.5 A self-contained chapter: Gauss, Riemann, and the Göttingen atmosphere

In his lectures on the history of mathematics (Klein 1926, 249) Felix Klein says that Riemann "could not have listened to many lectures of the by then

70-year-old Gauss, who lectured little anyway. The shy young student was certainly unable to establish human relations with Gauss; after all, Gauss taught very reluctantly, took little interest in most of his students, and was quite unapproachable. We must nevertheless call Riemann a student of Gauss, in fact, the only true student of Gauss who grasped his inner (!) ideas, as we are now gradually getting to know them in outline from the *Nachlass*" (Riemann "kann nicht viel bei dem damals schon 70jährigen Gauß, der ohnehin wenig las, gehört haben. Menschliche Beziehungen zu Gauß hat der junge, schüchterne Student sicher auch nicht anknüpfen können; Gauß lehrte nur widerwillig und brachte der Mehrzahl seiner Hörer wenig Interesse entgegen, war auch sonst recht unnahbar. Trotzdem müssen wir Riemann einen Schüler von Gauß nennen, ja er ist der einzige eigentliche Schüler von Gauß, der auf dessen innere (!) Ideen eingegangen ist, wie wir sie jetzt im Umriß allmählich aus dem Nachlaß kennen lernen").

Klein, who was one of the most enthusiastic propagators of Riemann's complex analysis, goes on to say: "Riemann's close relation to Gauss in his scientific ideas is quite wonderful and almost puzzling for us." ("Ganz wunderbar und fast rätselhaft für uns ist Riemann's nahe Beziehung zu Gauß in seinen wissenschaftlichen Ideen") (p. 249). "The only explanation for this is that the Göttingen atmosphere was then saturated with these geometric ideas, and exerted its uncontrollable but powerful pressure on the very gifted and receptive Riemann. It is of utmost importance what spiritual environment a person enters; it influences him far more strongly than the facts and concrete knowledge that are offered to him!" ("Wir können es gar nicht anders ansehen, als daß die Göttinger Atmosphäre damals mit diesen geometrischen Interessen gesättigt war und ihren unkontrollierbaren aber starken Zwang auf den sehr begabten und empfänglichen Riemann ausübte. Es ist eben viel wichtiger, in welche geistige Umgebung ein Mensch hineinkommt, die ihn viel stärker beeinflußt als Tatsachen und konkretes Wissen, das ihm geboten wird!") (p. 249). Then comes Klein's comment (p. 251) on the concurrence of their basic physical conceptions: "And finally there is that mystical, undeniable and yet not clearly understandable influence of the general atmosphere on a sensitive spirit." ("Es ist eben jener mystische, nicht abzuleugnende und doch auch nicht klar zu fassende Einfluß der allgemeinen Atmosphäre auf einen empfindlichen Geist").

The present conception of the history of science makes it impossible for us to be satisfied with Klein's necromancy. Klein had access in Göttingen to all the sources we have relied on; he made incomplete use of them.

Of course, when he was still a gymnasium student Riemann knew who Gauss was. We mentioned that he read Gauss' works beginning in the first semester of his university study. During his second semester, in the winter of 1846/47, he took Gauss' method of least squares, and in the summer of 1849, after returning from Berlin, his higher geodesy course. Years later Dedekind described Gauss' lectures, given in a workroom of the observatory; it is almost inconceivable that the Privy Councillor engaged in discussions with his listeners during or after the lectures (Dedekind *Werke* II, 293).

Today one asks a scientist for reprints of his papers, but at that time things were different. For example, Riemann tried in vain for a long time to obtain a copy of Gauss' paper on conformal mappings. On 30 March 1849 he wrote to his father that Dirichlet had secured for him access to the library at Berlin where he could read Gauss' otherwise unobtainable papers. And it was Dr. Galle of the observatory who finally provided him with a copy of the previously mentioned paper (see Neuenschwander 1981, 89).

Riemann's other teachers, Jacobi, Dirichlet, and W. Weber, who shared with Gauss some of the main contemporary research interests and maintained contact with him, exerted palpable rather than "atmospheric" influence on him. The claim of Göttingen mathematicians, in Klein's time as well as later, that "Extra Gottingam non est vita (mathematica)" was a self-generated myth. It seems that Berlin and Italy agreed better with the "sensitive" Riemann than did Göttingen — and not just meteorologically.

We discuss Riemann's intellectual attraction to Gauss in the chapters on complex analysis and on geometry but will touch on it here as well. The present context is largely that of real analysis. Klein writes (p. 249): "As the first instance of the inner contact between the two, for which there is no direct philological proof, we regard Riemann's papers on the hypergeometric function, which make use of a multitude of ideas not published by Gauss." ("Als einen ersten Beleg für den inneren Kontakt zwischen beiden, der sich philologisch nicht unmittelbar nachweisen läßt, wollen wir etwa Riemann's Arbeiten über die hypergeometrische Funktion ansehen, in denen eine Menge von Gauß nicht veröffentlicher Ideen benutzt werden"). True, here Riemann goes over into the complex domain, but in the paper published as early as 1857 (W. 67) he wrote that he was familiar with unpublished investigations of Gauss in which the work was done in the real domain.

On the basis of Dirichlet's letter we concluded previously that Gauss presumably made no further reference to his earlier reflections pertaining to trigonometric series. But here too there is no need to exploit the Göttingen atmosphere. The obvious common source is Fourier. Gauss was inspired by Poisson and Riemann by Dirichlet, and thus each of them, indirectly, by Fourier. The only difference is that Gauss was concerned with applications to physics — he touched on the mathematical question of representability in passing but did not pursue it — while Riemann emphasized the mathematical aspect; the issue was presumably completely settled in all cases that "occurred in nature."

In complex analysis, Gauss and Riemann, in contrast to Cauchy, made essential use of geometry. But in the real domain, both, like Cauchy, adhered without exception to the analytical mode of thought as the only rigorous one.

There is some *algebraic* thinking in young Gauss' number theory. In spite of the fact that Riemann knew quite early the work of Legendre as well as the *Disquisitiones arithmeticae*, he never used such reasoning in his lectures or in his research. His famous paper on number theory is pure Cauchy analysis. In this case the Gaussian atmosphere failed to work!

Did Riemann try to establish contact with Gauss after his return from Berlin and before being awarded the doctorate? Neuenschwander found Riemann's letter to Gauss, dated 12 July 1850, that deals with the so-called Pfaff problem. In letters dated 24 November 1851 and 28 September 1852 Riemann mentions visits to Gauss (Neuenschwander 1981, 89–90).

Johann Friedrich Pfaff (1765–1825) was close to the combinatorial Hindenburg school mentioned in Section 4.2 of the Introduction. He taught at Helmstedt, beginning in 1788, and at Halle, beginning in 1800. He was the doctoral adviser ("Doktorvater") of Gauss at the Braunschweig Land University in Helmstedt. The prince had insisted that the native of the *Land* whose career he promoted should at least obtain his doctorate at Helmstedt after having been permitted to study abroad. Gauss kept up contact with Pfaff.

Today a Pfaffian form is a linear differential expression

$$\Omega = \sum_{k=1}^{n} p_k \, dx_k,$$

where the p_k are functions of n variables x_i and the question about the (highest-dimensional) integral manifolds of such forms is known as Pfaff's problem. In his letter Riemann dealt with this problem and alluded to a relevant letter by Gauss from 1816 (Gauss, *Werke* III, 231–241).

This attempt of Riemann to establish contact with Gauss was particularly ill-chosen. After 35 years Gauss was most unlikely to take a great deal of interest in the problem. Moreover, his letter was not an original paper but a review, which began with the words: "Prof. Gauss is submitting to the Royal Society a manuscript by Prof. Pfaff at Halle titled Methodus generalis, aequationum differentiarum partialium ... complete integrandi." ("Der Königlichen Societät ist durch Hrn. Prof. Gauss eine handschriftliche Abhandlung des Hrn. Prof. Pfaff in Halle vorgelegt, überschrieben Methodus generalis, aequationum differentiarum partialium ... complete integrandi.") Actually, Pfaff's paper was not published by the Göttingen Society, but appeared in the *Abhandlungen der Königlichen Akademie der Wissenschaften Berlin* (1814–15), 76–136.

It seems that during the time before the habilitation there were hardly any contacts between Gauss and Riemann. The usual generous explanation for this state of affairs has been Gauss' poor health. Riemann's habilitation submission is dated 4 December 1853. Gauss was then 77. How poor was Gauss' health at the time? The answer is provided by excerpts from letters written by Gauss to Alexander von Humboldt (see Biermann 1990, 201–202). On 10 May 1853 he complained that he had suffered for six or seven years from congestion in the chest and throat, from shortness of breath, palpitation, and sleeplessness, so "that there are ever fewer hours fit for the elaboration of scientific results." (so "daß die zur Verarbeitung wissenschaftlicher Untersuchungen geeigneten Stunden immer seltener werden"). One year later he reported that since 1854 his condition had gradually deteriorated to the point where "sitting at desk even for a short time [is] extremely painful" ("das Sitzen am Schreibtisch selbst nur während einer kurzen Zeit ungemein sauer" werde).

Nevertheless, in the fall of 1853 Gauss was fully able to engage in extensive scientific discussions with Dirichlet on, besides other things, trigonometric series. For Riemann's habilitation paper all he mustered was the following laconic statement: "The tendered paper contains so many indications of refined knowledge, good judgment, and independent skillfulness that it suffices completely for its purpose. For the test lecture I would prefer the third of the proposals. I will conduct the colloquium at the appointed time" (Dekanatsakte Philos. Fakultät Nr.137, fol. 118 — not Nr.18, as M. Noether (N. 112) wrote; courtesy of Dr. U. Hunger, Universitätsarchiv Göttingen) ("Die vorgelegte Abhandlung enthält so viele Beweise von feinen Kenntnissen, guter Beurtheilung und selbstthätiger Geschicklichkeit, daß sie für ihren Zweck vollkommen ausreicht. Unter den Vorschlägen für die Probevorlesung würde ich Nro.3 vorziehen. Die Führung des Colloquiums will ich s.Z. übernehmen").

On this occasion we want to consider the question whether and how Riemann continued the traditions of his great model when he was offered Gauss' former professorship in 1859. In this connection one must keep in mind two things. For one thing, during half a century Gauss had taken on a great many tasks, some of which had to be continued . For another, Riemann's health was already greatly impaired. Given his tendency to carry everything out in a precise and thorough manner, he was obliged to turn down some of these tasks rather than try to carry them out unsatisfactorily.

Officially, the main area of Gauss' responsibility was *astronomy*. After his death, Wilhelm Weber was appointed the deputy director of the observatory. Neither Riemann nor Dirichlet ever spelled him.

Gauss lectured regularly on the method of least squares. Riemann announced a lecture with this title just once, namely in the summer of 1860 (N. 712). Dirichlet and Riemann lectured on mathematical methods of physics, one area of Gauss' work in which he never established a lecture tradition.

In conclusion let us recall once more Klein's raptures over the mystical influence of the Göttingen atmosphere. We see things very differently now. It is certainly true that in complex analysis Riemann went on from the work of Gauss. But while Gauss openly expressed the highest praise for Eisenstein, there is hardly any proof that he ever commended Riemann for carrying forward his, that is Gauss', work. The reason for this is clear. Eisenstein stayed completely within the frame of Gauss' image of mathematics, which was dominated above all by the algorithmically graspable, whereas Riemann's transition to the conceptual, to manifolds, to properties of functions rather than to functional expressions, made Gauss uncomfortable.

There is one more thing we must confront Klein with. Dedekind was exposed to the same Göttingen atmosphere as Riemann and yet his orientation turned out to be very different from Riemann's.

3. Geometry; Physics; Philosophy

Preliminary remarks: The central role of the habilitation lecture of 1854

Geometry, physics, and philosophy are, at least at first sight, different disciplines, but in Riemann's case they are closely related and thus belong in one chapter. Riemann discussed their mutual relations in his habilitation lecture "Über die Hypothesen die der Geometrie zu Grunde liegen" ("On the hypotheses which lie at the foundation of geometry") (W. 272–287), presented in the summer of 1854.

As is done even today, Riemann was supposed to submit titles of three lectures, from which the faculty of philosophy was to choose one. The first title was "Geschichte der Frage über die Darstellbarkeit einer Function durch eine trigonometrische Reihe" ("The history of the question of representability of a function as a trigonometric series"), and the second, "Über die Auflösung zweier Gleichungen zweiten Grades mit zwei unbekannten Grössen" ("Solution of two quadratic equations in two unknowns"). Of course, the third title was "Über die Hypothesen..." To this day it is expected that the candidate will propose topics that belong to very different areas and thus demonstrate the breadth of his professional knowledge. Accordingly, Riemann proposed topics in analysis, algebra, and geometry. The faculty's, in this case Gauss', choice was not, as Dedekind put it, a breach "of the usual convention" (W. 549) and his suggestion that Gauss "chose the third [topic] because he was curious to hear how such a young person would handle so difficult a topic" was pure guesswork. The first topic was largely taken care of by the dissertation, and the second, the determination of the points of intersection of two quadratic curves, must have struck Gauss as trivial. Even if we take seriously the suggestion that someone of Riemann's stature could change even unpromising material into something unusual, the faculty could not but reject the first two topics. Riemann could not have been naive enough not to know this, and there is every reason to think that he planned to lecture about the ideas that were particularly important to him.

There are statements in the literature that Riemann expressed great surprise at the faculty's choice of the third topic. He did not. By December 1853 he had probably worked out the details of the first two topics but not of the third (W. 547). No manuscripts bearing on the elaboration of any of the three have been preserved. Riemann had been assigned the third topic as early as December. He seems not to have been overly worried by it, for he postponed its elaboration until Easter of 1854 (W. 548). This seems to indicate that he knew very well what he wanted to say in his lecture. At that time he was always occupied with issues of physical and philosophical relations. Mathematical issues, the concept of a manifold and Gauss' geometry of surfaces, had been touched upon in his dissertation. It is impossible to tell when the idea of an n-dimensional generalization occurred to him, and whether this was the result of his preoccupation with Herbart. Completion of the computational part of his work was to cause him fewer difficulties than preparation of a polished presentation.

The development of the new "Riemannian" geometry is just one aspect of Riemann's lecture. Needless to say, it is, in and of itself, a very significant mathematical achievement. However, as has been pointed out by, for example, Gregory Nowak ("Riemann's Habilitationsvortrag and the synthetic a priori status of geometry," in Rowe and McCleary 1989, I. 17–46), the lecture must be viewed as more closely connected with physics and philosophy than has been indicated thus far in most accounts.

The notion of space plays a key role in geometry, physics, and philosophy. Today mathematics and physics may be said to depend on the naive identification of the so-called intuitive space with $\mathbb{R}^3$ and with physical space, the sort of thing that is done in school and in lectures for freshmen. This pragmatic conception relegates a prolonged philosophical development to the sidelines (M. Jammer, *Concepts of space*, Harvard University Press, Cambridge, Mass., 1954). It became dominant in Euler's time, and Euler had a strong influence on Kant. Riemann rethought the issue and elevated the discussion to new levels by pointing out new mathematical possibilities. In mathematics he spoke about manifolds parametrized by n real determinations of magnitude. A manifold can be provided with different metric determinations. The "space" to which all this can be applied always denotes the physical entity.

All we can say at the outset about the "space" is that it is 3-dimensional and that we know when two given points are infinitesimally close (topological space). The measure of distance, the metric determination, "is added to it from elsewhere" ("kommt anders woher hinzu") (W. 286). It is a field quantity much like electricity, magnetism, and gravity. The "basis of metric relations must be sought in the binding forces that act" on space (der "Grund der Massverhältnisse" muss in auf den Raum "wirkenden bindenden Kräften gesucht werden") (W. 286). The Euclidean character of a metric determination of space is not certain *a priori*. The contemporary philosophers, who followed Kant, must have regarded this conception as outrageous.

At the end of his lecture Riemann made it clear to the faculty that he could very well draw relevant physical consequences, but that this was not warranted by "the nature of today's occasion" ("die Natur der heutigen Veranstaltung"). But he had said earlier that

> "A decision on these questions can be found only by starting from the conception of phenomena that has been approved in experience hitherto, for which Newton laid the foundation, and by modifying this structure gradually under the compulsion of facts which it cannot explain. Such investigations as start out, like this present one, from general notions, can promote only the purpose that this task shall not be hindered by too restricted concepts, and that progress in perceiving the connection of things shall not be obstructed by the prejudices of tradition."
>
> ("Die Entscheidung dieser Fragen kann nur gefunden werden, indem man von der bisherigen durch die Erfahrung bewährten Auffassung der Erscheinungen, wozu Newton den Grund gelegt, ausgeht und diese durch Thatsachen, die sich aus ihr nicht erklären lassen, getrieben allmählich umarbeitet; solche Untersuchungen, welche, wie die hier geführte, von allgemeinen Begriffen ausgehen, können nur dazu dienen, dass diese Arbeit nicht durch die Beschränktheit der Begriffe gehindert und der Fortschritt im Erkennen des Zusammenhangs der Dinge nicht durch überlieferte Vorurtheile gehemmt wird.")

Riemannian metric determinations entered physics through Einstein, who was prompted by unexplained facts of experience, and their use fully reflected Riemann's field-theoretic views. We will discuss them in detail in 3.2.2. Like other field quantities, the metric satisfies a differential equation, and, in Einstein's work, the "binding forces" are the result of the linkage of the metric field with other fields. In Riemann's time it was impossible to anticipate that the unexplained fact of the independence of the speed of light of the state of motion of the observer would, in special relativity, result in the combining of space and time into a manifold that underlies reality. But this does not in any way affect the validity of Riemann's view of physics as a field theory. Beginning in 1912, another hitherto unexplained fact, the equivalence of inertia and gravity, was explained by Einstein in his general theory of relativity. While it cannot be claimed that Riemann prophetically predicted this particular development of physics, it can be claimed that it fits the scheme he constructed. It was through the mathematician Marcel Grossmann, in Zürich, that Einstein learned about Riemannian geometry and the tensor analysis that emerged from it.

Riemann praised Herbart and avoided a direct attack on Kant. One of the faculty members was the philosopher R.H. Lotze, a follower of the Kantian tradition. As late as 1871 he caused problems for Klein by declaring that all non-Euclidean geometry is nonsense (Klein 1926, 152).

In the lecture, Riemann pushed to extremes his tendency to use as few formulas as possible. Of course, one can use his text to carry out the relevant computations; we will soon use the notations that have become standard since Einstein's time. Riemann used similar notations in his Paris prize essay of 1861, but they have not been adopted.

Speiser (1927, 108) made an interesting conjecture concerning the origin of Riemann's geometric ideas. In the physical speculations of 1 March 1853 (to which we will come back later) Riemann considered deformations of a hypothetical liquid substance in 3-dimensional Euclidean space (W. 530). This led him to a quadratic differential form whose coefficients depended on three space coordinates as well as on time. He thus obtained a family of Riemannian metrics in 3-dimensional space, which he wanted to relate to the propagation of gravitation, light, and radiant heat (W. 532). This was a key idea for a unified field theory of gravitation and electromagnetism. Dedekind's quotation, taken from a letter of 28 December 1853 (W. 532), indicates that, after composing his habilitation paper, Riemann immediately resumed his investigation of the connection among electricity, galvanism, light, and gravity. He thought that he had made very good progress and thus believed that he could soon publish his results. Moreover, he had reason to believe that Gauss had for years pursued similar topics.

He decided against publication. But in a letter to his brother, dated 26 June 1854, he reported that he continued to be so engrossed in the connection among fundamental physical laws that he could not stop thinking about it, all the more because the geometric topic had been the topic of his habilitation lecture. At the very end of the lecture he pointed to physical considerations: the basis of the metric relations must be looked for in the forces acting on space (W. 286).

Thus it is conceivable that Weyl (N.741) was wrong to claim, as he did in his introduction to the new edition of the lecture, that Riemann's attempts to find a connection between light, electricity, magnetism, and gravitation, contemporaneous with the preparation of his lecture, not only were unrelated but actually interfered with one another "in his mind" ("in seinem Gehirn"). Of course, it was Einstein who first fully perceived the connection between geometry and gravitation, and it was he and many of his successors who, after Weyl and Th. Kaluza, tried, like Riemann, to include electromagnetism and later other fields as well.

It has always been assumed that Gauss' "agitation at the profundity of the thoughts presented by Riemann" ("Erregung über die Tiefe der von Riemann vorgetragenen Gedanken") (Dedekind, W. 549) was due to Riemann's modification of Euclidean geometry. But it is also possible that Gauss recognized the physical significance of Riemann's ideas.

In Riemann's (unfortunately few) surviving notes of March 1853 there are formulas for "Riemannian" metric determinations in a physical context. No record of his subsequent intense reflections, mentioned in the letters to his brother, has been found.

3.1 Geometry

3.1.1 From Euclid to Descartes and to non-Euclidean geometry

Euclid's *Elements*, written about 300 BC, summarized the previous development of mathematics and canonized geometry for almost two millennia. The merits of the *Elements* speak for themselves. But given these merits, it easily escaped notice that the narrow restriction to fixed constructions hardly benefited applications of mathematics. This is so in spite of the subsequent legitimization of special curves such as the conic sections and the Archimedean spiral. That ellipses sufficed for Kepler was, so to say, a lucky break — he needed only the mathematics he was familiar with.

In 1637 René Descartes (1596–1650) set aside the all-embracing domination of the *Elements* by introducing his method of coordinates. This method made possible the "analytic," i.e., computational, verification of geometric theorems. It gained acceptance rather slowly; Newton continued to publish in the Euclidean style. Analytic geometry was fully developed only in the 18th century, and in Riemann's time it was an obvious tool of the research mathematician. But n-dimensional spaces consisting of real n-tuples were used, at best, in the parametrization of problems of mechanics involving a number of degrees of freedom and in similar applications. At first, the extensive writings of the Stettin gymnasium teacher H. Grassmann (1809–1877) met with virtually no response. They could have been of little use to Riemann, for whom linear algebra was a trivial matter.

The development of non-Euclidean geometry was independent of analytic geometry and of applications. Its style was that of the *Elements*. Its source was the failure to prove Euclid's parallel postulate on the basis of his other assumptions, i.e., to prove that two straight lines cut by a third so that the sum of the resulting interior angles is less than two right angles must meet.

Riemann seems not to have taken notice of non-Euclidean geometry. But he must have known about the issue of parallel straight lines. He mentions Legendre, whose *Élémens de géométrie*, with editions running from 1794 to 1823, enjoyed wide circulation. The *Élémens* contains the following theorem, whose proof does not depend on the parallel postulate: if the sum of the angles in some triangle is less than two right angles, then it is less than two right angles in all triangles. It follows that the length of the side of an equilateral triangle is determined by the sum of its angles. Legendre thought that this was absurd and failed to open the door to non-Euclidean geometry. (That the angle sum cannot exceed two right angles follows from the assumption that two points determine a unique straight line; the only remaining possibility is that the angle sum is equal to two right angles, and this implies the existence and uniqueness of parallels.)

A readable account of the evolution of geometry is J. Gray 1979. We also mention K. Mainzer's *Geschichte der Geometrie*, Wissenschaftsverlag Mannheim 1980.

Did Riemann know about non-Euclidean geometry at the time of his habilitation lecture? He does not mention it. Had he known about Gauss' interest in its development he would have at least hinted at it. Long before that, Gauss had systematically developed a plane geometry (and to a large extent also a space geometry) in which he had replaced the Euclidean parallel axiom by the assumption that through a given point in a plane, not on a given straight line in that plane, it is possible to pass more than one straight line that does not meet the given straight line. He spoke of this geometry, which he referred to as non-Euclidean, only to friends, for, as he put it in a letter to Bessel dated 27 January 1829, he feared "the screaming of the Boeotians." That this fear was not unfounded is shown by, say, Lotze's later utterance against Felix Klein. Gauss' thoughts on this subject became known only after the publication, between 1860 and 1863, of his correspondence with Schumacher.

Of course, there were the publications by N.I. Lobachevsky (1793–1856) and János Bólyai (1802–1860). Lobachevsky's essay "Géométrie imaginaire" in *Crelle* 17 (1837), 295–320, and his booklet *Geometrische Untersuchungen zur Theorie der Parallellinien* (*Geometric investigations on the theory of parallels*), published in Berlin in 1840, were available in the Göttingen university library and Riemann could have consulted them. On the basis of the borrowing list compiled by Neuenschwander, E. Scholz (1982, 220, 221, footnotes) concluded that Riemann had borrowed the volume of the journal on 15 February 1854, but that neither he nor any other mathematician had borrowed the booklet in the first half of the 1850s.

Even if we admit the, by no means certain, possibility that Riemann borrowed the journal because of Lobachevsky's essay, it could not have awakened his interest because it contained just computations, without an explicit presentation of foundations and objectives, and was reminiscent of Lambert's earlier investigations of trigonometry on an imaginary sphere. The notes from Riemann's *Nachlass*, published by Scholz (loc. cit.), show that, unlike Gauss, he was not interested in the foundations of elementary geometry. Rather, as in his use of the plane and of Riemann surfaces in the theory of functions, he wanted to develop and use his geometric ideas on n-dimensional manifolds as an aid to analysis, algebra, and physics.

Riemann's significance for geometry was very aptly described by Hermite in the introduction to *Œuvres de Riemann* (Paris 1898, X): "L'Auteur dépasse infiniment la question du postulatum d'Euclid ... C'est tout un monde inconnu interessant à la fois le philosophe et le géomètre ... "

Indeed, by introducing the concept of a manifold Riemann went far beyond the intellectual scheme of both Euclidean and non-Euclidean geometry. He began with a continuum without a geometric structure, one in which it was not initially possible to speak of, say, straight lines, distances between points, and angles. In this connection he referred to Gauss (W. 273) and, more specifically, to the brief hints which the latter

> "has given on the subject in his second essay on biquadratic residues, in the Göttingen learned notices, and in his Jubilee booklet ..."
>
> ("in der zweiten Abhandlung über die biquadratischen Reste, in den Göttingenschen gelehrten Anzeigen und in seiner Jubiläumsschrift dar-über gegeben hat ...")

Gauss' allusions are connected with the complex plane. We quote one of them. It is found in his paper *Beiträge zur Theorie der algebraischen Gleichungen* (*Contributions to the theory of algebraic equations*) (*Werke* III), written in 1849 in connection with the fiftieth anniversary of the awarding of his doctoral degree: "The wording of the proof is taken from the geometry of position, for in this way it gains maximal intuitive appeal and simplicity. Strictly speaking, the essential content of the whole argument belongs to a higher, space-independent, domain of the general abstract science of magnitude that investigates combinations of magnitudes held together by continuity. At present, this domain is poorly developed, and one cannot move in it without the use of language borrowed from spatial images." ("Ich werde die Beweisführung in einer der Geometrie der Lage entnommenen Einkleidung darstellen, weil jene dadurch die grösste Anschaulichkeit und Einfachheit gewinnt. Im Grunde gehört aber der eigentliche Inhalt der ganzen Argumentation einem höheren von Räumlichem unabhängigen Gebiete der allgemeinen abstracten Grössenlehre an, dessen Gegenstand die nach der Stetigkeit zusammenhängenden Grössencombinationen sind, einem Gebiete, welches zur Zeit noch wenig angebaut ist, und in welchem man sich auch nicht bewegen kann ohne eine von räumlichen Bildern entlehnte Sprache.")

This can certainly be regarded as an allusion to manifolds coordinatized by continua of n-tuples of numbers, as well as to the use of geometric language in a nongeometric context. Gauss considered the points of the plane given by real coordinates t, u and introduced an "algebraic structure," that of the complex numbers. Riemann was to introduce real n-tuples and to investigate a "metric structure."

3.1.2 *Gauss' theory of surfaces*

Viewing the plane as a surface in 3-dimensional space suggests the possibility of a significant analytic generalization of plane Euclidean geometry to a geometry on a curved surface in space. Riemann referred to "the famous essay of Privy Councillor Gauss on curved surfaces" (W. 276), the first systematic presentation of so-called differential geometry. Before Gauss there were only isolated results on particular classes of surfaces and on applications of analysis to geometry. The preferred way of describing a surface in space furnished with Cartesian coordinates x, y, z was by means of an equation $F(x, y, z) = 0$ or by an equation $z = f(x, y)$.

Imitating the use of latitude and longitude in geography, Gauss described a surface by means of three functions $x = x(p, q)$, $y = y(p, q)$, $z = z(p, q)$ of two parameters p and q. Today we prefer to use the position vector r and to number the parameters, i.e., we describe a surface by means of an equation

$$r = r(u_1, u_2)\,. \tag{1}$$

The arclength of a curve $r = r(t) = r(u_1(t), u_2(t))$ is given by

$$\int ds = \int |\dot{r}| dt\,. \tag{2}$$

Denoting the partial derivative of r in (1) with respect to u_k by r_k, we can write briefly

$$ds^2 = \dot{r}^2 dt^2 = \left(\sum_i r_i\, du_i,\ \sum_k r_k\, du_k\right) = \sum_{i,k} g_{ik}\, du_i\, du_k \tag{3}$$

with

$$g_{ik}(u_1, u_2) = r_i r_k\,. \tag{4}$$

Here the sums run from 1 to 2. We assume that all derivatives exist and are continuous and that the tangent vectors r_1 and r_2 to the parameter lines are linearly independent. Then $g_{11} > 0$, $g_{22} > 0$, $g_{12} = g_{21}$, and $g = g_{11}g_{22} - g_{12}g_{21} > 0$. As a normal vector to the surface we can take $n = r_1 \times r_2/\sqrt{g}$.

Suppose we want to investigate the geometry of a surface without reference to its embedding in space. Then, given the parameter system u_1, u_2, for measuring distances we need only know the three functions g_{ik}. Moreover, knowledge of these functions makes possible the measurement of angles and areas, the latter by means of $\iint \sqrt{g}\, du_1\, du_2$. That a curve on a surface linking two of its points has least length is an assertion that belongs to the "intrinsic geometry" of that surface. It leads to the notion of a geodesic, which is the analog of a straight line in the plane.

Bending of a plane piece of surface does not change distances on it. The plane piece and its bent version are differently embedded in space, but their intrinsic geometries are the same. What is different is their curvatures in space. In Euler's time it was known that the curvature of a surface can be characterized by the curvatures of curves on it in the following manner. At a fixed point on the surface consider the curvatures of its normal sections, i.e., of the curves of intersection of the surface with the planes containing the normal to the surface at the point in question. Let k_1 and k_2 be the extreme values of these curvatures; the corresponding curves turn out to be orthogonal. The sign of the normal curvature $k = nr''$ can be negative. (r'' is the curvature vector of the curve, i.e., the second derivative of its position vector with respect to the arclength s.)

To obtain measures of the curvature of the surface at the point in question one forms the mean curvature $H = (k_1 + k_2)/2$ and the Gaussian curvature $K = k_1k_2$. For plane surfaces we have $k_1 = k_2 = H = K = 0$. For a cylindrical surface obtained by bending a plane surface we have $K = 0$, but in general $H \neq 0$. In general, bending affects the individual values of k_1 and k_2.

The most important discovery in Gauss' theory of surfaces, one that he himself referred to as *Theorema egregium* (extraordinary theorem), is that K can be expressed in terms of the g_{ik}. This means that it belongs to the intrinsic geometry of the surface, and therefore can be computed by measurements of lengths alone. Gauss' proof is purely computational. In the context of this result, geometric intuition seems of little relevance. Gauss gave yet another geometric interpretation of K. When $r(u_1, u_2)$ ranges over a piece of the surface, $n = n(u_1, u_2)$, the corresponding normals to the surface, determine on the unit sphere a set of points called its spherical normal image. Gauss showed that the value of K at a point of the surface is the limit of the quotient of the area of the spherical image of a piece of the surface containing this point by the area of the piece of the surface as the latter shrinks to the point. It is easy to see that $K = 0$ for planes, cylinders, cones, and tangential developables. Indeed, in the case of these surfaces the spherical image is either a point or a piece of a curve, and thus has zero area. All these surfaces can be obtained by bending pieces of a plane. According to Gauss, these are the only surfaces for which $K = 0$.

Consider a piece of a sphere of radius r. Here $k_1 = k_2 = H = 1/r$ and $K = 1/r^2$. The geodesics are parts of great circles. It turns out that the spherical excess of a geodesic triangle on a sphere is K times the area of the triangle. This is a special case of a more general result proved by Gauss: the angular excess of a geodesic triangle on a surface is its total curvature, i.e., the surface integral of the Gaussian curvature over the triangle. If $K < 0$, then the angular excess is negative, and we speak of an angular defect. According to Gauss, we can determine K by measurements on the surface!

Gauss' summary of surface theory (*Werke* IV, 344–345) is already a program of study of 2-dimensional "Riemannian" geometries: "These theorems suggest consideration of the theory of curved surfaces from a different viewpoint, one that opens to research an extensive and as yet quite underdeveloped field. If one views surfaces not as boundaries of solids but as solids with one vanishing dimension, and as bendable but not stretchable, then one becomes aware of the need to distinguish two essentially different kinds of relations, namely, those that assume a particular form of the surface in space and those that are independent of the different forms which a surface can take on. It is the latter we speak of here .... The starting point of all such investigations must be that the nature of a curved surface in itself is given by the expression of an indeterminate line element in the form

$$\sqrt{Edp^2 + 2Fdpdq + Gdq^2} \ldots \text{"}$$

("Diese Sätze führen dahin, die Theorie der krummen Flächen aus einem neuen Gesichtspunkt zu betrachten, wo sich der Untersuchung ein weites noch ganz unangebautes Feld öffnet. Wenn man die Flächen nicht als Grenzen von Körpern, sondern als Körper, deren eine Dimension verschwindet, und zugleich als biegsam, aber nicht als dehnbar betrachtet, so begreift man, dass zweierlei wesentlich verschiedene Relationen zu unterscheiden sind, theils nemlich solche, die eine bestimmte Form der Fläche im Raum voraussetzen, theils solche, welche von den verschiedenen Formen, die die Fläche annehmen kann, unabhängig sind. Die letztern sind es wovon hier die Rede ist .... Alle solchen Untersuchungen müssen davon ausgehen, dass die Natur der krummen Fläche an sich durch den Ausdruck eines unbestimmten Linearelements in der form

$\sqrt{Edp^2 + 2Fdpdq + Gdq^2}$ gegeben ist ... ").

After 1818 Gauss was deeply involved in the survey of the kingdom of Hanover; the 10-mark bills issued in 1991 show a grid of survey triangles. There is no doubt that this practical activity inspired his surface theory, and especially its aspect of intrinsic geometry. For further details see W. K. Bühler 1981, Chapter 9.

The *Theorema egregium* motivated the search for additional interpretations of the Gaussian curvature K associated with intrinsic geometry. Fix a point on a surface or on a 2-dimensional Riemannian manifold. The geodesics issuing from this point provide a schlicht covering of a neighborhood of it. Following Gauss, we can introduce in this neighborhood polar coordinates r, φ. Here r is the distance to the origin measured along a geodesic, and φ is the angle, with vertex at the origin, that this geodesic forms with a fixed one of the lines issuing from the origin. Let K_0 be the Gaussian curvature of the surface at the origin. Then the expression for the line element is

$$ds^2 = dr^2 + \left(r^2 - \frac{K_0}{3}r^4 + \cdots\right)d\varphi^2 .$$

The lines r=const are called geodesic circles. It is not difficult to see that the circumference of a circle with radius r (i.e., with dr=0) is

$$U(r) = \int_{\varphi=0}^{2\pi} ds = 2\pi r - \frac{\pi}{3}K_0 r^3 + \cdots$$

and its area is

$$F(r) = \int_0^r U dr = \pi r^2 - \frac{\pi}{12}K_0 r^4 + \cdots .$$

This shows that K_0 can be obtained by measurements on infinitely small circles as the limit when $r \to +0$ of either

$$\frac{3}{\pi}\frac{2\pi r - U(r)}{r^3}$$

or

$$\frac{12}{\pi}\frac{\pi r^2 - F(r)}{r^4} .$$

These interpretations, due, respectively, to Bertrand and Puiseux and to Diguet, appeared in the 5th edition (1850) of *Application de l'analyse à la géométrie*, published by the students of G. Monge, and were easily accessible.

Riemann started (W. 281) from Gauss' interpretation involving the angular excess in infinitely small geodesic triangles.

All intuitive interpretations of K show that, from the viewpoint of intrinsic geometry, the name "curvature" was not aptly chosen. $K > 0$ states that the circumferences and areas of circles are less than in the Euclidean case and that K measures the deviation from the Euclidean behavior. Sometimes the terms "shrinking" and "inflation" were used in connection with the cases $K > 0$ and $K < 0$ respectively. The name "curvature" suggests embedding in a higher-dimensional space in which the surface, or the Riemannian space, seems curved. But the "curvature" of the space of the theory of relativity, say, does not assume such an embedding of the world in a higher-dimensional space!

Gauss obtained the *Theorema egregium* by means of involved computations, and the definitions of the curvature K make essential use of the embedding of the surface in space — more specifically, of the normals to the surface. Riemann's proof does not rely on the embedding. This alone was nothing new, for the French papers just mentioned achieved the same thing. For Riemann this theorem was the key to a new approach to geometry, whereas for the Frenchmen it was one result among many.

For Riemann's differential geometry see K. Reich 1973.

3.1.3 The n-fold extended manifold

Without, as yet, possessing a perfected terminology, Riemann introduced what we would call today a topological space, with the additional property that each point has a neighborhood homeomorphic to a region of $\mathbb{R}^n$ for a given n. He called it an n-fold extended magnitude and observed (W. 286, footnote) that what was involved was

> "the preliminary study for contributions to analysis situs." Discrete manifolds occur frequently, but "on the other hand there are in everyday life such infrequent occasions to form concepts whose modes of determination form a continuous manifold, that the positions of objects of sense, and the colors, are probably the only simple concepts whose modes of determination form a multiply extended manifold."
>
> ("die Vorarbeit für Beiträge zur Analysis situs." Diskrete Mannigfaltigkeiten träten häufig auf, "dagegen sind die Veranlassungen zur Bildung von Begriffen, deren Bestimmungsweisen eine stetige Mannigfaltigkeit bilden, im gemeinen Leben so selten, das die Orte der Sinnengegenstände und die Farben wohl die einzigen einfachen Begriffe sind, deren Bestimmungsweisen eine mehrfach ausgedehnte Mannigfaltigkeit bilden.")

In mathematics one is more frequently led to such concepts. He mentions "Riemann" surfaces

> "for the treatment of many-valued analytic functions" and thinks highly of his invention of these surfaces: the lack of this concept "is likely a principal reason why the celebrated theorem of Abel and the contributions of Lagrange, Pfaff, and Jacobi to the general theory of differential equations have remained so long unfruitful."
>
> ("für die Behandlung der mehrwärtigen analytischen Functionen" und schätzt seine Erfindung derselben hoch ein: der Mangel an diesem Begriff "ist wohl eine Hauptursache, dass der berühmte Abel'sche Satz und die Leistungen von Lagrange, Pfaff, Jacobi für die allgemeine Theorie der Differentialgleichungen so lange unfruchtbar geblieben sind") (W. 274).

He tried to explain in words the description of a manifold by means of *n* real coordinates but noted that the dimension

> "need not be a finite number, rather, it may call for either an infinite series [we say: a countable sequence] or a continuous manifold of numerical determinations [after Cantor, an uncountable index set]"
>
> ("nicht eine endliche Zahl, sondern entweder eine unendliche Reihe [bei uns: eine abzählbare Folge] oder eine stetige Mannigfaltigkeit von Größenbestimmungen [nach Cantor eine überabzählbare Indexmenge] erfordern [könne]").

Relevant examples are functions on a given region or the possible shapes of a space figure.

When Riemann used the term "*analysis situs*" in his *Theorie der Abel'schen Functionen* of 1857, he cited Leibniz (W. 91). He wanted the term to designate the part of the study of continuous magnitudes

> "that does not view the magnitudes as existing independently of position and measurable in terms of each other, but makes the subject of investigation solely their relations of position and region, while entirely ignoring metric relations."
>
> ("welcher die Größen nicht als unabhängig von der Lage existierend und durcheinander messbar betrachtet, sondern von den Massverhältnissen ganz absehend, nur ihre Orts- und Gebietsverhältnisse der Untersuchung unterwirft.")

In this work he considered, in the 2-dimensional case, not only what later came to be known as "set-theoretic" topology but also connectivity relations in the sense of combinatorial topology. The latter played no role in the habilitation lecture. On the other hand, he used the required differentiability properties without mentioning them; he dealt with what were later called "differentiable manifolds."

For Riemann, who always thought in analytic terms, this concept was the most natural generalization of Euclidean geometry. At the very beginning of his lecture (W. 272) he stated that geometry assumes as given the concept of space and the fundamental concepts for constructions in space (he probably had in mind points, lines, angles, and so on), and provides just nominal definitions for them. He probably had in mind Euclid's statements that a point is that which has no part, a line is length without breadth, and an angle is the inclination to one another of two lines.

> "The relation of these presuppositions is left in the dark ... From Euclid to Legendre ... this darkness has been lifted neither by the mathematicians nor by the philosophers who have labored upon it. The reason for this lay perhaps in the fact that the general concept of multiply extended magnitudes, in which spatial magnitudes are comprehended, has not been elaborated at all."
>
> ("Das Verhältnis dieser Voraussetzungen bleibt dabei im Dunkeln ... Diese Dunkelheit wurde auch von Euklid bis auf Legendre ... weder von den Mathematikern, noch von den Philosophen, welche sich damit beschäftigten, gehoben. Es hatte dies seinen Grund wohl darin, dass der allgemeine Begriff mehrfach ausgedehnter Grössen, unter welchem die Raumgrössen enthalten sind, ganz unbearbeitet blieb.")

Riemann thought it possible to characterize the multiply extended magnitudes, nowadays called spaces or (differentiable) manifolds, by representing their points as number tuplets. He cited (W. 273) Gauss and the philosopher Herbart. It was the set-theoretic way of thinking which later made it possible to conceive of a space as simply a set of points endowed with a structure and, in particular, to think of a metric space as a set of points with a distance function. In general, such spaces cannot be dealt with by analytic means.

Riemann's reduction of geometry to analysis could not go unchallenged. However, his geometries turned out to be particularly useful for applications in physics. This is not true to the same extent of the variants of Euclidean geometry constructed on axiomatic foundations and known since Hilbert's time as "Foundations of geometry." (But see in this connection the books by W. Benz, *Geometrische Transformationen*, Mannheim 1992, and *Real Geometries*, Mannheim 1994, both published by B.-I. Wissenschaftsverlag.)

Using modern terms we can say that Riemann starts from manifolds with an analytic structure, complex in the case of Riemann surfaces and differentiable in the case of differential geometry. Riemann himself managed quite well without an explicit formulation of such distinctions, but, as can be seen with the advantage of hindsight, the subsequent development was hampered by the lack of clear conceptual locutions. These were formulated in the 20th century, beginning with H. Weyl's book *Die Idee der Riemannschen Fläche* of 1913, which is of historical significance. The new version of this book, *The Concept of a Riemann Surface*, prepared by Weyl in the last year of his life (1955) and published in English by Addison-Wesley, is for us more readable.

Nowadays, our starting point is a (connected) topological space M. In addition, we take a topological vector space V; in the case of differential geometry $V = \mathbb{R}^n$. Let $\{U_\mu\}$ be a covering of M by means of open sets U_μ such that every U_μ is homeomorphic to an open subset $V_\mu \subseteq V$. Every V_μ is called a local chart of M and $\{V_\mu\}$ is called an atlas. In this way coordinates are introduced in M. In order to be able to calculate with the coordinates one imposes on them differentiability conditions: if $U_\mu \cap U_\nu \neq \emptyset$, then the transition from V_μ-coordinates to V_ν-coordinates is to be given by a function that is differentiable sufficiently many times. Another atlas is admissible if its union with the previous one is again an atlas. For details consult a textbook of differential geometry such as Laugwitz 1977. The totality of admissible atlases yields a differentiable structure on M. For the Riemann surfaces of the theory of functions $V = \mathbb{R}^2$, and the functions are not just differentiable functions but orientation-preserving conformal mappings (a conformal structure). Another option is $V = \mathbb{C}$ with only holomorphic functions (a complex structure). If the surface covers the complex plane or the complex sphere, then the relevant structure can be transferred to it in an obvious way. The branch points are an exception. At these points one makes use of a local uniformizing variable.

This cartographic terminology is related to Gauss' theory of surfaces. In the case of such surfaces one can speak of normal vectors and tangent planes. However, in the general case no embedding in a Euclidean space is available and a substitute must be created using M alone. If $x(t)$ and $y(p)$ are differentiable coordinate representations of curves with $x(0) = y(0) = P$, and if $dx/dt = dy/dp$ for $t = p = 0$, then we say that the two curves are equivalent. It is easy to see that this definition is independent of the coordinate system, that one can form equivalence classes, and that these classes form a vector space isomorphic to V. We call it the tangent space $T(P)$. In this way we associate with every $P \in M$ a $T(P)$. Coordinates were needed as a resource and are of no geometric interest. For dealing with the vector spaces $T(P)$ we have at our disposal the complete apparatus of linear algebra. In particular, symmetric bilinear forms lead to Riemannian metrics.

3.1.4 *The metric determinations*

Let $x(t)$, t a real parameter, be a given curve on a manifold with coordinate representation $x = (x_1, \ldots, x_n)$. A metric determination fixes the length of the piece of the curve between $x(t_0)$ and $x(t_1)$. The formula for Riemann's corresponding general statement is

$$\int_{t_0}^{t_1} F(x, \dot{x})\, dt,$$

or briefly $ds = F(x, dx)$. F must be positive. Moreover, if the length of the curve is to be independent of the direction of traversal and of the choice of the parameter t, then we must stipulate that $F(x, r\xi) = |r| F(x, \xi)$. These requirements are satisfied by positive definite quadratic forms

$$F^2(x, \xi) = \sum_{i,k} g_{ik}(x)\xi_i \xi_k, \quad g_{ik} = g_{ki}\,.$$

Riemann restricted himself from the first to these metric determinations but noted the possibility of more general functions F. In this connection he said (W. 278):

> "The next case in order of simplicity would probably contain the manifolds in which the line element can be expressed by the fourth root of a differential expression of the fourth degree." This would be rather time-consuming and would give little insight, "particularly since the results cannot be expressed geometrically."
>
> ("Der nächst einfache Fall würde wohl die Mannigfaltigkeiten umfassen, in welchen sich das Linienelement durch die vierte Wurzel aus einem Differentialausdrucke vierten Grades ausdrücken lässt." Das würde ziemlich zeitraubend sein und wenig Einsicht geben, "zumal da sich die Resultate nicht geometrisch ausdrücken lassen.")

This justification of the restriction to quadratic forms is not very convincing, and we will find it necessary to come back to Helmholtz' apt characterization of this special case as well as to the developments, related to the work of Finsler, involving arbitrary $ds = F(x, dx)$. On the other hand, it must be admitted that when he restricted himself to the metric determinations later named after him, Riemann seized upon the "correct" generalization of Gauss' intrinsic geometry of surfaces.

The method of coordinates makes possible the analytic treatment of geometry. However, the geometric propositions must be independent of the choice of coordinates. In particular, if we introduce new coordinates $\bar{x}$ as functions of the coordinates x, then it must be true that

$$F(x, dx) = ds = \bar{F}(\bar{x}, d\bar{x}),$$

and the coefficients of the quadratic form, which may always be supposed to be symmetric, must satisfy the condition

$$\bar{g}_{jl} = \sum_{i,k} g_{ik} x_{i,j}\, x_{k,l} \quad \text{where} \quad x_{i,j} = \frac{\partial x_i}{\partial \bar{x}_j}.$$

Riemann wrote (W. 278):

> "One can transform such an expression [i.e., a quadratic differential expression] into another similar one by substituting for the n independent variables functions of n new independent variables."
>
> ("Man kann einen solchen Ausdruck [d.h. einen Differentialausdruck zweiten Grades] in einen anderen ähnlichen transformieren, indem man für die n unabhängigen Veränderlichen Functionen von n neuen unabhängigen Veränderlichen setzt.")

Before continuing our examination of Riemann's text we interpolate an observation pertaining to the case of constant $g_{ik}(x) = g_{ik}$. We know from linear algebra that positive definiteness ensures the existence of a linear coordinate transformation such that, in terms of the new coordinates $\bar{x}$, the $\bar{g}_{jl}$ form the identity matrix. This means that the metric is that of Euclidean geometry. Since we can also make use of nonlinear coordinate transformations, it is conceivable that the same result can be obtained in the case of nonconstant g_{ik}. If this were the case, then, for a suitable choice of coordinates, all Riemannian geometries would coincide with Euclidean geometry. We know that this is false in the case of the theory of surfaces, i.e., for $n = 2$. Riemann adduced the following argument for an arbitrary value of n:

Because of symmetry, there are $\frac{n(n+1)}{2}$ coefficients g_{ik}

> "that are arbitrary functions of the independent variables; but by introducing new variables one can satisfy only n relations and so can make only n of the coefficients equal to given quantities."
>
> ("welche willkürliche Functionen der unabhängigen Veränderlichen sind; durch Einführung neuer Veränderlicher wird man aber nur n Relationen genügen können und also nur n der Coefficienten gegebenen Größen gleich machen können.")

The number of leftover functions is $\frac{n(n-1)}{2}$, i.e., just one in the case $n = 2$. This counting argument makes it at least plausible that "the line element cannot be reduced to the form $\sum dx^2$" in all Riemannian spaces. Riemann calls the manifolds in which this happens "flat."

Riemann uses the following fact without providing an explicit justification. An m-dimensional submanifold, given by $x = x(u)$, $x = (x_1, \dots, x_n)$, $u = (u_1, \dots, u_m)$, $x_{i,\alpha} = \frac{\partial x_i}{\partial u_\alpha}$, carries an induced metric

$$ds^2 = \sum_{i,k} g_{ik}\, dx_i\, dx_k = \sum_{\alpha,\beta} \gamma_{\alpha\beta}\, du_\alpha\, du_\beta\,;$$

$$\text{since} \quad dx_i = \sum_\alpha x_{i,\alpha}\, du_\alpha, \quad \text{we have} \quad \gamma_{\alpha\beta} = \sum_{i,k} g_{ik}\, dx_{i,\alpha}\, dx_{k,\beta}\,.$$

3.1.5 Curvature

> "To give a tangible meaning to the measure of curvature of an n-dimensional manifold at a given point and in a surface direction passing through that point, it is necessary to start out from the principle that a shortest line, originating in a point, is fully determined when its initial direction is given. According to this, a determinate surface is obtained when one prolongs into shortest lines all the initial directions going out from a point and lying in the given surface element; and this surface has at the given point a determinate measure of curvature, which is also the measure of curvature of the n-dimensional manifold in the given point and the given surface direction."
>
> ("Um dem Krümmungsmass einer nfach ausgedehnten Mannigfaltigkeit in einem gegebenen Punkte und einer gegebenen durch ihn gelegten Flächenrichtung eine greifbare Bedeutung zu geben, muss man davon ausgehen, dass eine von einem Punkte ausgehende kürzeste Linie völlig bestimmt ist, wenn ihre Anfangsrichtung gegeben ist. Hiernach wird man eine bestimmte Strecke erhalten wenn man sämtliche von dem gegebenen ausgehenden und in dem gegebenen Flächenelement liegenden Anfangsrichtungen zu kürzesten Linien verlängert, und diese Fläche hat in dem gegebenen Punkte ein bestimmtes Krümmungsmass, welches zugleich das Krümmungsmass der nfach ausgedehnten Mannigfaltigkeit in dem gegebenen Pukte und der gegebenen Flächenrichtung ist") (W. 281).

This is Riemann's definition of the geometric magnitude now referred to as *sectional curvature.* Riemann considered the Gaussian curvature K of the 2-dimensional manifolds containing the point P_0 and the geodesics tangent to them at P_0. K depends not only on P_0 but also on the given surface direction. In the case of "flat" manifolds we always have $K = 0$. More generally, there are manifolds with constant K. In the latter it is possible to choose the coordinates so that

$$ds^2 = \frac{1}{1 + \frac{K}{4} \sum x_i^2} \sum dx_j^2$$

(W. 314). This is the only formula in the whole essay!

> "The common character of those manifolds [of constant curvature] can also be expressed thus: that the figures lying in them can be moved without stretching."
>
> ("Der gemeinsame Charakter dieser Mannigfaltigkeiten [K constant]) ... kann auch so ausgedrückt werden, dass sich die Figuren in ihnen ohne Dehnung bewegen lassen.")

At this point it seems appropriate to note that spaces of constant curvature admit a constructive buildup of a geometry with congruence theorems. However, the concept of parallelism has no meaning if $K \neq 0$. It is reasonable to assume that, had he been familiar with the works of Lobachevsky and Bólyai, Riemann would have indicated this, however briefly.

Later we will discuss the computations associated with Riemann's assertions. In the lecture he indicated (W. 279) that he was using normal coordinates, analogous to the Cartesian coordinates in the case $K = 0$. The n coordinate axes form with one another $\frac{n(n-1)}{2}$ 2-dimensional surface directions, and Riemann concluded (W. 279/280) that

> "A while ago, $n\frac{n-1}{2}$ functions of position were found to be necessary for the determination of the metric relations of an n-fold extended ... manifold; it follows, that if the measure of curvature is given at every point in $n\frac{n-1}{2}$ surface directions, then we can determine from these [data] the metric relations of the manifold ... "
>
> ("Zur Bestimmung der Massverhältnisse einer nfach ausgedehnten ... Mannigfaltigkeit wurden vorhin $n\frac{n-1}{2}$ Functionen des Orts nöthig gefunden; wenn also das Krümmungsmass in jedem Punkte in $n\frac{n-1}{2}$ Flächen-richtungen gegeben wird, so werden daraus die Massverhältnisse der Mannigfaltigkeit sich bestimmen lassen ... ")

He said that the metric relations can be expressed in this way regardless of the choice of coordinates. If so, then clearly the metric is determined if we know the curvature K as a function of position and of surface direction. At least this is the case in general; in fact, earlier (W. 280), Riemann stated a restriction. (Actually, if $K = 0$, then the metric is Euclidean but it is determined only up to an affinity. Besides, given K, there is no simple way to determine the g_{ik} even in the case $n = 2$.)

How can one tell if the line element of an n-dimensional manifold is flat, i.e., if there exist coordinates $\bar{x}$ such that $ds^2 = \sum d\bar{x}_i^2$? Riemann's hints, quoted earlier, suggest that this is the case precisely when the measure of curvature vanishes at every point in $n\frac{n-1}{2}$ surface directions. But this is hardly a verifiable condition. What is needed is a formula involving the g_{ik} that yields the result through calculation.

Riemann carried out this computation in the second half of a paper that he submitted to the Académie des Sciences on 1 July 1861. This was his answer to the prize question on heat distribution posed by the Academy in 1858. The paper (W. 401–404) remained unknown until 1876, the year of publication of the first edition of Riemann's collected works. (A German translation of the Latin original is found in Böhm-Reichardt 1984.) His notations are few. They are appropriate but have not been adopted. The symbolism below has been in use since Einstein's time. Its two main features are that the coordinates are provided with superscripts and that one sums over an index that appears once as a superscript and once as a subscript. Summation signs are left out.

Consider an element of arclength in two coordinate systems x^j and $\bar{x}^i$. We have:

$$ds^2 = g_{jl}dx^j dx^l = \bar{g}_{ik}d\bar{x}^i d\bar{x}^k = \bar{g}_{ik}\bar{x}^i{}_{,j}\,\bar{x}^k{}_{,l}\,dx^j dx^l$$

($f_{,j}$ stands for $\partial f/\partial x^j$). The symmetry of the g and the $\bar{g}$ implies that

$$g_{jl} = \bar{g}_{ik}\bar{x}^i{}_{,j}\,\bar{x}^k{}_{,l}\ .$$

Suppose the manifold is flat, $ds^2 = \sum (d\bar{x}^i)^2$, i.e., $\bar{g}_{ik} = \delta_{ik}$ (the Kronecker symbol, the unit matrix). Differentiation with respect to x^r yields

$$g_{jl,r} = \delta_{ik}\bar{x}^i{}_{,j\,,r}\,\bar{x}^k{}_{,l} + \delta_{ik}\bar{x}^i{}_{,j}\,\bar{x}^k{}_{,l\,,r}\ .$$

By rewriting, we obtain from this

$$\delta_{ik}\bar{x}^i{}_{,j_1\,,j_2}\,\bar{x}^k{}_{,j_0} = \frac{1}{2}(g_{j_0 j_1\,,j_2} - g_{j_1 j_2\,,j_0} + g_{j_2 j_0\,,j_1})\ .$$

The right side of the latter expression is referred to as the Christoffel symbol $\Gamma_{j_1 j_2 | j_0}$. By further differentiation and substitution of the preceding equations Riemann obtained

$$\begin{aligned} &g_{j_0 j_2\,,j_1\,,j_3} + g_{j_1 j_3\,,j_0\,,j_2} - g_{j_0 j_3\,,j_1\,,j_2} - g_{j_1 j_2\,,j_0\,,j_3} \\ &\quad + 2g^{lm}(\Gamma_{l j_1 | j_3}\Gamma_{m j_0 | j_2} - \Gamma_{l j_0 | j_3}\Gamma_{m j_1 | j_2}) = 0. \end{aligned} \tag{I}$$

The matrix (g^{lm}) is the inverse of the matrix (g_{ik}). Riemann denoted the left side of (I) by the abbreviation $(j_0 j_1,\ j_2 j_3)$. This expression, multiplied by -2, came to be known as the Riemann curvature tensor and to be denoted by $R_{j_0 j_1 j_2 j_3}$.

The $\bar{x}$ do not appear in the equation (I) and one has necessary conditions for the g_{ik} to belong to a flat metric determination. These conditions are also sufficient. Riemann indicated this in a vague way; the first proofs were given by Christoffel and Lipschitz 1869–1870 in *Crelle's Journal* 70 and 72.

3.1.6 Effects on geometry and physics in the first fifty years after Riemann's death

Immediately after the appearance of Riemann's lecture Hermann Helmholtz (1821–1894; von H. as of 1882) picked up his ideas and noted, first of all, that he himself had also thought of the idea of an n-dimensional manifold and had actually jotted it down. He wanted to base geometry on facts, and adopted the free mobility of rigid bodies as a physically determined fact. Like Riemann's "mobility without stretching," this requirement characterizes the spaces with constant curvature. What characterizes the Riemannian metric determinations within the class of all metric determinations $ds = F(x, dx)$ (in the sense of 3.1.4) is the requirement of free mobility of just infinitesimal rigid bodies (Helmholtz 1868). Today we can see this, in the sense of Riemann's motto, "almost without computation," as follows.

We require only freedom of rotation about a fixed point (x_0), and put $F(x_0, dx) = f(dx)$. We describe the state of affairs in the infinitesimal neighborhood of the point (Section 3.1.3) by means of the tangent space $T(x_0)$ and the function $f(\xi)$ for the vectors $\xi \in T(x_0)$. Freedom of rotation without change of length means, in particular, that for any two vectors $\xi, \eta \in T$ with $f(\xi) = f(\eta)$ there is an endomorphism D of T such that $D\xi = \eta$ and $f(D\zeta) = f(\zeta)$ for all $\zeta \in T$. We are to show that under these conditions f is necessarily a square root of a positive definite quadratic form. Consider the unit ball $K = \{\xi;\ \ f(\xi) \leq 1\} \subseteq T$. It is not difficult to see that there is a unique ellipsoid E with minimal volume, centered at 0, such that $E \supseteq K$. We associate with it the quadratic form with value 1 at the boundary of E and denote by $g(\xi)$ the root of this form. In view of the minimality of E, there is at least one ξ with $f(\xi) = g(\xi) = 1$. Let η be arbitrary with $f(\eta) = 1$, and let D with $D(\xi) = \eta$ be the corresponding automorphism. Clearly, $DK = K$, so that D is volume-preserving. DE is an ellipsoid whose volume is the same as that of E, and $DE \supseteq DK = K$. In view of the uniqueness of E, $DE = E$, and therefore $g(\eta) = g(D\xi) = g(\xi)$. The boundary of K coincides with that of E, $f(\zeta) = g(\zeta)$ for all ζ, and, as asserted, f is a root of a quadratic form. *The assumption of freedom of rotation of infinitesimal rigid bodies about arbitrary points implies the Riemannian nature of a space* (e.g., Laugwitz 1977, 143 ff.).

Helmholtz was one of the most eminent natural scientists of his time, a medical doctor, a physicist, and a physiologist. The mathematical considerations of such people often tend to be mistrusted by rigorous mathematicians. Moreover, Helmholtz's approach to the problem was somewhat different from Riemann's. His starting point was not a metric determination $ds = F(x, dx)$ but the motions of the space. In that case, he should have also required a monodromy axiom, which asserts that after a full turn a body is superimposed on itself without enlargement. Sophus Lie (1842–1899), who developed his continuous transformation groups soon after, rushed into the Helmholtz problem and changed it from a characterization of Riemannian metrics into a characterization of the orthogonal group.

In this connection Hilbert commented: "It seems to me that Lie always introduces into the issue a preconceived onesidedly analytic viewpoint and forgets completely the principal task of non-Euclidean geometry, that of characterizing the various possible geometries by the successive introduction of *elementary* axioms, concluding with the only remaining Euclidean geometry. Given this approach, it is not surprising that the assertions of other mathematicians do not fit his rigid analytical framework" (Hilbert, letter to Klein, dated 15 November 1893; *The Hilbert-Klein correspondence*, published by G. Frei, Göttingen 1985, 101). ("Lie trägt, wie es mir scheint, immer einen vorgefassten einseitig analytischen Standpunkt in die Sache hinein und lässt dabei ganz die Hauptaufgabe der Nichteuklidischen Geometrie, durch successive Einführung *elementarer* Axiome die verschiedenen möglichen Geometrieen bis zum schliesslichen Aufbau der allein noch übrigen Euklidischen Geometrie ganz ausser Augen. Bei dieser Verfahrensweise ist es dann nicht zu verwundern, dass die Behauptungen anderer Mathematiker in sein festes analytisches Fachwerk nicht hineinpassen") (Hilbert, Brief an Klein vom 15 November 1893; *Briefwechsel Hilbert-Klein*, herausg. v. G. Frei, Göttingen 1985, 101).

Helmholtz regarded free mobility of finite, or at least infinitesimal bodies, as a physical fact. He thought that the existence of rigid, 3-dimensional, extended measuring rods is a necessary precondition for every measurement of length. Later Einstein recognized the physically problematic nature of the concept of a rigid body. Actually, tightly stretched (1-dimensional) threads suffice for the measurement of length. But even then our reasoning yields a characterization of Riemannian spaces in terms of a homogeneity postulate: if at a point (x_0) there is to be no distinguished direction, then, since there is at least one ξ with $f(\xi) = g(\xi)$, this must hold for all such ξ. But now, conversely, it turns out that a simple homogeneity requirement enforces the freedom of rotation of infinitesimal rigid bodies and the existence of the orthogonal group!

In 1886 Fritz Schur characterized the spaces with constant curvature by means of a stronger homogeneity property. In general, the sectional curvature K depends on the point of the Riemannian space and on the 2-dimensional "sectional direction" at this point. But if the space is "locally" homogeneous, in the sense that K does not depend on the sectional direction at the given point, then K is also independent of the point, i.e., it is constant (F. Schur, "Räume konstanten Krümmungsmaßes II," *Math. Annalen* 27 (1886), 537–567; Laugwitz 1977, 110).

Riemann himself did not repudiate (W. 278) the more general $ds = F(x, dx)$.

> "The investigation of this more general species would not call for essentially different principles, but would be considerably time-consuming..."
>
> ("Die Untersuchung dieser allgemeinern Gattung würde zwar keine wesentlich andere Principien erfordern, aber ziemlich zeitraubend sein...")

The more general metric determinations were first studied in *Über Kurven und Flächen in allgemeinen Räumen* (1918; Nachdruck Birkhäuser, Basel 1951), the Göttingen dissertation of Paul Finsler, suggested by Carathéodory. It soon turned out that in such spaces we can write

$$ds^2 = \sum g_{ik}\, dx^i\, dx^k, \quad \text{with}$$

$$g_{ik} = g_{ik}(x;\xi) = \frac{1}{2}\frac{\partial^2 F^2(x;\xi)}{\partial\xi^i\partial\xi^k},$$

and that indeed no essentially different principles are required. In this case, all magnitudes depend not only on the point (x) but also on the tangent vectors (ξ). It is reasonable to make a restricting convexity assumption about $f(\xi)$. Finsler geometry has become an extensive research area, and attempts have been made to apply it. Its foundations are presented in Laugwitz 1977, 140–156.

But now we return once more to the 19th century. W.K. Clifford (1845–1879) translated Riemann into English and took seriously his suggestions about the interrelationship between physics and geometry. He saw in matter and its motion a manifestation of a time-dependent behavior of curvature. We quote:

> I hold in fact: (1) That small portions of space are of a nature analogous to little hills on a surface which is on the average flat. (2) That this property of being curved or distorted is continually being passed on from one portion of space to another after the manner of a wave. (3) That this variation of the curvature of space is really

> what happens in that phenomenon which we call the motion of matter whether ponderable or ethereal. (4) That in this physical world nothing else takes place but this variation, subject, possibly, to the law of continuity. (Clifford. *On the Spatial Structure of Matter*, Cambridge Philosophical Society, February 21, 1870; *Proceedings*, 1876.)

In his time, Clifford was regarded as a visionary.

In the first fifty years after Riemann's death, for reasons to be discussed later, Riemannian metrics played hardly any role in geometry or in physics.

3.1.7 The algorithmic developments

The computations at the end of Section 3.1.5 give one a foretaste of the way in which computing methods associated with Riemannian manifolds were at first bound to develop. Viewed from today's vantage point they strike one as prolix and opaque. Above all, what turned out to be a major obstacle was that the development of linear algebra failed for a long time to keep pace with the progress of analysis. For example, it was only after 1900 that vector notation was adopted in textbooks of mathematical physics, and it was not until the appearance, in 1918, of Hermann Weyl's *Space–Time–Matter* that "abstract" vector spaces, introduced axiomatically, began to appear in textbooks.

It is best to begin — ahistorically — with the tangent vector spaces $T(P)$ associated with a manifold (see the end of Section 3.1.3). We consider a definite one of them, $T = T(P)$. The linear forms over T form the dual space T'. In traditional terminology, the elements of T are called contravariant vectors, and those of T', covariant vectors. Then there are tensors, i.e., multilinear forms over T and T'. A bilinear form $g(\xi, \eta)$, $\xi, \eta \in T$, is called a twofold covariant tensor; in terms of coordinates, $g(\xi, \eta) = g_{ik}\, \xi^i\, \eta^k$. If g is symmetric and positive definite, then we have a scalar product in T. If we have such a form for every $T(P)$, then we have a Riemannian metric; in terms of coordinates, $ds^2 = g_{ik}(x)\, dx^i\, dx^k$. One speaks of a tensor field. With Riemannian curvature there is associated a fourfold covariant tensor field. Its coordinate representation is $R_{ijkl}(x)$.

In differential geometry we must frequently resort to the use of coordinate systems for computational reasons. But we are primarily interested in results whose significance is independent of a particular choice of coordinates, i.e., in results that are "invariant." A similar statement holds for analysis on manifolds. One needs, as auxiliary means, relations that describe the transition from one coordinate system to another. A prototype of a contravariant vector is a tangent vector to a curve. For the transition from an x-system to an $\bar{x}$-system we have

$$\frac{dx^i}{dt} = \frac{\partial x^i}{\partial \bar{x}^k}\frac{d\bar{x}^k}{dt} . \tag{a}$$

For a scalar field $M \to \mathbb{R}$ we have $\varphi(x) = \bar{\varphi}(\bar{x})$ and

$$\varphi_{,i} = \frac{\partial \varphi}{\partial x^i} = \frac{\partial \bar{\varphi}}{\partial \bar{x}^j}\frac{\partial \bar{x}^j}{\partial x^i} = \bar{\varphi}_{,j}\frac{\partial \bar{x}^j}{\partial x^i} , \tag{b}$$

and $d\varphi$ is a linear form over T:

$$d\varphi = \varphi_{,i}\, dx^i = \bar{\varphi}_j \frac{\partial \bar{x}^j}{\partial x^i}\frac{\partial x^i}{\partial \bar{x}^k} d\bar{x}^k = \bar{\varphi}_{,j}\, \delta^j_k d\bar{x}^k = \bar{\varphi}_{,j}\, d\bar{x}^j = d\bar{\varphi} .$$

$\varphi_{,i}$ represents a covariant vector. Equations (a) and (b) are the transformation laws for (the coordinate representations) of contravariant and covariant vectors respectively. For a simply covariant and simply contravariant tensor we have

$$t^i_l = \frac{\partial x^i}{\partial \bar{x}^k}\, \bar{t}^k_j\, \frac{\partial \bar{x}^j}{\partial x^l} , \tag{c}$$

and so on. We are interested in invariants, in scalar functions on M that are independent of the choice of a coordinate system. A simple example is obtained from (c) by putting i=l and summing, as required by the convention:

$$t^i_i = \frac{\partial x^i}{\partial \bar{x}^k}\frac{\partial \bar{x}^j}{\partial x^i}\, \bar{t}^k_j = \delta^j_k\, \bar{t}^k_j = \bar{t}^j_j .$$

Trace-forming, or contraction, makes an invariant out of a tensor. In the case of tensors of higher order this can hardly be realized without the use of coordinates. The tensor represented by (t^i_k) can also be regarded as an endomorphism of the vector space T. Its eigenvalues, and in particular their elementary symmetric functions, are certainly invariants because the coordinate system is not involved in their definition. As for the trace, it is the sum of the eigenvalues. One can also work with pairs or triples of tensors and form invariants such as $g_{ik}h^{ik}$, $g_{ik}t^{ik}_j v^j$, and $g^{il}g^{jk}R_{ijkl}$. In the second half of the 19th century there developed a flourishing theory of algebraic invariants whose aim was to find the totality of independent (algebraic) invariants.

In 1894 Klein wrote the following about Riemann (Klein *Werke* III, 494–495): "[With his Riemannian geometry] he prepared the ground for a new chapter of the differential calculus, namely, the study of quadratic differential expressions of arbitrary variables, and, more specifically, of the invariants of such differential expressions with respect to arbitrary transformations of the variables." ("Er hat dabei [mit seiner Riemannschen Geometrie] den Grund zu einem neuen Kapitel der Differentialrechnung gelegt: Zur Lehre von den quadratischen Differentialausdrücken beliebiger Variabler, bzw. von den

Invarianten, welche diese Differentialausdrücke gegenüber beliebigen Transformationen der Variablen besitzen.") This is an excellent description of the development of Riemannian geometry in the decades after Riemann's death.

The algebraic invariants discussed so far belong to the linear algebra of the vector space T. More interesting were differential invariants, which added tensor analysis to tensor algebra. After Christoffel, tensor analysis flourished primarily in Italy, owing to the contributions of Beltrami, Bianchi, Ricci, and Levi-Cività.

Is it possible to differentiate tensor fields? If in (c) we differentiate with respect to x^r then, clearly, the result does not again conform to the transformation law of a tensor because of the appearance of second derivatives of the coordinate transformation. Beginning in 1887 Ricci developed covariant differentiation. Let

$$\Gamma^i_{jk} = g^{ir}\Gamma_{jk|r}$$

be the Christoffel symbols of a Riemannian metric. If u_j, v^i, and t^i_k are representations of tensors, then so too are

$$u_j;_k = u_j,_k - \Gamma^i_{jk}u_i$$

and

$$v^i;_k = v^i,_k + \Gamma^i_{jk}v^j, \quad t^i_k;_j = t^i_k,_j - \Gamma^r_{lj}t^i_r + \Gamma^i_{lj}t^l_k\,.$$

The divergence of a vector field yields $v^i;_i$, the divergence of the gradient of a scalar field yields $(g^{ik}\varphi,_k);_i$, and the curvature tensor is related to the twofold covariant derivatives:

$$v^i;_j;_k - v^i;_k;_j = R^i_{hjk}v^h\,,$$

$$R^i_{hjk} = \Gamma^i_{hj},_k - \Gamma^i_{hk},_j + \Gamma^r_{hj}\Gamma^i_{rk} - \Gamma^r_{hk}\Gamma^i_{rj}\,.$$

This yields also formulas for Euclidean spaces in curvilinear coordinates. There are intimations of some of these matters in Riemann's Paris prize essay.

Tensor analysis flourished as a result of general relativity. The growing flood of indices was finally collected in an encyclopedic work, J.A. Schouten's *Ricci-Calculus, An Introduction to Tensor Analysis and its Geometrical Applications*, Springer, Berlin etc. 1954, with a bibliography of some 1200 titles. The book appeared exactly 100 years after Riemann's habilitation lecture. Here tensors are still defined by transformation laws, which play a secondary role in our presentation. We regard this formalism as an expedient, as a necessary evil. It is a consequence of the properties of multilinear forms.

We witness here a development similar to that in complex analysis. Algorithmic thinking was rampant until well into the 20th century and seemed to have supplanted Riemann's thinking in terms of manifolds with differentiable or complex structure. But in the second half of the 20th century Riemann's thinking again came to occupy a leading position. Of course, for many computational problems, especially those involving applications, it is expedient to use power series expansions in the case of complex functions and index representations in the case of tensors. But today, from a purely mathematical viewpoint, it seems inappropriate to regard such representations as intrinsic definitions of these concepts. This is equally true of Weierstrass' definition of an analytic function as a certain collection of power series and of the definition of a tensor as a collection of number-tuples with certain transformation laws. "Definitions" such as the following (Schouten, *Ricci-Calculus*, 6) are understandable only if we think of multilinear forms (E_n is our T, $A_\kappa^{\kappa'} = \partial\bar{x}^{\kappa'}/\partial x^\kappa$, for "quantity" read "tensor," say): "A *quantity* in E_n is a correspondence between the rectilinear coordinate systems and the ordered sets of N numbers satisfying the conditions: 1. To every coordinate system (κ) there corresponds one and only one set of N numbers; 2. if ϕ_λ; $\lambda = 1, ..., N$ corresponds to (κ) and $\phi_{\lambda'}$; $\lambda' = 1, ..., N$ to (κ'), then the $\phi_{\lambda'}$ are functions of the ϕ_λ and the $A_\kappa^{\kappa'}$ are homogeneous linear in ϕ_λ and homogeneous algebraic in $A_\kappa^{\kappa'}$. The ϕ_λ are called the components of the quantity with respect to (κ)."

In 1917 Levi-Cività introduced into the calculus the genuinely geometric idea of parallel displacement ("Nozione di parallelismo in una varietà qualunque," *Rendiconti del Circolo matematico di Palermo* 42). The underlying idea can be explained even within the framework of the Gaussian theory of surfaces. If a vector tangent to a surface at a point is moved parallel to itself to an infinitely close point of the surface, then the new vector need no longer be tangent to the surface, but its orthogonal projection on the tangent plane to the surface at the new point is again tangent to the surface. The projection vector is said to be the result of parallel displacement (in the sense of Levi-Cività) of the initial tangent vector. The change in the vector coordinates v^i due to the parallel displacement from (x) to $(x + dx)$ is given by

$$dv^i = -\Gamma^i_{jk} v^j dx^k ,$$

and this can be used to define parallel displacement in any Riemannian space. A vector can be displaced along a curve. It turns out that this does not affect its length. In general, parallel displacement of a vector along a closed curve yields a vector different from the initial one. An exception is parallel displacement in spaces with constant 0 curvature, but (in general) only along nullhomotopic curves. The unit tangent vectors at points on a geodesic are obtained from one another by parallel displacement along that geodesic. Thus geodesics are the "straightest" curves. Parallel displacement of vectors around infinitesimal surface elements gives a new geometric interpretation of Riemannian (Gaussian) curvature. In this connection H. Weyl noted that parallel transport does not call for a Riemannian metric. All one requires is gamma quantities. The latter need not be Christoffel symbols but must be related by an appropriate transformation law. In this way Weyl obtained the so-called spaces with affine connection. Their relation to Riemannian spaces is analogous to that of affine spaces to Euclidean metric spaces. This development also resulted in an enrichment of the formalism.

3.1.8 The influence of Felix Klein

During the first fifty years after Riemann's death, tensor analysis developed slowly and was far less important than invariant theory. We will describe the latter in greater detail in the sequel, partly in order to correct from a modern viewpoint one of its historiographers, namely Felix Klein. In the second volume of his lectures in the years 1915–1917 (Klein 1927) Klein discusses Riemannian geometry under the heading "Groups of analytic point transformations with a quadratic differential form taken as a basis" (pp. 136–204) ("Gruppen analytischer Punkttransformationen bei Zugrundelegung einer quadratischen Differentialform").

Klein not only witnessed developments in complex analysis and in geometry in the last decades of the 19th century but also helped to shape them through his papers and lectures. Hence the need for a deeper appreciation of him and of his role.

In 1872, a few years after Riemann's death and after the publication of his lecture on geometry, there appeared a paper by Klein that was to become highly influential and to be known as the *Erlangen Program* (*Erlangen Programm*). It had the effect of determining for many decades what was to be meant by the term "geometry." In particular, it marginalized for a time Riemannian geometry. It was only as a result of an external influence, namely that of general relativity, that Klein's horizons were extended, and, as a result, Riemann's geometry seemed acceptable to him. This was in stark contrast to complex analysis, where Klein appreciated and used Riemann's work from the very beginning.

Klein studied in Bonn under Plücker and was his assistant. Plücker taught physics as well as mathematics. In mathematics he worked primarily in the area of line geometry, the then "newer geometry." In 1868 Klein obtained his doctoral degree for work in this area. After Plücker's death (in 1868) Klein worked with Lipschitz, who was obviously unwilling to familiarize him with Riemannian geometry (Klein 1927, 168). In the winter of 1868/1869 Klein continued his education in the seminars of Kummer and Weierstrass at Berlin,

32: Felix Klein

and he habilitated in 1871 in Göttingen. He was to return to Göttingen for good in 1886. Erlangen, the Technical University in Munich, and Leipzig were for him, so to say, intermediate way stations.

Klein was a dominant personality. Unlike Weierstrass, he saw to it that his lectures were quickly published, and he was concerned about *Mathematische Annalen*, educational policies, and about the history and applications of mathematics. By bringing Hilbert and Minkowski to Göttingen he made it again, for the first time after Riemann's death, a mathematical center on a par with Berlin.

The *Erlangen Program* came into being when the mercurial young Rhinelander merged two active currents of thought. One was the view of geometry summarized in Arthur Cayley's (1821–1895) dictum "projective geometry contains all geometry." The other was group theory as a method of demonstrated efficacy, particularly in algebra. Using modern terminology, one could summarize Klein's synthesis as follows: a "geometry" is a "space" (a set of points) with a "structure," and the one-to-one structure-preserving mappings of the space onto itself form a group that can be called its group of automorphisms. In the case of the projective plane the group elements are the collineations that map straight lines onto straight lines. By singling out a definite straight line one can obtain the subgroup of collineations that map this line onto itself. This yields affine geometry. The fixed straight line can be viewed as "the line at infinity." It now makes sense to talk about parallel straight lines: if the point of intersection of two straight lines lies on the line at infinity, the same is true of their images. The subgroups of similarity and congruence mappings belong to Euclidean geometry. The first great success of this viewpoint, beyond the classification of the known "geometries," was the Cayley-Klein model of the hyperbolic plane. The associated group consists of the collineations that map a nondegenerate conic onto itself.

According to Klein, to study geometry is to investigate the invariant theory of an appropriate group. For example, in Euclidean geometry two points are a certain distance apart, and this distance is an invariant. It is unchanged by a congruence: the distance between the images of two points is the same as that between their preimages.

In affine geometry the simplest invariant is the inner ratio of three collinear points, and in projective geometry it is the cross ratio of four collinear points. The latter can be used to define the distance between two points in hyperbolic geometry; the necessary quadruple of points is obtained by adding to the two given points the two points in which the straight line joining them meets the distinguished conic.

There seems to be no place in the Klein scheme for Riemannian geometry, successfully proclaimed before 1872 by Helmholtz, Lipschitz, and Christoffel. An automorphism of a Riemannian space is a length-preserving mapping of this space onto itself, and it can happen that its only automorphism is the identity mapping. According to Riemann and Helmholtz, only spaces of constant curvature can have "sufficiently many" automorphisms. Accordingly, elliptic geometry, a geometry with constant positive curvature, was for a long time referred to as Riemannian geometry (as indeed it continues to be called in biographical accounts of Klein), in much the same way that hyperbolic geometry is variously referred to as Bolyaian, Lobachevskian, or Gaussian (in this connection see Klein 1927, 22).

The extent to which Klein was steeped in group representations is illustrated by his postcard to Hilbert of 28 August 1894. This was Klein's response to Hilbert's "On the straight line as the shortest connection between two points" ("Über die gerade Linie als kürzeste Verbindung zweier Punkte"). Here Hilbert referred to the Cayley-Klein model of hyperbolic geometry in the interior of an ellipse and observed that one could work with the interior of a convex curve or surface. If distances were determined in a way analogous to that in the Cayley-Klein model, then straight lines would again yield the shortest connections between points. Klein wrote: "This pleases me all the more because for once there is *no* group here" ("Mir ist die Sache um so sympathischer als da nun einmal *keine* Gruppe vorliegt") (G. Frei, *Briefwechsel Hilbert-Klein*, Göttingen 1985, 110). We see that for Klein a geometry without a group was something of a surprise. At the time when he was busy with his lecture on Riemann and his importance for the development of mathematics, he seems not to have been aware that Riemannian spaces had for a long time supplied such examples.

At first Klein's program was intended as a classification principle for existing geometries, and since the young man did not as yet know everything in existence, this classification helped him. But later the principle became a dogma, and a geometry became the invariant theory of a group. We know of similar developments. In the middle of the 20th century Bourbaki formulated a classification principle intended for the whole of mathematics. Klein brought together groups and geometry, and Bourbaki algebraic and topological structures. We realize the fruitfulness of bringing conceptually distinct things under one roof but also the dangers of dogmatism.

Klein was always interested in physics, and he could easily fit the relativity theory of 1905 to his scheme: the Lorentz group and its invariant theory could be regarded as an adequate description of electromagnetism. But then came general relativity, whose physical justification was obvious to him, and he had to accept Riemannian geometry. He even became quite proficient in it. But how was invariant theory of groups to be fitted to it?

The answer to this question is found in the second part of his lectures (Klein 1927), subtitled "Die Grundbegriffe der Invariantentheorie und ihr Eindringen in die mathematische Physik" ("The fundamental concepts of invariant theory and their penetration of mathematical physics"). Even this subtitle is a biased description of the historical events. Klein used a trick to bend his Erlangen Program so as to make it seem to fit Riemannian geometry. His argument turned on the ambiguity of the term "transformation," which even today is a source of tribulation for all beginning students of linear algebra. "Transformation" can denote a (usually bijective) mapping of a set onto itself or onto another set, or a coordinate transformation. While the starting point of the original conception of Klein's Erlangen Program seemed to be the first of these two meanings, he now embraced the second one.

In the chapter titled "Gruppen analytischer Punkttransformationen..." ("Groups of analytic point transformations...") we read (pp. 136/137) that he envisaged groups of arbitrary substitutions

$$x_j = \varphi_j(y_1, ..., y_n), \quad j = 1, 2, ..., n. \tag{1}$$

"We usually view $x_1, ..., x_n$ as point coordinates in an n-fold extended space and simply speak of point transformations. Moreover, one is at liberty to regard the coordinate system as fixed and the space, or the structures in it, as transformable, or, if this happens to be more convenient, to regard R_n as fixed and the coordinate system as changeable (in which case the equations (1) represent a 'coordinate transformation')." ("Indem wir die $x_1, ..., x_n$ zumeist als Punktkoordinaten in einem n-fach ausgedehnten Raume auffassen, sprechen wir kurzweg von Punkttransformationen. Es steht dabei frei, sich entweder das Koordinaten als fest und den Raum bzw. die in ihm gelegenen Gebilde als transformierbar anzusehen, oder, wie es zunächst bequemer ist, den R_n als fest und das Koordinatensystem als veränderlich (so dass also die (1) eine 'Koordinatentransformation' vorstellen)").

Actually, one is not all that free. If we interpret (1) as a coordinate transformation, then we must bear in mind that such a transformation is only locally meaningful. Some fragmentary group properties can be saved by radical means in the form of pseudogroups (Veblen and Whitehead 1932). But this is of no use whatever to differential geometry and we need not go into it. As for mappings of the space or of "the structures in it" — in Klein's case this is a new restriction or extension — we have already noted that they are significant only if they are automorphisms, and of these there are too few. This was clear to the differential geometers. In fact, rather than speak of invariance, the Italian geometers preferred to speak of absoluteness, i.e., of independence in relation to the coordinate system. In Klein's *Mathematische Annalen* 54 (1901), 125–201, there appeared the paper of G. Ricci and T. Levi-Cività titled "Méthodes de calcul différentiel absolu et leurs applications."

The desirable objective is not the invariant theory of a group of (coordinate) transformations but the coordinate-free development of geometry. In the case of affine and Euclidean geometries the solution was in the offing; one can look it up in the early sections of H. Weyl's *Space–Time–Matter*. In the case of differentiable manifolds there is no such easily accessible possibility for the beginner.

Riemann worked within a far older geometric tradition than did Klein in the Erlangen Program. There are neither coordinates nor motions in Euclid and there are none in Hilbert's modern (1899) formulation of the foundations of geometry. One does not verify the congruence of triangles by going to the trouble of moving the whole infinite plane — a procedure inconceivable to Euclid. What one needs is comparisons of distances, and Riemann introduced length measurement of curves. Coordinates have been used since Descartes (1637), but at first they were not necessarily rectangular. Rather, they were *ad hoc*, dictated by the requirements of particular problems. We saw that Riemann operated in much the same way, in the sense that he computed exclusively in terms of specially adapted, normal coordinates, inevitably derived from the geometry. The existence of a congruence-preserving group of motions in the cases of Euclid and Hilbert is a quasi-accidental, *a posteriori* consequence.

Coordinates and mappings play a secondary role in geometry. But in Klein's case they become the primary definition of its essence. We cannot rule out the possibility that by "transformation" Klein meant not the mapping but the transition to a different coordinate system, but in that case he made a virtue out of a necessity. Since it is possible to approach the issue in a direct, or "absolute," manner, the invariant theory of transformation groups is geometrically superfluous. In this way the Erlangen Program promoted a mode of thought that was further from the origins of geometry than was Riemann's formulation.

In his *The Genesis of the Abstract Group Concept* (The MIT Press, 1984, 192/193), H. Wussing also discusses the aged Klein's flawed assessment of the Erlangen Program. Klein tried very hard to prove the significance of the program as a preliminary stage of the (special) theory of relativity, and used a kind of play on words for this purpose.

It is impossible to decide whether or not the Erlangen Program initially hindered the further development of Riemann's differential geometry. But it is striking that notable advances were made in Italy, outside the domain of Klein's influence, while little happened in Germany after 1872 (R. Beez, F. Schur, A. Voss). To gauge the distance that separated Klein from Riemann's intentions it is enough to recall that Riemann's primary argument for the restriction to quadratic differential forms was that the investigation of the more general case would be relatively time-consuming (W. 277–278). True, Riemann was very busy at the time, but his followers would have had time enough. Instead of developing Riemann's ideas people preferred to exploit the existing algorithm of invariant theory. Moreover, what Riemann had intimated failed to come through in Klein's work.

Klein was also far from taking abstract vector spaces seriously (only then would the concept of a differentiable manifold, with its tangent vector spaces in our sense of the term, have been accessible). In fact, he scoffed at the concept of a vector space already used by contemporary physicists (Klein 1927, 48): " ... that ... to a greater or lesser extent vector theory is viewed as something that is set against traditional analytic geometry, i.e., coordinate geometry, and must be developed independently of it, and should perhaps be even regarded as the foundation of coordinate geometry. This is precisely the reverse of the invariant-theoretic treatment of geometry ... that links the use of coordinates to the mobility of our notion of space through the medium of a group of substitutions. The contradiction in question arose in England, where the 'poor old Cartesian with his axes' plays a standard role in polemical representations." ("... daß... mehr oder minder die Vektorlehre als etwas aufgefaßt wird, was der traditionellen analytischen Geometrie, d. h. der Koordinatengeometrie, entgegengesetzt ist und unabhängig von ihr entwickelt werden muß, vielleicht gar als Grundlage der Koordinatengeometrie zu betrachten ist. Das ist also genau das Gegenteil von der invariantentheoretischen Behandlung der Geometrie … , die den Gebrauch der Koordinaten durch das Intermedium der Substitutionsgruppe mit der Beweglichkeit unserer Raumanschauung verbindet. Der Gegensatz um den es sich handelt, ist in England entstanden, wo der 'poor old Cartesian with his axes' in den polemischen Darstellungen eine stehende Rolle spielt.") He added, in a footnote: "It is curious that the gentlemen who forbid coordinate axes nevertheless leave unchanged the preferred status of the origin." ("Merkwürdig, daß die Herren welche die Achsen verpönen, trotzdem dem Koordinatenanfangspunkt eine bevorzugte Stellung belassen.") Klein overlooked that what was involved was the nullvector and not the nullpoint.

The adherents of each of the several algorithmic trends in differential geometry regarded themselves as Riemann's followers. (We think that we rather do justice to Riemann by emphasizing his tendency to prove with virtually no computations.) A clear determination of the merits of these claims is impossible, because his texts contain too few completed computations in this area. It is different in complex analysis. Here, whenever possible, Riemann proved results "almost without computations," but whenever necessary he made use of every available algorithmic tool without dogmatically tying himself down. On the other hand, the respective algorithms and calculi of his followers were marked by exclusivist tendencies.

3.1.9 Dedekind: Analytic investigations related to Bernhard Riemann's paper on the hypotheses which lie at the foundations of geometry

In his remarks on Riemann's Paris prize essay (W. 405–423) Heinrich Weber tried to supply the missing computations for this paper and thereby also for the habilitation lecture. He mentioned that this was done "on the basis of an older (unpublished) investigation by R. Dedekind" ("auf Grund einer älteren (ungedruckten) Untersuchung von R. Dedekind") (W. 405). Dedekind's manuscript, whose title is that of this section, became generally accessible only in 1966, a century after Riemann's death (Göttingen, Cod. Ms. Dedekind). A transcription with a French translation and a commentary has been published: M.-A. Sinaceur, "Dedekind et le programme de Riemann," *Rev. Hist. Sci.* 43 (1990), 221–294.

When she published Dedekind's collected works, Emmy Noether could not take this manuscript into consideration. Owing to Ludwig Bierbach, it ended up in the Göttingen collection.

The only person influenced by the manuscript was Weber. It contains no surprising reflections for modern readers. Nevertheless, we propose to discuss it briefly because it sheds light on Dedekind's interest in Riemann's ideas and on the difference in the modes of thought of the two friends. In particular, it makes evident Dedekind's more algebraic orientation.

Dedekind set down a program of 22 sections, some just titles, and wrote in the margin: "This outline has been realized in the text only in part and in a different order" ("Dieser Entwurf ist im Text nur theilweise, und in anderer Folge, ausgeführt.")

As a sample, we reproduce the entire
"§1. Object of study: $ds^2 = \sum a_i^{i'} dx_i dx_{i'}$, a positive form. Lines. Direction of an element from x to $x + dx$ determined by the n values

$$\frac{dx_1}{ds}, \ldots, \frac{dx_n}{ds}$$

where ds is positive."
("§1. Gegenstand der Betrachtung: $ds^2 = \sum a_i^{i'} dx_i dx_{i'}$ positive Form. Linien. Richtung eines Elementes von x zu $x + dx$ bestimmt durch die n Werthe

$$\frac{dx_1}{ds}, \ldots, \frac{dx_n}{ds}$$

wo ds positiv.")
In the relevant detailed part of the paper Dedekind does linear algebra and employs clever notations such as

$$(u, u') = \sum a_i^{i'} u_i u'_{i'}$$

and

$$(d, d') = \sum a_i^{i'} dx_i d'x_{i'} = \sum b_i^{i'} dy_i d'y_{i'},$$

where a and b are the coefficients of the form with respect to the coordinates x and y. Here he introduces a coordinate-free representation of an inner product. (He speaks of a bilinear function.) He develops a substantial chunk of multilinear algebra in n dimensions, including the "Gram" determinant and its relation to linear dependence and alternating forms. In some of these formulations one can discern features of the calculus developed later by E. Cartan. He repeatedly expresses delight at attaining independence of coordinates. Here Dedekind is far ahead of his contemporaries with their analytic-geometric orientation. But he remains essentially in the realm of linear algebra (tensor algebra) of a single tangent vector space and does not get to "covariant" differentiation. However, the more he advances, the closer he comes to the Einstein summation convention; close to the end of the manuscript he sums over an index precisely when it occurs once as a subscript and once as a superscript!

In §§7–8 Dedekind goes beyond purely algebraic arguments. Specifically, very much in the spirit of Riemann, he brings together two subjects treated by the latter. One is Riemannian geometry and the other is what Dedekind calls investigations of potential. He extends the Laplace operator Δ to n-dimensional Riemannian spaces, much as Beltrami was to do later, and notes the independence of the choice of coordinates as a consequence of the Green integral formula, an application for which he provided a justification. He closes §7 with a remark that, unfortunately, he fails to pursue: "Finally, one perceives the possibility of extending the Dirichlet Principle to spaces of the kind considered here." ("Man erkennt endlich die Möglichkeit der Übertragung des Dirichletschen Prinzips auf Räume der hier betrachteten Art.")

§8 is the last section worked out in some detail. It is devoted to variational problems. In addition, Dedekind obtains a definition of the (first) curvature and principal normal of a curve in an n-dimensional Riemannian space. He indicates that he intends to investigate variational problems dealing with the content of m-dimensional spaces in n-dimensional ones. "But this is so far removed from the actual aim of this essay that I defer its communication to another occasion." (Das läge "jedoch dem eigentlichen Zweck dieser Abhandlung so fern dass ich die Mittheilung derselben mir für eine andere Gelegenheit verspare.")

This was never realized. But it is interesting to note that during this early phase (probably around 1868) Dedekind definitely intended to concern himself further with Riemannian geometry.

3.2 Physics

3.2.1 The interest in physics

Riemann had an abiding interest in physics and at times devoted himself more to that subject than to mathematics. Four of the nine papers that he himself managed to publish must be viewed as belonging to physics. About half of the lectures announced by him are devoted to methods of mathematical physics. While he did not always have a deft touch in his first attempts at physics, a few eminent contemporary physicists took him seriously. His contributions of permanent interest are, above all, mathematical, but they also include his papers of 1860 and 1861, devoted to shock waves and to the motion of a homogeneous liquid ellipsoid respectively. A short time ago, some experts (P. D. Lax, N. 807–810; S. Chandrasekhar, N. Lebovitz, N. 811–820) praised Riemann's work in these two areas. This being so, we can dispense with a discussion of the details in these papers that are not very relevant to the image of Riemann as a mathematician.

What did Riemann himself view as essential? The following note (W. 507) probably belongs to the period of his habilitation:

> "My principal task is a new interpretation of the well-known laws of nature — their expression by means of other fundamental concepts — that would make possible the utilization of experimental data on the interaction of heat, light, magnetism, and electricity for the investigation of their correlations."
>
> ("Meine Hauptarbeit betrifft eine neue Auffassung der bekannten Naturgesetze — Ausdruck derselben mittelst anderer Grundbegriffe — wodurch die Benutzung der experimentellen Data über die Wechselwirkung zwischen Wärme, Licht, Magnetismus und Elektricität zur Erforschung ihres Zusammenhangs möglich wurde.")

He mentions this "principal task" in addition to the further realization of Section 20 of his dissertation (introduction of the imaginary into analysis) and the development of

> "new methods of integration of partial differential equations" which he "already successfully applied to several physical topics."
>
> ("neue Methoden zur Integration partieller Differentialgleichungen" welche er "bereits auf mehrere physikalische Gegenstände mit Erfolg angewandt habe.")

Thus at the time, in addition to Section 20, Riemann had a physical program that concerned the very phenomena which were later to be brought together under the umbrella of electromagnetism. It was to this "principal task" that he referred in the letter of 29 December 1853 to his brother, in which he also mentioned its import and stated that the matter was almost ready for publication. This program fitted the kind of unified field theory that the great 20th-century physicists were to try in vain to construct. But for the moment Riemann published nothing, and what we find in the *Nachlass* is rather disappointing (W. 528–538). Nevertheless, we will discuss this material (in Section 3.3.3) because it is an original speculation. His note of 1858, "Ein Beitrag zur Elektrodynamik" ("A contribution to electrodynamics") is more substantial. We will discuss it at the end of Section 3.2.2.

We know from Riemann's biography that as a student he also concerned himself with physics and was to continue the tradition of his teachers Gauss, Dirichlet, Jacobi, and later Weber. The many-sidedness of his interests is also attested by a list, mentioned by M. Noether, of outlines of theses for a doctoral disputation (N. 710–711):

"1. There are no magnetic fluids.
2. Faraday's 'induction in curved lines' is not tenable.
3. We can premise analysis with the differential calculus without loss of generality.
4. The reversion pendulum is not the best tool for the determination of the length of pendulums.
5. The theory of conservation of energy is not yet sufficiently established experimentally.
6. The concept of electrical potential has not yet been understood in the science of electricity with the necessary precision."

("1. Es existieren keine magnetische Fluida.
2. Faradays 'induction in curved lines' ist nicht haltbar.
3. Man kann ohne der Allgemeinheit zu schaden die Differentialrechnung der Analysis vorausschicken.
4. Das Reversionspendel ist nicht das geeignetste Mittel um die Pendellängen zu bestimmen.
5. Die Lehre von der Erhaltung der Kraft ist experimentell noch nicht genügend erwiesen.
6. Der Begriff Spannung ist bisher in der Elektrizitätslehre noch nicht mit der gehörigen Schärfe aufgefaßt worden.")

The following remark belongs with item 5. On 23 July 1847, during Riemann's first semester at Berlin, Helmholtz presented his famous lecture "Über die Erhaltung der Kraft" ("On the conservation of energy") to the physical society in Berlin. It seems that the audience received the lecture with a great deal of skepticism. This was also Dirichlet's reaction. Helmholtz reported later that of all the members of the Berlin Academy only the mathematician Jacobi espoused his cause (H. Helmholtz, *Über die Erhaltung der Kraft*, Ostwalds Klassiker Nr. 1, Leipzig 1915, p. 59). It is possible that Riemann came in contact with these problems already at that time. The idea of a conservation law also turns up in Riemann's speculations about masses of spirit (W. 511).

Upon his return to Göttingen, Riemann formed a close association with Weber. It may be that this turn to physics was partly motivated by practical thoughts about the future. The only professional objective other than the financially very unreliable position of university instructor (Privatdozent) was that of a gymnasium teacher. For the latter one had to be familiar with more than mathematics. Riemann's participation in the pedagogical seminar lends support to the guess that he gave very serious thought to such possibilities. Of course, he studied not just for the sake of examinations. In his case, mathematics, physics, and psychology combined in a single scientific entity that can be sensed in his later work.

3.2.2 *Physics as a field theory*

In Riemann's time the mathematical treatment of physics was dominated by partial differential equations. This approach had grown out of particle mechanics. In the equation of motion $m\ddot{x} = -F(x)$, $x = (x_1, x_2, ...)$, where the dots above x indicate differentiation with respect to time t, one assumed the existence of forces acting directly at a distance. This thought-scheme was also adopted as the foundation in the mechanics of continua, in elastic forces in solids, and in the cohesion of fluids. The problem was to determine the forces between the particles, the "atoms." The ether too was thought of in mechanical terms as a kind of matter. Descartes and Huygens thought that the ether led the planets in their courses; and Huygens and Fresnel thought that it was the medium for the propagation of light. Around 1820 physicists distinguished ordinary (ponderable) matter, the heat substance *Caloricum*, various electrical *Fluida*, and ether. Paris, then the center of mathematical-physical research and learning, boasted luminaries such as Laplace, Fourier, Ampère, Poisson, and Cauchy. In England, basing himself on his experiments and eschewing mathematical formulations, Michael Faraday (1791–1867) thought of electromagnetic phenomena in a new way which was destined to turn the ether concept into that of a field. On this view a physical event is a change of the state of the field described by means of functions $\psi(x, t)$, and the task of physics is to find the appropriate field equations, i.e., the partial differential equations for the field quantities ψ that describe the course of the event on the basis of its present state. According to this conception, the partial differential equation determines the change in the field quantities at a point x from the field quantities at this point and in its immediate (infinitesimal) neighborhood. Thus local action in field theory took the place of action at a distance in the old scheme of particle mechanics.

Around 1810 Laplace set down the following summary: All said and done, natural phenomena must be reduced to actions "*ad distans*" from particle to particle, and the study of these actions must serve as the basis for every mathematical theory of the phenomena. The nonempirical character of this principle of Laplace raised problems: the laws of the forces between particles were unknown and, in fact, not measurable; and in what way were the actions transmitted? Fourier pushed all these issues aside; he referred to heat, whereas Laplace initially had optics in mind. But both wanted to include all of physics

33: Jean-Baptiste Joseph Fourier

within a single program. "We need not set up uncertain hypotheses about the nature of heat; knowledge of the mathematical laws that control its effects is independent of any hypothesis; it calls only for careful testing of the main facts of ordinary observations and confirmation through precise experiments" (Fourier, *Théorie analytique de la chaleur*, 1822, §22). One can observe temperature as a function of place and time, and the Fourier differential equation for this function is confirmed by experiments which he himself carried out with all necessary thoroughness. Fourier's book is the prototype of a field theory.

The agreement with observations was decisive for the success of Fourier's conception, for it is not convincing in terms of the theory of particles as centers of force. This conception implies that an "atom" is subject to actions of just its immediate neighbors, but not those of the neighbors of neighbors. Occasionally, people had taken advantage of this already in the 18th century in connection with one-dimensional vibrations. If the field quantity u is the deviation, then one obtains the correct differential equation $u_{tt} = c^2 u_{xx}$ if one assumes that the acceleration is proportional to the resultant of $u(x+dx)-u(x)$ and $u(x-dx)-u(x)$ and neglects $u(x \pm 2dx)$. Since that time, infinitesimal arguments for the "derivation" of differential equations of physics have followed this pattern. Gauss, Dirichlet, and Wilhelm Weber also regarded partial differential equations as the foundation of physics, and Riemann assimilated this conception, which was to dominate all of his teaching activity. He began his lectures in the winter semester of 1854/1855 with "Die Theorie der Integration der partiellen Differentialgleichungen nebst Anwendung derselben auf verschiedene Probleme der Physik." ("The theory of integration of partial differential equations and their application to various problems of physics.") After that no academic year passed without his announcing lectures of this kind, sometimes with a special topic such as elasticity or gravity, or electricity and magnetism. These lectures were the basis of the books by K. Hattendorff 1869, 1876. The first of these books was later continued by Heinrich Weber, and for generations "Riemann–Weber" was the standard work on mathematical physics for students and for professionals alike. Of course, the Hattendorff revisions were not the same as the authentic lectures, but it is undoubtedly Riemann's merit that his mode of thought and methodology spread throughout the German-speaking world. After 1925, P. Frank and R. von Mises consciously continued this tradition.

In the fall of 1854 Riemann wrote down fundamental remarks on mathematical physics, possibly intended as an introduction to the lectures on partial differential equations and their application to physics that he gave in the winter semester of 1854/1855. Hattendorff used these notes verbatim as an introduction to Riemann 1869. Neither he nor his contemporaries and fellow students regarded the part of the book other than the introduction as a faithful account of Riemann's lectures. Below we reproduce excerpts from Riemann's text.

> "It is well known that scientific physics has existed only since the invention of the differential calculus."
>
> ("Eine wissenschaftliche Physik existiert bekanntlich erst seit der Erfindung der Differentialrechnung.")

The concept of an accelerating force as a simple cause of motion is due to Galileo. To this was added the concept of a simple cause of force in the form of Newton's idea of a center of attraction.

> With these "two concepts physics builds even today. ... No new forward step has been taken since Newton; so far, all attempts to go beyond these fundamental concepts in order to penetrate into the core of nature have failed; the influence of later philosophical systems, where it has manifested itself in the physical literature, has had the sole success of distorting Newton's original conception and of introducing inconsistencies into it."
>
> (Mit diesen "beiden Begriffen konstruiert die Physik noch heute. ... Es ist seit Newton kein neuer Fortschritt gemacht; alle Versuche, über diese Grundbegriffe hinaus ins Innere der Natur zu dringen, sind bis jetzt mißglückt; der Einfluß der späteren philosophischen Systeme, wo er sich in der physikalischen Literatur geltend gemacht, hat nur den Erfolg gehabt, die ursprüngliche Auffassung Newtons zu verunstalten und Inkonsequenzen in dieselbe einzuführen.")

Newton published his principal work in 1687. Sixty years later, in 1747, d'Alembert published his treatment of the vibrating-string problem, involving a partial differential equation. After another 60 years, on 21 December 1807, Fourier submitted to the Paris Academy his first paper on the theory of heat. Since that time, said Riemann, physics had been dominated by partial differential equations.

> "Thus in all these physical theories, which are explained by the assumption of molecular forces, partial differential equations now form the actual foundation ascertained by experience. The same is true of phenomena that are explained by means of the assumption of forces of attraction and repulsion whose action is inversely proportional to the square of the distance: of magnetism, electricity, and gravitation, provided that we are dealing with extended bodies.
>
> What has been shown to be a fact by induction, (namely) that differential equations form the actual foundation of mathematical physics, can also be shown *a priori*. True elementary laws can operate only in the infinitely small, and apply to points in space and time. But such laws are in general partial differential equations, and the derivation of laws for extended bodies and time periods requires their integration. Thus we need methods for deriving from laws in the infinitely small such laws in the finite, and, more precisely, (laws) that one derives with complete rigor, without permitting oneself omissions. Because it is only then that we can test them by way of experience."
>
> ("In allen diesen physikalischen Theorien, welche man durch Annahme von Molekularkräften erklärt, bilden also jetzt partielle Differentialgleichungen die eigentliche durch die Erfahrung festgestellte Grundlage. Dasselbe gilt auch von den Erscheinungen, welche man durch Annahme von Anziehungs- und Abstoßungskräften erklärt, die umgekehrt proportional dem Quadrat der Entfernung wirken: von dem Magnetismus, der Elektrizität und von der Gravitation, sobald man es dabei mit ausgedehnten Körpern zu tun hat.
>
> Was sich eben auf dem Wege der Induktion als Tatsache ergeben hat, daß die eigentlichen Grundlagen der mathematischen Physik Differentialgleichungen sind, ergibt sich auch a priori. Wahre Elementargesetze können nur im Unendlichkleinen, für Raum- und Zeitpunkte stattfinden. Solche Gesetze werden aber im allgemeinen partielle Differentialgleichungen sein, und die Ableitung der Gesetze für ausgedehnte Körper und Zeiträume aus ihnen erfordert die Integration derselben. Es sind also Methoden nötig, durch welche man

> aus den Gesetzen im Unendlichkleinen diese Gesetze im endlichen ableitet, und zwar in aller Strenge ableitet ohne sich Vernachlässigungen zu erlauben. Denn nur dann kann man sie an der Erfahrung prüfen.")

Here we must consider somewhat more carefully a number of interesting viewpoints.

For Riemann, the only scientific physics was mathematical physics. More precisely, and as we would put it today, for him scientific physics was a model to be described by differential equations. It did not include the statics of Archimedes. Galileo, who had died less than a year before Newton's birth, just barely made it as the founder of dynamics. What Riemann goes on to describe is nowadays called field physics: the cause of motion is the accelerating force, the cause of force is "the attracting or repelling center," the central force fields are, purely and simply, *the* foundation of physics.

Experience, i.e., induction, has shown that partial differential equations are the appropriate mathematical tool. But this could have been expected *a priori* (surely a strange sentiment from someone who has just rejected philosophical speculations). This claim is connected with the rejection of all action at a distance, expressed in the assertion that the true laws of nature are inherent in the infinitely small. Once this is granted, we are, of course, down to differential equations — partial differential equations, since there are several variables for space and time.

We recall the formulations in the habilitation lecture presented a few months earlier, and the historical outline in Riemann's habilitation essay in which d'Alembert and Fourier are already viewed as the physicists who had substantially advanced the mathematical treatment of physics. All this enables us to see clearly the philosophical core of Riemann's activities in 1854.

Wilhelm Weber propounded an atomic conception of electricity according to which positive and negative particles interact instantaneously. On the other hand, Riemann, like Maxwell after him, had in mind a theory of local action with finite propagation velocity. Both Weber and Riemann stated laws of the force that two moving charged particles exert on one another. The short note "Contribution to electrodynamics" ("Ein Beitrag zur Elektrodynamik"), presented by Riemann in Göttingen on 10 February 1858, opens with the following formulation (W. 288–293):

> "I take the liberty of communicating to the Royal Society an observation that brings closely together the theory of electricity and magnetism and that of light and radiant heat. I have found that the electrodynamic actions of galvanic currents can be explained if one

assumes that the action of one electrical mass on others is not instantaneous but propagates itself towards them with constant speed (equal, within the limits of observational errors, to that of the speed of light). Under this assumption, the differential equation of the electrical force is the same as that for the propagation of light and radiant heat."

("Der Königlichen Societät erlaube ich mir eine Bemerkung mitzutheilen, welche die Theorie der Elektricität und des Magnetismus mit der des Lichts und der strahlenden Wärme in einen nahen Zusammenhang bringt. Ich habe gefunden, dass die elektrodynamischen Wirkungen galvanischer Ströme sich erklären lassen, wenn man annimmt, dass die Wirkung einer elektrischen Masse auf die übrigen nicht momentan geschieht, sondern sich mit einer constanten (der Lichtgeschwindigkeit innerhalb der Grenzen der Beobachtungsfehler gleichen) Geschwindigkeit zu ihnen fortpflanzt. Die Differentialgleichung der elektrischen Kraft wird bei dieser Annahme dieselbe, wie die für die Fortpflanzung des Lichts und der strahlenden Wärme.")

Today what matters is Riemann's insight, which preceded Maxwell's decisive treatise by 15 years, rather than the fuzzy justification which follows it. He quickly withdrew the note. It appeared posthumously in *Poggendorffs Annalen der Physik und Chemie* 131, 1867. In Vol. 135, 606, Clausius criticized Riemann's unjustified interchange of the order of two integrations and, like Heinrich Weber (W. 293), assumed that this flaw may have been the reason for Riemann's backing down. But far more important than all this is Riemann's introduction of retarding potentials and his intimation that light, radiant heat, electricity, and magnetism exhibit similar behavior.

He dealt with this issue in his lectures, but here we have only Hattendorff 1876 to go by. After the appearance of this book, Clausius used physical arguments to criticize Wilhelm Weber's as well as Riemann's force laws ("Über die Ableitung eines neuen elektrodynamischen Grundgesetzes," *Crelle* 82 (1877), 85 ff.)

We see that the fate of Riemann's "principal task" was not a happy one. In the letter to his brother at the end of 1853 he thought that he had brought it to the point where he could publish it without reservations. But more than four years later he lacked the courage to publish even a partial result (without inclusion of gravity).

What suggested to him the idea of putting light and electromagnetism under the same roof? In the above letter he mentioned the guess that Gauss had long concerned himself with this idea and had presumably told this in confidence to Weber, who seems to have intimated as much to Riemann. In fact, there is a letter from Gauss to Weber written in 1845 (Gauss, *Werke* V, 629). At about the same time, without any solid basis, Faraday said in effect that there may be a connection between optical and electrical phenomena (*Philos. Magazine* 28 (1846), 345). This influenced Maxwell. In Germany, in the mid 1850s, Weber and Kohlrausch realized that a certain constant c in Weber's law of force is numerically equal to $\sqrt{2}$ times the speed of light, and Riemann also mentions this with reference to his own law (W. 293).

For Riemann, local action and finite speed of propagation belonged together. In this respect he felt he was a follower of Newton. Speaking about Riemann in the summer of 1857, Dedekind reported (W. 553) that Riemann was deeply impressed by Newton's pronouncement about the impossibility of direct action at a distance. In his draft on gravity and light (W. 534, footnote) Riemann quoted him verbatim:

> "Newton says: 'That gravity should be innate, inherent, and essential to matter, so that one body can act upon another at a distance through a vacuum, without the mediation of anything else, by and through which their action and force may be conveyed from one to another, is to me so great an absurdity, that I believe no man who has in philosophical matters a competent faculty of thinking can ever fall into it.' See the third letter to Bentley."

Here Riemann could find in Newton's own words a denial of the widespread view that their author thought of gravity as action at a distance — a view that suggests itself strongly and yet naively. Actually, Newton never took a clear stand on this issue in any of his works. Rather, he refused to make hypothetical pronouncements going beyond the mathematical description of the acting forces. Riemann's quotation comes from Newton's correspondence with Bentley, begun in 1692. The clergyman Richard Bentley was to present public lectures in London in order to combat atheism. He approached Newton in Cambridge and asked him for help in composing a generally intelligible account of natural philosophy. To do this Newton had to leave out mathematical descriptions. This explains the origin of the pronouncements cited by Riemann.

3.2.3 Mathematical methods for physics

This heading could serve as a motto for Riemann's lifework. To lend substance to this generality we will devote a large part of this section to a discussion of a few of Riemann's papers in which he starts from physical problems and develops mathematical methods using concrete examples. After 1857, i.e., after he had finished the two great papers on Abelian functions and hypergeometric

series, Riemann concentrated on groups of such problems. During that time he had a heavy teaching load and his lectures were largely devoted to mathematical physics. In the winter of 1858/59, when he substituted for the ill Dirichlet, he lectured for 12 hours a week, on Mondays, Tuesdays, Thursdays, and Fridays. He taught the following three courses:
– Theoretical physics, in particular the theory of elasticity and the mechanical theory of heat (from 9 to 10),
– Integration of partial differential equations with applications to physical questions (from 12 to 1),
– On functions of a variable complex magnitude, in particular, on hypergeometric series and related transcendental functions (from 4 to 5).
Originally, the second of these courses had been announced by Dirichlet (information supplied by Neuenschwander in N. 865).

The intensive work on the zeta function must have also been done at that time.

The beginning of Riemann's own review of his paper on air waves (W. 176), published in 1859, is characteristic of his approach:

> "This investigation does not claim to provide useful results for experimental research; the author would like it to be regarded as just a contribution to the theory of nonlinear partial differential equations. Just as the most fruitful methods for the integration of linear partial differential equations have been obtained not by developing the general idea of this problem but rather by treating particular physical problems, so too the theory of nonlinear partial differential equations seems to be best advanced by a circumstantial treatment of particular physical problems which takes into consideration all constraints. In fact, the solution of the very special problem that is the subject of this paper has yielded new methods and approaches that will probably also play a role in more general problems."
>
> ("Diese Untersuchung macht nicht darauf Anspruch, der experimentellen Forschung nützliche Ergebnisse zu liefern; der Verfasser wünscht sie nur als einen Beitrag zur Theorie der nicht linearen partiellen Differentialgleichungen betrachtet zu sehen. Wie für die Integration der linearen partiellen Differentialgleichungen die fruchtbarsten Methoden nicht durch Entwicklung des allgemeinen Begriffs dieser Aufgabe gefunden worden, sondern vielmehr aus der Behandlung specieller physikalischer Probleme hervorgegangen sind, so scheint auch die Theorie der nichtlinearen partiellen Differentialgleichungen durch eine eingehende, alle Nebenbedingungen berück-

sichtigende, Behandlung specieller physikalischer Probleme am meisten gefördert zu werden, und in der Tat hat die Lösung der ganz speciellen Aufgabe, welche den Gegenstand dieser Abhandlung bildet, neue Methoden und Auffassungen erfordert, welche wahrscheinlich auch bei allgemeineren Aufgaben eine Rolle spielen werden.")

So far, mathematical physicists had dealt almost exclusively with waves of infinitely small amplitude. The resulting linear differential equations could be treated by the method of separation of variables and by series or integrals à la Fourier. But Riemann ventured to treat the nonlinear differential equation associated with "airwaves of finite amplitude" (W. 156–175, published in 1860). By means of a clever change of variables he managed to obtain a linear differential equation with, of course, nonconstant coefficients. He developed the integration method named after him, which transferred Green's procedure from the elliptic to the hyperbolic case. This method has been generalized in many ways and is still relevant today. But from the standpoint of physics he obtained more than he promised at the beginning of his review: he predicted the occurrence of "condensation impulses" ("Verdichtungsstöße") — in modern parlance, shock waves. For a detailed appreciation we refer the reader to Peter Lax (N. 807–810).

A short time later, following directly the work of Dirichlet, Riemann worked on another question of the mechanics of continua, which became the subject of the paper "Ein Beitrag zu den Untersuchungen über die Bewegung eines flüssigen gleichartigen Ellipsoids" ("A contribution to the investigations on the motion of a liquid homogeneous ellipsoid"). Investigations of the equilibrium shapes of homogeneous rotating fluids, which are of obvious interest because of the shapes of the celestial bodies, date back to Newton and were continued by MacLaurin (1742) and Lagrange (1811), besides others. In 1834 Jacobi made the surprising discovery that if the angular momentum exceeds a certain critical value, then ellipsoids without rotational symmetry can also appear as equilibrium shapes. Beginning in the winter of 1856/57 Dirichlet investigated what happens when the critical value is exceeded, and discussed the issue in his lectures; Dedekind continued this last work of his teacher and got it published. In modern terms, the problems involved are stability and bifurcation problems.

We quote just one of Riemann's results in this area (W. 204):

"Thus this investigation shows that in the case of rotation of a flattened ellipsoid of revolution about its smaller axis, well known since the time of MacLaurin, the stability of the state of motion is necessarily labile as soon as the ratio of the smallest axis to the others is less than .303327 ...; in this case, in the event of the slightest difference between the other two (axes), the liquid mass would completely change its shape and state of motion and a continual oscillation would set in about the state that corresponds to the minimum of the [energy] function G. The latter (state) consists in a uniform rotation of an ellipsoid with different axes about the smallest one, combined with an equidirectional inner motion in which the particles move along similar ellipses perpendicular to the axis of rotation. Moreover, the period of the motion is equal to the period of the rotation, so that already after half a revolution of the ellipsoid each particle is back to its original position."

("Aus dieser Untersuchung ergibt sich also, dass in dem schon seit Maclaurin bekannten Falle der Rotation eines abgeplatteten Umdrehungsellipsoids um seine kleinere Axe die Beständigkeit des Bewegungszustandes nur labil ist, sobald das Verhältniss der kleinern Axe zu den andern kleiner ist als .303327 ... ; bei der geringsten Verschiedenheit der beiden andern würde in diesem Falle die flüssige Masse Form und Bewegungszustand völlig ändern und ein fortwährendes Schwanken um den Zustand eintreten, welcher dem Minimum der [Energie-] Function G entspricht. Dieser besteht in einer gleichförmigen Umdrehung eines ungleichaxigen Ellipsoids um seine kleinste Axe verbunden mit einer gleichgerichteten innern Bewegung, bei welcher die Theilchen sich in einander ähnlichen zur Umdrehungsaxe senkrechten Ellipsen bewegen. Die Umlaufszeit ist dabei der Umdrehungszeit gleich, so dass jedes Theilchen schon nach einer halben Umdrehung des Ellipsoids in seine Anfangslage zurückkehrt.")

For further relevant developments see Chandrasekhar and Lebovitz (N. 811–820).

A paper of lesser physical significance written by Riemann around that time is his "Commentatio mathematica..." (W. 391–404; we follow the German translation in Böhm-Reichardt 1984), a prize essay presented to the Paris Academy on 1 July 1861. Its content is described by its full title: "A mathematical essay which is an attempt to answer the following question proposed by the august Paris Academy: To determine what must be the heat condition of an unbounded homogeneous solid so that a system of isotherms, known at a given time, remains such after an arbitrary time, and in such a way that the temperature at a point can be expressed as a function of time and of two further independent variables." ("Mathematische Abhandlung, durch die versucht wird, auf die von der hochberühmten Pariser Akademie vorgelegte Frage zu antworten:

Zu bestimmen, welches der Wärmezustand eines unbegrenzten homogenen festen Körpers sein muß, damit ein zu einem vorgegebenen Zeitpunkt vorliegendes System von Isothermen nach einer beliebigen Zeit ein ebensolches bleibt derart, daß die Temperatur in einem Punkte als Funktion der Zeit und zweier weiterer unabhängiger Variabler ausgedrückt werden kann.")

The paper is mathematically important. Using a very efficient notation, Riemann here developed much of the method later known as the tensor calculus. In particular, he introduced the curvature tensor and showed that its vanishing is necessary and sufficient for the associated quadratic differential form to have constant coefficients in a suitably chosen coordinate system.

He must have been familiar with such computations at the time of his habilitation lecture of 1854. For one thing, he could even give up the assumption of the homogeneity of the medium. Of course, the various special cases that had to be considered, like it or not, for the complete solution of the Paris problem must have bored him. The editors note (W. 391, footnote) that his illness prevented him from providing a detailed treatment of the problem. Be that as it may, no prize was awarded and the prize problem was withdrawn in 1868. Weyl described Weber's annotations (W. 405–423) as excellent (N. 754). We discussed Dedekind's pertinent reflections in 3.1.9. By the time Riemann's collected works appeared in 1876 the principal mathematical result had already been rediscovered by Christoffel and by Lipschitz.

The papers written between 1859 and 1861 and the lectures presented during that time show Riemann to have been a master of mathematical analysis who successfully devoted himself to concrete questions of physics and of number theory. During this time he did not withdraw into himself, and nothing indicates that he continued his lonely speculations on natural philosophy. As well, the theory of functions of a complex variable now came second as a research area.

Riemann's lectures from this time, as presented in the Hattendorf publications of 1869 and 1876 and their "successor books" (H. Weber, Frank-Mises), helped to propagate the field-theoretic approach to physics and permanently influenced the mathematical methods of physics. Neuenschwander (N. 860) has discovered additional drafts whose examination is bound to be a rewarding task. Those familiar with Riemann's work will easily spot certain flaws in the Hattendorf accounts. The two papers by E. Gödecker, subtitled "Based on the lecture of B. Riemann," are far closer to Riemann's style and spirit. One paper is called "Die Vertheilung der Wärme in der Kugel" ("Distribution of heat in a ball"). Programm des Johanneums zu Lüneburg, Ostern 1872. The other is called "Die Bewegung eines kreisförmigen Ringes in einer unendlichen incompressibeln Flüssigkeit" ("Motion of a circular ring in an infinite incompressible fluid"). Programm, Gymnasium und Realschule zu Göttingen, 1879. These papers should definitely be included in a new edition of Riemann's collected works. Gödecker took notes in Riemann's course in 1860/1861.

For the sake of completeness we must mention some of Riemann's reflections on physics written long before his intellectual prime, when he was an assistant in Wilhelm Weber's physical seminar. We have in mind the papers II, III, and XX.

On 21 September 1854 the newly qualified *Privatdozent* delivered a lecture at a meeting of scientists at Göttingen. Its title was: "Über die Gesetze der Vertheilung der Spannungselectricität in ponderabeln Körpern, wenn diese nicht als vollkommene Leiter oder Nichtleiter, sondern als dem Enthalten von Spannungselectricität mit endlicher Kraft widerstrebend betrachtet werden" ("On the laws of distribution of potential electricity in ponderable bodies when the latter are regarded not as perfect conductors or nonconductors but as opposing the retention of potential electricity with finite force"). He followed the views of R.H.A. Kohlrausch (1809–1857) on the so-called residue in Leyden bottles. This lecture, as well as the two subsequent papers, contributed to the contemporary discussion on the foundations of electricity. The two subsequent papers were: "Zur Theorie der Nobili'schen Farbenringe" ("On the theory of Nobili's color rings"), published on 28 March 1855 (W. 87–162), and "Neue Theorie des Rückstandes in electrischen Bindungsapparaten" ("New theory of the remainder in electrical absorption devices") (dated 1854: W. 366–378). The first of these is of some mathematical interest. Here Riemann used semiconvergent (asymptotic) series, and this motivated the subsequent works of Hankel and H. Weber (W. 65). Riemann withdrew the second paper. The editors speculated that this was due to his unwillingness to make a change suggested by the journal. Another, very plausible, guess is suggested in Weber's footnote in W. 371: Riemann did not want to give up his brief presentation on the foundations of electricity (§3, W. 370–371) but feared that "he might then scandalize the physicists" ("bei den Physikern damals Anstoss zu erregen"). His use of elliptic functions is of some mathematical interest.

Two other physical papers (W. 431–444), prepared by Weber from the *Nachlass*, were computations relating to examples in the lectures.

As stated at the beginning of this section, we wanted to find out how physics stimulated Riemann to develop mathematical methods and what physical impulses promoted his work in complex analysis and geometry. In the next section we will consider the physicists' view of Riemann's physical ideas.

3.2.4 *Riemann's electrodynamics from the viewpoint of the physicists*

For 20th-century physicists Riemann's name stands just for the name of a supplier of mathematical tools. But at the dawn of the 20th century, in 1905, before the appearance of the fundamental papers of Einstein and Planck, prominent physicists took Riemann seriously as one of their own. This comes through with particular clarity in the articles in the second half of the fifth volume (physics) of the *Encyklopädie der Mathematischen Wissenschaften*. Issue 1 appeared on 16 June 1904 and was devoted to electricity. The article V. 12, by R. Reiff and A. Sommerfeld, presented an historical account titled "Standpunkt der Fernwirkung. Die Elementargesetze" ("The standpoint of action at a distance. The elementary laws") (pp. 3–62). Already its outward features point to the importance the authors attribute to Riemann. They devote to him seven printed pages, and the Hattendorff 1876 version of his lectures on gravity, electricity, and magnetism is one of the six listed textbooks. We summarize the article below and repeat certain points made earlier.

The subsection devoted to Gauss and Riemann begins with the words (p. 45): "While Weber consistently championed the standpoint of action at a distance, diametrically opposite tendencies were put forward in his immediate milieu by his teacher Gauss and by his student Riemann" ("Während Weber in consequenter Weise den Standpunkt der Fernwirkungen verfocht, machten sich in seiner unmittelbaren Umgebung, bei seinem Lehrer Gauss und seinem Schüler Riemann, Strömungen entgegengesetzter Richtung geltend"). In this connection we must mention first of all the letter that Gauss wrote to Weber in 1845 (*Werke* V, 627). The letter speaks of the still-missing link with the electrostatic forces between electrical elements, forces which "ensue when they are in motion with respect to one another." Gauss says specifically that these forces are based on "an action that is *not instantaneous*, but propagates itself in time (like light)" ("*nicht instantanen*, sondern (auf ähnliche Weise wie beim Licht) in der Zeit sich fortpflanzenden Wirkung").

We know that Riemann found out about Gauss' position indirectly, through Weber. Be that as it may, these circumstances are included in the part of Riemann's letter of 28 December 1853 (W. 547) to his brother, devoted to the connection between electricity, magnetism, light, and gravitation, an issue that Gauss seems never to have expressed himself on. Riemann had the impression that Gauss had been working on this issue for a number of years. This may have been due to Weber's account but need not have been correct. All that was found in Gauss' *Nachlass* was a manuscript, dated 1835, with a formula that did not correspond to a finite speed of propagation (Gauss, *Werke* V, 616).

Reiff and Sommerfeld (p. 46) emphasized that the paper "Ein Beitrag zur Elektrodynamik," presented by Riemann to the Göttingen Society of Sciences on 10 Febrary 58 (W. 288–293), made him a forerunner of Maxwell, and that the "recent electron theory" had, in a certain sense, led back to Riemann's form of a (retarded) elementary potential.

In fact, Riemann was the first to formulate the differential equation

$$\frac{1}{c^2}\frac{\partial^2 U}{\partial t^2} = \Delta U + 4\pi\rho$$

for the potential U and the charge density ρ (W. 290) that was later deduced from Maxwell's theory, and to observe that his results agreed with experience if c was taken to be the velocity of light.

> "I have found that the electrodynamic actions of galvanic currents can be explained if one assumes that the action of one electrical mass on others is not instantaneous but propagates itself towards them with constant speed (equal, within the limits of observational errors, to that of the speed of light). Under this assumption, the differential equation of the electrical force is the same as that for the propagation of light and radiant heat."
>
> ("Ich habe gefunden, dass die elektrodynamischen Wirkungen galvanischer Ströme sich erklären lassen, wenn man annimmt, dass die Wirkung einer elektrischen Masse auf die übrigen nicht momentan geschieht, sondern sich mit einer constanten (der Lichtgeschwindigkeit innerhalb der Grenzen der Beobachtungsfehler gleichen) Geschwindigkeit zu ihnen fortpflanzt. Die Differentialgleichung der elektrischen Kraft wird bei dieser Annahme dieselbe, wie die für die Fortpflanzung des Lichts und der strahlenden Wärme") (W. 288).

That the propagation of waves in space is governed by the equation $\frac{1}{c^2}\frac{\partial^2\phi}{\partial t^2} - \Delta\phi = 0$ had been known for almost a century; Euler and Lagrange had used it to study sound. Poisson had used the equation $\Delta U = -4\pi\rho$ for the potential U in electrostatics. It was therefore highly plausible, from a theoretical viewpoint, that Riemann's equation governed electrodynamics. For this one had to assume, however, that electrical actions are propagated with a fixed finite velocity c. But Riemann followed his repeatedly stated conviction that a theory has to be based on experience. He therefore immediately tried to bolster his differential equation by means of known phenomena. Given the contemporary state of techniques of measurement, this was far from simple. Reiff and Sommerfeld summarized his efforts in these words (p. 46): "Riemann sums this elementary potential [the fundamental solution of his wave equation] over all electrical masses e and e' of two current-carrying conductors, and finds by means of computations, whose admissibility had been contested by R. Clausius, a value that coincides with the [Franz] Neumann potential if c is taken to be equal to the velocity of light." ("Riemann summiert nun dieses Elementarpotential [die Grundlösung seiner Wellengleichung]) über alle elektrischen Massen e und e' zweier stromführender Leiter und findet durch Rechnungen, deren Zulässigkeit von R. Clausius angefochten worden ist, einen Wert, der sich mit dem [Franz] Neumannschen Potential deckt, wenn c gleich der Lichtgeschwindigkeit angenommen wird.")

Neumann's potential was not quite reliable and Riemann's computation was not convincing. In fact, Riemann had left out contributions whose order of magnitude was equal to that of the parts he had kept. Moreover, he had made an inadmissible interchange of the order of two integrations. A careful analysis is given by T. Archibald in Rowe and McCleary 1989 II, p. 54 ff. All we can say is that Riemann was convinced of the validity of the wave equation for electrodynamics. But he must have realized that his attempt to substantiate it on the basis of experience was feeble and he withdrew it. It was published posthumously (*Poggendorffs Annalen der Physik und Chemie* 131 (1867), 237 ff.). In the meantime Maxwell's far more comprehensive paper, "A Dynamical Theory of the Electromagnetic Field," appeared in the *Transactions of the Royal Society of London* 155 (1865), 459 ff.

Reiff and Sommerfeld produced a very detailed appreciation of the merits of Riemann's account of the science of electricity as set forth in the lectures published by Hattendorff in 1876. In the simplest case of two moving particles with charges e and e', Riemann set down a balance of energy that included the kinetic energy T, the potential energy U of electrostatic origin, and a potential energy V of an electrodynamic nature. He put the latter equal to

$$V = -\frac{ee'u^2}{2rc^2},$$

with r the distance between the particles and u their relative velocity. Then he took from mechanics the idea of Hamilton's principle and derived from it a law of force.

The discussion of different laws of force and electrodynamic potentials lasted for more than 50 years. It began with Grassmann (1845), continued with Franz Neumann, W. Weber, Riemann, and Carl Neumann, and ended with Clausius. The confrontation pitted a dualistic conception, which assumed that positive and negative electricity in a conductor move in opposite directions, against a unitary one which assumed that one kind of electricity in a conductor moves while the other remains at rest. Riemann favored the first of these conceptions while Clausius regarded the second as the more natural. (The term "electron" was introduced, by Johnstone Stoney, only in 1891!) In 1858 Riemann made no reference to an ether, but its existence was generally accepted as a necessary medium for the propagation of action. At the end of their report, Reiff and Sommerfeld discussed the possible experiments that could have determined the validity of one of the laws of force. At that time, Riemann was regarded as one of the competitors on the basis of his treatment of the concept of potential in his lectures, and was included among the competent physicists in this area. A few years later the dispute became irrelevant. Soon Einstein's electrodynamics of moving bodies, i.e., the special theory of relativity, became the theoretically persuasive foundation of electrodynamics, which included Maxwell's equations.

In retrospect, we can regard it as an irony of history that Riemann's wave equation, combined with the principle that the velocity of light is independent of the frame of reference, is already relativistically invariant. One could have inferred this by formulating Riemann's geometry for indefinite metric determinations and appropriately transferring the differential operators from the Paris prize essay. But at that time observational facts, such as the Michelson experiment, were not yet available.

3.2.5 Riemannian geometry in the physics of the twentieth century: Einstein and Weyl

Sixty years after Riemann's habilitation lecture, Albert Einstein's (1879–1955) general relativity theory effected a breakthrough in physics that changed the role of Riemann's geometry from wallflower to prima ballerina. At the same time Riemann's dream of a general field theory was realized so far as gravitation was concerned. Einstein's special theory of relativity of 1905 had dealt with electromagnetism. To this day, many physicists aspire to a unified theory of all physical fields. The first attempts in this direction, directly connected with the work of Einstein, were due to H. Weyl and Th. Kaluza. We will provide a brief description of these attempts in the sequel. All of these developments significantly advanced differential geometry. Just as in Riemann's case, the pattern was that of an interplay between physics and mathematics.

The requirements of his special theory of 1905 forced Einstein to make two inessential modifications to Riemann's geometry: the square of the arclength element is no longer a positive definite quadratic form, and the dimension of the manifold is four rather than three — time being added. In this theory we have the metric

$$ds^2 = -dx_0^2 + dx_1^2 + dx_2^2 + dx_3^2,$$

with $x_0 = ct$ (c is the velocity of light and t is time), sometimes referred to as the Minkowskian metric. The space of the general theory is a 4-dimensional manifold in which every tangent space is equipped with a Minkowskian metric. In contradistinction to this, the tangent spaces of the actual Riemannian spaces were Euclidean in suitably chosen coordinate systems, i.e., $ds^2 = \sum dx_i^2$. Einstein's world is locally Minkowskian while Riemann's spaces are locally Euclidean. What counts for the calculus is that in both cases we have nondegenerate quadratic forms. It is this that allows application of the machinery of Riemannian geometry and of tensor analysis; for g_{ik} there is g^{ik}.

At the end of his lecture (W. 286) Riemann had said that the basis for the metric relations in space must be sought in the binding forces acting on it. Einstein found these in general gravitation. He started with the observational fact that the inertial mass m of a body, the mass that appears as a factor in Newton's equations of motion, in which force is a product of mass and acceleration, cannot be distinguished from its weight, or gravitational mass m_w, the "gravitational charge" ("Gravitationsladung") responsible for motion in the gravitational field. The ratio $m : m_w$ is a universal constant that can be taken to be 1. Conceptually, the two masses must be distinguished; the experiential fact of their equivalence calls for theoretical understanding.

In his Princeton lectures of 1921 Einstein offered the following explanation. According to the principle of inertia, a particle of matter not acted upon by a force moves uniformly along a straight line. This is also the case in the 4-dimensional continuum of the special theory. The most natural, i.e., the simplest, generalization of a straight line in Riemannian geometry is the "straightest" line, i.e., a geodesic. Therefore, in accordance with the principle of equivalence, we must assume that the motion of a material particle under the sole influence of inertia and gravitation must be given by the equation

$$\frac{d^2x^i}{ds^2} + \Gamma^i_{jk}\frac{dx^j}{ds}\frac{dx^k}{ds} = 0.$$

When all the components Γ^i_{jk} of the gravitational field vanish, this equation reduces to an equation of a straight line. (Here and in the sequel we follow Einstein's *The Meaning of Relativity*, 5th ed., Methuen, London 1951.)

The Γ-quantities are formed from the derivatives of the g_{ik}, which means that this tensor must play the role of the gravitational potential. In the Newtonian theory, the (scalar) gravitational potential ϕ is controlled by the Poisson equation

$$\Delta\phi = 4\pi K\rho$$

with matter density ρ, i.e., by a linear second-order differential equation. This means that we must look for a differential equation for the tensor (g_{ik}) which has no higher derivatives than the second and is linear in them. Then there is also the requirement that the divergence of the left side of the new differential equation vanish, a requirement suggested by the conservation theorem for the energy impulse tensor of the special theory. From all this, Einstein was able to obtain the field equation of his general theory:

$$R_{ik} - (1/2)g_{ik}R = -\kappa T_{ik}.$$

Here R_{ik} and R are obtained from the Riemannian curvature tensor by contraction, κ is a number related to the gravitation constant K, and T_{ik} is the energy tensor of matter.

Einstein began to publish papers on general relativity in 1915. His Zürich associate Marcel Grossman discussed tensor analysis with him. It is noteworthy that as a student and a member of the student circle "The Olympia Society" in Zürich, Einstein became familiar with Riemann's habilitation lecture and with the writings of Clifford. (In this connection see *The Collected Papers of Albert Einstein*, Vol. 2, Princeton University Press, 1989, p. xxv.)

The starting point of the theory was the observational fact of the equivalence of the inertial and gravitational masses. Further confirmation came from otherwise unexplainable physical phenomena (the deviation of the orbit of Mercury from an ellipse, the bending of light rays in the gravitational field of the sun). Moreover, Riemannian geometry gave satisfying models; for example, it admitted the model of a closed universe of finite volume. What was disappointing was the failure to obtain a unified field theory of gravitation and electromagnetism, the kind of theory Riemann had had in mind. It was this problem that Einstein wrestled with for the rest of his life. The first attempt was due to Hermann Weyl, with whom Einstein had contact in Zürich in 1913. In the summer of 1917 Weyl lectured for the first time on the theory of relativity at the ETH in Zürich, and these lectures gave rise to the book *Space–Time–Matter*, which has appeared in many editions beginning in 1918. In the meantime Einstein had gone to Berlin.

We have come across connections between the work of Weyl and of Riemann on numerous occasions. No mathematician of the first half of the 20th century cared as much as did Weyl for greater depth of thought in analysis, geometry, mathematical physics, and philosophy. Like Riemann, Hermann Weyl (1885–1955) came from northern Germany. He was born in Elmshorn in Holstein. He obtained his doctorate under Hilbert in 1908 with a paper on singular integral equations, became a *Privatdozent* at Göttingen in 1910, a professor at the University of Zürich in 1913, and Hilbert's successor at Göttingen in 1930.

He left Göttingen in 1933 and stayed at The Institute for Advanced Study in Princeton until 1951. In the last years of his life he returned to Europe and resided mostly in Zürich.

In his book of 1913 Weyl had presented rigorously, for purposes of complex analysis, Riemann's ideas on continuous manifolds, and could now take part in developing from the very beginning the mathematics of the theory of relativity. As we know, his attempt to incorporate electromagnetism was not a success.

In 1918 Weyl came up with the idea that one could introduce different gauges of length in different parts of the universe, and that the metric was therefore determined only to within a variable gauge factor. This added to the Γ^i_{jk} an expression $\delta^i_j u_k + \delta^i_k u_j - g_{jk} u^i$. The covariant vector (u_k) is identified with the electromagnetic potential.

Beginning in 1918, Theodor Kaluza (1885–1954; the same age as Weyl to the day) followed a different road. He added a fifth dimension to the 4-dimensional universe by putting at the foundation the product space of a 4-dimensional manifold by a circle of very small radius, which makes the g_{ik} periodic functions of this fifth coordinate. Only four of the five new g_{ik} are needed for the electromagnetic potential. Later, when one tried to incorporate yet other fields, one was glad to have the superfluous function, which was initially viewed as a nuisance. Beginning in 1926, O. Klein tried to bring in quantum-theoretic arguments. Since Kaluza's approach was not limited to dimension 5, attempts have subsequently been made to work with yet higher dimensions. For recent developments, and for Kaluza's discussions with Einstein, see the following two articles in *Jahrbuch Überblicke Mathematik* 1986: E.W. Mielke, "Kaluza-Klein Theorien," pp. 127–138; D. Laugwitz, "Theodor Kaluza," pp. 179–187.

We see that Riemann's geometry allows much flexibility in the building of mathematical models. As for choosing among such models, Riemann himself had remarked (W. 286) that one must bring in hitherto unexplainable facts. For both of Einstein's theories one had the principles of the constancy of the speed of light and of the equivalence of the gravitational and inertial masses. On the other hand, all attempts to construct a unified field theory have suffered from the unavailability of equally convincing facts of experience. Einstein remarked (*loc. cit.* 187) in connection with the ideas of Kaluza: "It often seems to me that the magnetic field of the earth depends on an as yet unknown connection between gravitation and electromagnetism ... "

Another of Weyl's contributions, which he himself regarded as very important, bears on the extension of the Helmholtz space problem, treated in Section 3.1.6, to indefinite Riemannian metrics. In terms of the notations in 3.1.6 we can say that, in general, the function F in $ds = F(x, dx)$ need no longer be positive or exclusively real at all points (x) and in all space directions (dx). It is

34: Hermann Weyl

natural to ask what requirements characterize the case when F^2 is a quadratic form in the dx. Weyl treated this problem in 1923 and a very readable account is found in the paper Weyl 1988, 36–38, written in 1925. Only in the case of quadratic forms is there a sufficiently large group of linear congruence mappings of the tangent space onto itself, provided that the dimension is at least 3. As Weyl himself put it: "To the extent to which we can judge, the group-theoretic viewpoint was foreign to Riemann's geometric reflections" ("Der gruppentheoretische Standpunkt ist jedoch, soviel wir beurteilen können, Riemann's geometrischen Betrachtungen fremd geblieben") (Weyl 1988, 31). For this reason we think it more appropriate to characterize indefinite metrics by homogeneity properties alone, as was done in 3.1.6 for the definite case. This can be achieved in a manner similar to that in Laugwitz 1977, 147. Then the existence of a sufficiently large group is again a consequence rather than an assumption.

In the mathematical treatment of physics, the basic difference between Riemann and Weyl is that the latter relied on group-theoretic methods. Weyl's approach is a reflection of the tenets in his book *Gruppentheorie und Quantenmechanik*: if one knows of a physical system that it has certain symmetry properties, then we can obtain information from the relevant group.

For the sake of completeness we mention the book of Monastyrsky 1987 in which Riemann and topology are looked at from the viewpoint of a physicist.

The book's style is lively but the details, unfortunately, are unreliable. (*Translator's note:* The second edition, 1999, has corrected many of the errors of the first edition.)

3.3 On philosophy

3.3.1 Preliminary remarks

Freudenthal wrote (1975, 448) enthusiastically about Riemann: "one of the most profound and imaginative mathematicians of all time, he had a strong inclination to philosophy, indeed was a great philosopher. Had he lived and worked longer, philosophers would acknowledge him as one of them." Unfortunately, Freudenthal failed to provide more detailed support for this emphatic opinion. What is more, he wrote that (p. 454): "Riemann left many philosophical fragments — which, however, do not constitute a philosophy." True, he did go on to mention the philosophical aura of the habilitation lecture. E. Scholz, and before him H. Wussing (*Carl Friedrich Gauß*, Leipzig 1973; p. 84 of the 5th edition of 1989), emphasized the "consistent materialistic standpoint of the mathematician and natural scientist Riemann" (einen "konsequenten materialistischen Standpunkt des Matematikers und Naturforschers Riemann").

The number of philosophical fragments left behind by Riemann is not all that great. They take up 17 pages in the collected works (W. 509–525), to which, following Dedekind and Weber, we could perhaps add his reflections on the foundations of physics (*Naturphilosophie*, W. 526–538). E. Scholz (1982, 432–433) found some supplementary notes on Herbart. Riemann's manuscript on the mechanics of the ear (W. 338–350) and his habilitation lecture must also be considered when we think of him as a philosopher. His significance for what we now call the foundations of mathematics will be taken up in Chapter 4.

At the universities of Riemann's time, the boundaries between philosophy on the one hand and psychology, physiology, and pedagogy (in the sense of the art or science of teaching) on the other were not sharply drawn, and this must be kept in mind here. J. F. Herbart (1776–1841) taught at Göttingen from 1802 to 1809 and again as a full professor (Ordinarius) beginning in 1833. He must also be classified as a psychologist and pedagogue (in the above sense). In 1844 his position was taken over by R. H. Lotze (1817–1881) who had also been qualified to lecture in physiology. We will see that, conversely, full-time physiologists were active in philosophy.

Nothing discussed in this part (3.3) of Chapter 3 was submitted for publication by Riemann himself. When it comes to the fragments, it is not always possible to decide when they were written, nor is it clear in all cases whether they represent his own, possibly provisional, views or are to be regarded as just excerpts from, or commentaries on, the writings of others. We must be aware that every interpretation is risky and that the available writings admit different explanations.

The primary reason for risk-taking must be the one which impelled the editors to include the fragments in the edition of Riemann's collected works: "philosophical speculations... attended (Riemann) throughout a major part of his life" (W. 507). As Riemann's friend, Dedekind was qualified to put the matter thus. The editors refrained from interpreting the fragments; but today, given the historical distance, it may be easier to take the risk. Our starting assumption is that Riemann strove for a unified worldview. In terms of physical phenomena, we conjectured that he wanted to justify a unified field theory. We suppose that the field conception, replacing the older mechanistic viewpoint, was to include the combined element of spirit and soul; as Riemann put it (W. 529), substance travels from the material into the spiritual world, spiritual substance is formed in the latter, and the spiritual world intervenes in the material world.

One is tempted to sense in Riemann echoes of the so-called romantic natural philosophy which spread in Germany following Friedrich Schelling (1775–1854; taught last in Berlin, until 1846). Schelling's relevant writings had appeared at the turn of the century, but they continued to influence physicians, biologists, and the physicist J.W. Ritter, the discoverer of ultraviolet radiation. This philosophy saw the whole of nature as a large organism endowed with a living soul, and sought a unified worldview that would encompass spirit and nature. Goethe, whose methodological and scientific conceptions were akin to Schelling's, mentioned him frequently as an ally, for example in connection with the theory of colors. The two were personally well acquainted. At Goethe's instigation, Schelling became a professor at Jena, where he taught from 1798 to 1803. They corresponded until 1827.

It is possible that Riemann got to know the writings of the natural philosophers already in his youth. But we regard as more reliable conjectures about the influences that may have come from his Göttingen milieu, and it is these that we propose to discuss next. Whereas romantic natural philosophy regarded the *spirit* as dominant in the all-inclusive unity, Riemann spoke of the *substance* of the spirit (Geistes*substanz*) and of *masses* of spirit (Geistes*massen*). We will see that such an outlook, interpreted as a turn toward materialism, was then *au courant.*

3.3.2 The spiritual atmosphere in 1853/54: The debate over materialism

Around 1853, at the time when Riemann set down the most important of his notes on natural philosophy, there raged a debate, particularly intense at Göttingen, that has entered the history of philosophy under the name of the debate over materialism (Materialismusstreit). Today we can give the whole matter a condescending smile. Indeed, it was a passing phenomenon, which F.A. Lange described in his important *Geschichte des Materialismus* of 1866 as settled and replaced by the deeper dispute over Darwin's theory of evolution. Nevertheless, we will give reasons for assuming that this conflict had an effect on Riemann and on his philosophical reflections. The key issue was the relation of Christian faith to contemporary natural science.

We recall that on 21 September 1854 Riemann presented a lecture at the Göttingen meeting of natural scientists and physicians. The principal lecture, titled "Human creation and soul substance" ("Menschenschöpfung und Seelensubstanz") was given on September 18. The lecturer was the famous Göttingen physiologist Rudolf Wagner (1805–1864), who is of some significance for the history of mathematics on account of his *Gespräche mit Carl Friedrich Gauß in den letzten Monaten seines Lebens* (*Conversations with Carl Friedrich Gauß during the last months of his life*) (published in 1975, edited by Heinrich Rubner, Nachr. Akad. Wiss. Göttingen, philolog.-hist. Kl., Heft 6/1975). Wagner, who for many years had been involved in arguments with the materialists, wanted to use his lecture to restate and justify the standpoint of theology and the church in physiological terms. The opposing side revived the physiological component of 18th-century French materialism with writings such as *Der Kreislauf des Lebens* (*The Cycle of Life*) (by J. Moleschott, Mainz 1852). Wagner's principal opponent was the physiologist C. Vogt of Geneva, author of the paper *Köhlerglaube und Wissenschaft* (*Blind Faith and Science*), widely disseminated in a number of editions (Gießen from 1854 on). The polemic soon ceased to be objective, but the discussion took place among serious scientists as well. According to Wagner's note dated 24 December 1854 (*loc. cit.* p. 25), Gauss had told him that he shared his view concerning the soul substance, the speed of whose progression he took to be comparable to that of a galvanic current rather than to that of light, for, what was being conveyed, while imponderable, was nevertheless material. Following its transposition after death to another world, the soul had to be provided with another body.

Obviously, one must not connect Gauss with the gross materialistic speculations about animal magnetism and soul migration that were revived after 1850.

In order to form a picture of that time we mention a few more curiosities.

Beyond astrology, which prospers at all times, the realm of the occult is subject to the dictates of fashion. Given Riemann's receptiveness to psychological and physiological problems, such phenomena could not pass him unnoticed, even though he lived a withdrawn life. We can take it for granted that he could tell wheat from chaff and could see through quackery. But to understand the atmosphere of the time we must not ignore its fashionable currents, especially when we realize the extreme fluidity of the boundaries of contemporary serious science.

Biermann (1990, 29 and 201f.) tells us that in 1853 table-tipping or -turning, the ostensible method for establishing contact with the souls of the dead, spread in Europe like an epidemic. Gauss referred to the practice as table-pressing (Tischdrücken), and thus pointed to a mechanical explanation of the phenomenon. At Heidelberg, members of the faculty of law ran "like crazy" ("wie wahnsinnig") after a rotating table. On 10 May 1853 Gauss wrote to Alexander von Humboldt: "I have been able to contemplate the inanities of the day with a good deal of equanimity, nay, could laugh heartily at some conversation pieces like the table-pressing attempts of the members of the Heidelberg law faculty" ("Die jetztigen Tagesthorheiten habe ich ziemlich mit Gleichmuth betrachten, ja über einige Genrebilder, wie die Versuche der Heidelberger Juristenfakultät mit Tischdrücken, herzlich lachen können.") And on 24 April 1853 Hirst, whom we mentioned in the Introduction, wrote in his Berlin diary: "Wednesday evening we spent with Dirichlet ... During the evening Prof. Hensel (husband of the late Fanny Hensel, another sister of Mendelssohn's) came. It was proposed that we should make the experiment of moving the table (Tisch rücken) which is now the subject of fashionable twaddle ... The scientific men in Berlin almost all ridicule the idea. It deserves, however, a closer experiment" (Gardner and Wilson, *Amer. Math. Monthly* 100 (1993), 625).

This is what was taking place in the salons, including the salons of scientists!

The very time of these professorial polemics and table-tipping was also the time of Riemann's reflections that he dated as "Found on 1 March 1853" (W. 258) and provided with the ambitious title "Neue mathematische Principien der Naturphilosophie" ("New mathematical principles of natural philosophy") in clear imitation of Newton. These reflections must have been very important to him, for he focussed on them rather than on his habilitation paper. In Dedekind's words (W. 547): "Clearly, the beginning of 1853 was a time of virtually exclusive preoccupation with natural philosophy ..." ("Offenbar fällt in den Anfang des Jahres 1853 eine fast ausschliessliche Beschäftigung mit der Naturphilosophie ...")

On the scientific side of the ledger we must mention a man whom Riemann quoted on a number of occasions. This was Gustav Theodor Fechner (1802–1887), the founder of psychophysics, an early stage of experimental psychology, who held a chair in physics at Leipzig until 1839. He developed the theory that all things are endowed with living souls. In 1836 he published *Das Büchlein vom Leben nach dem Tode* (*The booklet about life after death*), and in 1848 *Nanna oder über das Seelenleben der Pflanzen* (*Nanna, or the soul-life of plants*). Riemann also referred to Fechner's *Zend-Avesta*:

> "While in his *Nanna* Fechner tries to demonstrate that flowers are endowed with living souls, so in the *Zend-Avesta* the starting point of his reflections is the doctrine that the stars are endowed with living souls."
>
> ("Wie Fechner in seiner *Nanna* die Beseeltheit von Pflanzen darzuthun sucht, so ist der Ausgangspunkt seiner Betrachtungen im *Zend-Avesta* die Lehre von der Beseeltheit der Gestirne.")

The fact that Riemann not only took note of Fechner's publications but devoted to them a detailed review, which was almost ready to go to press (but in fact did not appear), shows that he followed the matter with a measure of sympathy. What interested him in addition to Fechner's worldview was his methodology. There are areas of knowledge in which the method of the physicists, which consists in bolstering theoretical speculations by experiments, fails:

> "The method that he [Fechner] uses is not abstraction of general laws by means of induction and their use and testing in the explanation of nature, but analogy."
>
> ("Die Methode deren er [Fechner] sich bedient, ist nicht die Abstraction allgemeiner Gesetze durch die Induction und die Anwendung und Prüfung derselben in der Naturerklärung, sondern die Analogie.")

While remaining critical, Riemann found that Fechner's presentation had "persuasive power" (sei von "überzeugender Kraft") (W. 515).

3.3.3 New mathematical principles of natural philosophy

Riemann's draft "found on 1 March 1853" constituted a direct link with Newton's *Philosophiae naturalis principia mathematica* of 1687 (W. 528–532). We recall Riemann's fundamental observations in his lectures, described in 3.2.2, to the effect that attempts to go beyond Newton's fundamental concepts in order to "penetrate into the interior of nature" had failed, and that later philosophical systems had even distorted Newton's ideas. His aim is to accept the foundations of astronomy and physics discovered by Galileo and Newton and to provide for them mathematically comprehensible justifications. He wants to go beyond an empirically secured *description* of nature, of the kind first consistently arrived at by means of partial differential equations by Fourier in 1822 in the form of his theory of heat, to an *explanation* of nature.

Since his "speculation" cannot be of immediate practical use to astronomy, Riemann asks for understanding on the part of the prospective "readers of this page," a turn of phrase which shows that while composing his draft he fully intended to publish it.

The basis for Newton's general laws of motion is presumably provided by the "inner state of the ponderables" (in dem "inneren Zustand der Ponderabilien"). This is Riemann's basic thesis. This inner state cannot be accurately perceived by direct observation. But, in addition to the observations of the outside world, we have yet another experience: "We observe a continual activity of our soul." ("Wir beobachten eine stetige Thätigkeit unserer Seele.") We cannot draw conclusions about the inner state of ponderables by means of "induction," but we can do so by means of "analogy" with "our own inner perception" (mit "unserer eigenen inneren Wahrnehmung.")

From the observation of the activity of the soul Riemann derives what he regards as a

> "Fact: thus something permanent underlies every act of our soul, which enters our soul together with this act but disappears completely from the world of phenomena at that very moment."
>
> ("Thatsache: Jedem Act unserer Seele liegt also etwas Bleibendes zu Grunde, welches mit diesem Act in unsere Seele eintritt, aber in demselben Augenblick aus der Erscheinungswelt völlig verschwindet.")

In order to be able to follow Riemann further, we must accept that this is for him an irrefutable observational fact of absolute certainty. The permanent element that has entered the soul proclaims itself through the memory on special occasions but ceases to exert any influence on the phenomena. Like Herbart, and the psychology developed after him, Riemann seems to believe that the soul itself has no substantiality but that every single idea formed in us has. When he speaks elsewhere of masses of spirit (W. 509 ff.), this is for him apparently always consistent with the notion of the nonsubstantiality of the soul itself. As parts of the soul, the "compact masses of spirit" (die..."compacten Geistesmassen") which come into being during its life continue to exist after death, but their isolated continued existence does not last. While these speculations hold no further interest for us, they demonstrate Riemann's consistent conviction of the validity of the law of preservation of substance.

Reasoning by analogy, Riemann derives from the "fact" of inner perception a "hypothesis" pertaining to the world of phenomena, or the material world:

> "Space is filled with a substance that constantly flows into the ponderable atoms, and disappears into them from the world of phenomena."
>
> ("Der Weltraum ist mit einem Stoff erfüllt, welcher fortwährend in die ponderablen Atome strömt und dort aus der Erscheinungswelt verschwindet.")

The two can be combined in a single

> "General hypothesis: In all ponderable atoms, substance from the material world constantly enters the world of the spirit."
>
> ("Generalhypothese: In allen ponderablen Atomen tritt beständig Stoff aus der Körperwelt in die Geisteswelt ein.")

Just before that, spirit substance formed there and is regarded as the cause of the vanishing of substance. The spirit world intervenes in this way in the material world at the locations of ponderable bodies. Riemann does not talk of a transformation of the substance that has entered into the spirit substance. Rather, the coming into being of the spirit substance is the cause of the vanishing of substance from the material world. Riemann explains in a footnote that the entering amount of substance is proportional to the force of gravitation.

We see how he disassociates himself from his materialistic contemporaries. He does speak of spirit substance, but this is not another physical form of substance of the material world on which it, so to say, feeds. It is qualitatively different. Here it is important to note the time-order of cause and effect: spirit substance is formed and immediately thereafter substance vanishes from the material world, the world of phenomena.

Let us focus, however, on Riemann's intention to justify new "mathematical" principles. He wants to express

> "the action of general gravitation on a ponderable atom in terms of the pressure of the space-filling substance in its immediate neighborhood,"
>
> ("die Wirkung der allgemeinen Gravitation auf ein ponderables Atom durch den Druck des raumerfüllenden Stoffes in der unmittelbaren Umgebung desselben")

i.e., he wants to justify a theory of local action. Although his formal statements do not lead very far, the investigation of the deformation of infinitesimal particles of matter suggests that the idea of a Riemannian metric was at hand. Riemann says elsewhere (W. 533) that matter can be represented as a physical space whose points move in a geometric space. Gravitation and propagation of light must be explained in terms of the motion of this matter. He tries to obtain differential equations for this scheme, and finally formulates a variational principle (W. 538). All this did not lead to tangible results that agreed with experience, but we can discern in it a prototype of the idea of a unified field theory of gravitation and electromagnetism (light) resumed in the 20th century.

Until recently one was bound to form the impression, unaffected by Dedekind's cautious allusions, that Riemann kept his reflections on natural philosophy to himself. This impression was corrected by the reprinting of Schering's memorial address of 1866 in N. 828–847. Ernst Schering (1833–1897) attended Riemann's lectures. He obtained his doctorate in 1858, became an associate professor of mathematics in 1860, and a full professor and successor of Gauss in his role as astronomer in 1869. In 1873 he published three articles in the *Göttinger Nachrichten* whose content was clearly related to Riemann's geometric and gravitational ideas.

Schering claimed that Riemann passed on to him results in natural philosophy. We quote him verbatim (N. 838). "First he eliminates from all laws of interaction all data that relate to action at a distance, for such data always depend on the nature of the surrounding space, and therefore a law pronounced in it is bound to involve the implicit space structure. He obtains the necessary transformation for the action of masses, of free electricity, and of closed galvanic currents by considering the forces as the physical expression of certain motion formulas of a medium that fills, on the whole uniformly, the threefold extended space. The points in space at which the acting bodies are located are viewed as infinitely dense spots of the medium, or, more intuitively, as places to which the medium goes from the definite threefold extended space to the multiply extended space that surrounds it everywhere." ("Zunächst eliminirt er aus allen Gesetzen für Wechselwirkungen alle die jenigen Bestimmungsstücke, die sich auf Distanzwirkungen beziehen, weil solche immer abhängig von der Beschaffenheit des umgebenden Raumes sind und deshalb ein darin ausgesprochenes Gesetz schon die nicht mit ausgesprochene Raumconstruction involviret. Für die Wirkung der Massen, der freien Electricität und der geschlossenen galvanischen Ströme erreicht er die nöthige Transformation der bekannten Gesetze, indem er die Kräfte als den physischen Ausdruck gewisser Bewegungsformeln eines den dreifach ausgedehnten Raum im Allgemeinen gleichmässig erfüllenden Mediums betrachtet. Die Punkte des Raumes, an welchen sich die wirkenden Körper befinden, werden dabei als unendlich verdichtete Stellen des Mediums betrachtet oder anschaulicher als Orte, an welche das Medium aus dem bestimmten dreifach ausgedehnten Raume in den ihn überall umgebenden mehrfach ausgedehnten Raum austritt.")

We witness here a certain continued development of the reflections of 1853. It is unlikely that Riemann talked about this issue with a student so early, so that what he communicated to Schering should probably be assigned a later date. This is also suggested by the fact that Riemann mentioned the paper of 1858 to Schering and expressed satisfaction that it had not been published at the time, for in the meantime he managed to fit it into the general principles for the other fundamental laws (N. 839). Moreover, he said that he had a general minimality principle, analogous to Gauss' principle of least constraint, that included all laws, namely, those that govern light phenomena as well as those, mentioned earlier, that govern electricity, closed galvanic currents (magnetism), and gravitation. It is clear that this goes well beyond the fragment on gravitation and light (W. 523–538), where it is mentioned just as an objective.

The one rather implausible part of Schering's communication is the reference to a higher-dimensional space that surrounds 3-dimensional space everywhere and appears as a *deus ex machina* that absorbs all substance no longer needed for the building of spirit masses (Geistesmassen). Or perhaps not.

In view of the absence of relevant notes by Riemann, it is best to leave these conjectures and to turn instead to the question of how he may have arrived at his system of natural philosophy. Unfortunately, here we are also hindered by a paucity of clues. But at least we have his own statement (W. 507) to the effect that what led him to the new conception of the laws of nature was

> "primarily the study of the works of Newton, of Euler, and – on the other hand – those of Herbart"
>
> (dass er "hauptsächlich durch das Studium der Werke Newton's, Euler's und – andererseits – Herbart's geführt" worden sei)

There is no doubt about Newton's influence. In addition to Newton's *Principia*, Riemann was familiar with David Brewster's *Memoirs of Sir Isaac Newton*, which also included some of Newton's letters. We mentioned this already at the end of Section 3.2.2.

It is not clear which of Euler's works Riemann had in mind. Even if we had at our disposal Neuenschwander's list of the items borrowed by Riemann from the university library this would not be of great help, for it is one thing to borrow an item from a library and quite another thing to have its content in one's mind while writing. And then there were other sources Riemann had access to. We may well accept Andreas Speiser's supposition (1927) that he was familiar with the second volume of the Brussels 1838 edition of Euler's works. This volume contains not only the famous "Briefe an eine Deutsche Prinzessin" ("Letters to a German princess") (published in 1768), in which Euler explained his views on natural philosophy, but also other works in which he set up the hypothesis of an infinitely fluid substance (called ether), present in the universe, which was to explain gravitation, light, electricity, and magnetism. The "Letters" were so widespread that Riemann could not have overlooked them. Moreover, David Speiser quotes Euler's posthumous *Anleitung zur Natur-Lehre* (*Guide to the study of nature*) and thinks that Riemann may have been influenced by it (D. Speiser, *Eulers Schriften zur Optik, zur Elektrizität und zum Magnetismus*; Leonhard Euler 1983, 226).

Be that as it may, Euler described a field theory dealing with the very same phenomena Riemann wanted to bring together in such a theory. He was able to explain successfully many facts of experience, especially those involving magnetism. Propagation of light in ether could be regarded as analogous to the propagation of sound in air, and this gave one the right partial differential equation. To model the interaction of ether and matter, Euler assumed that every material body is permeated in all directions by little pores that are normally filled with ether, and that the electrical properties of a body depend on the ease with which the pores open, so that ether can be forced into or out of them. This was to account for the two possible signs of electrical charge.

Riemann refrained from using such baroque models and relied instead on his vague and unspecified inner states of ponderables. But his field conception was otherwise very close to Euler's. Before Riemann, Faraday had been influenced by Euler's field conception, though only by its nonmathematical aspects. He had also learned about it from Euler's "Letters" (D. Speiser, p. 224).

Compared with Euler, did Riemann suggest anything new in addition to his attempt at explanation by analogy that relied on inner states of matter? Speiser (see pp. 106/107) notes that Euler encountered two difficulties. One was that the notion of the tremendous pressure the ether was presumably under could not be easily reconciled with the infinitude of space. Another was the question of what was happening to the ether that was presumably always flowing into the atoms. Riemann could point to finite spaces in his geometry, and, as we saw, he had an explanation for the vanishing of the ether substance. Other than this, there is a seamless connection between Euler and Riemann. Euler's idea of a field theory was well developed, and in this respect Riemann did not go much further.

Before exploring (in the next section) Herbart's influence on Riemann, we want to consider the question of how Riemann himself later evaluated, and possibly continued, his reflections on natural philosophy dating back to the 1850s. Schering's recollections, mentioned earlier, shed some light on both issues.

Riemann continued to value these matters highly. But he realized that for speculative formulations to gain recognition it was necessary to develop them in a concrete context. This is the position from which we look at Riemann's last work, his *Mechanik des Ohres* (*The mechanics of the ear*) (W. 338–350). Since he knew that he had little time left to live, Riemann concentrated in mathematics on rounding out his principal function-theoretic work by working on the theta functions (W. 212–224) and on advising his students, mainly Prym and Roch in addition to Hattendorff. He devoted all of his remaining energy to his last work. Its first section, "Über die in der Physiologie der feineren Sinnesorgane anzuwendende Methode" ("On the method to be used in the physiology of the finer sense organs"), is of permanent scientific value.

While it is true that Riemann developed his methodological considerations using the special case of the ear, they are transferable to other organs, or, more generally, to objects not fully accessible to direct observation. The other objects he may have had in mind were the "inner states of the ponderabilia."

Riemann wanted to infer an organ's structure and function from its performance, or, as he put it, to deduce (unobservable) causes from (observable) actions. This called for three steps: the finding of an hypothesis sufficient for the explanation of the performance; investigation of the extent to which the hypothesis is needed for the explanation; and confrontation with experience. When constructing an hypothesis one should not reject the use of analogy and teleology, but when using them one should avoid arbitrariness.

We see that Riemann disassociated himself from the positivist attitude as spelled out in detail by, say, Fourier (see Section 3.2.2). We can guess that, had Riemann's procedure been successful in the concrete example of the ear, such success would have given him the possibility of applying it to natural philosophy as well. This guess is supported by the fact that Riemann picked an example in which the dynamics of fluids played an essential role. Schering and Henle, the editors of the fragment, write (W. 338, footnote) that Riemann regarded the associated mathematical problem as a problem in hydraulics.

3.3.4 The role of Herbart's philosophy

Riemann himself admitted on numerous occasions that he owed essential insights and impulses to Herbart's philosophy and psychology. In his 1854 habilitation lecture he mentioned Herbart in the same breath with Gauss and stated that the two men had provided the groundwork for his concept of a manifold. In a note (W. 507) he observed that he had been led to his principal work on the laws of nature largely through the study of Newton, Euler, and Herbart. And in his last work, devoted to the mechanics of the ear, he named Newton and Herbart as his models in methodological studies. Thus Riemann himself put Herbart next to his greatest authorities in mathematics and physics, and so we have every reason not to underestimate the influence of Herbart's writings on Riemann's scientific development.

Johann Friedrich Herbart (1776–1841) came from North Germany. Between 1797 and 1800 he stayed in Bern and worked as a private tutor. While there he met Pestalozzi. In 1802 he obtained his doctoral degree in philosophy and pedagogy at Göttingen. In 1809 he went to Königsberg as Kant's successor, and he returned to Göttingen in 1833 to take over the post of a professor of philosophy. Until relatively recently he was regarded as a scientist only by experts in pedagogy, but lately, in view of his approach to philosophy, he has been seen as a precursor of the modern theory of knowledge. For him the task of philosophy in the individual disciplines is to provide a systematic procedure which Herbart himself defines as the "elaboration of concepts" (die "Bearbeitung der Begriffe"). Special mention must be made of his *Psychologie als Wissenschaft, neu gegründet auf Erfahrung, Metaphysik und Mathematik* (*Psychology as a science, newly based on experience, metaphysics, and mathematics*) I/II, Königsberg 1824/25. Before going into details we wish to emphasize that we can view Riemann's work in mathematics and physics completely in the sense of Herbart's conception, according to which the effect of philosophy in the individual sciences takes the form of a tendency to systematic activity and to thinking in concepts. Here we must leave undecided the question of whether Riemann's way of doing science was the result of Herbart's influence, or whether, after studying Herbart, he realized that he had come across a mature conception that accorded with his own nature, one that could offer him strong support. An argument in favor of the latter supposition is that already in his youthful work, long before studying Herbart's, Riemann had shown a preference for the conceptual over the algorithmic.

Dedekind described (W. 545) the period after Riemann's return to Göttingen in 1849 as follows: "The first germs of his ideas in the philosophy of science must have developed at that time, simultaneously with his involvement in philosophical studies, largely oriented toward Herbart." ("In dieser Zeit müssen bei gleichzeitiger Beschäftigung mit philosophischen Studien, welche sich namentlich auf Herbart richteten, die ersten Keime seiner naturphilosophischen Ideen sich entwickelt haben."). There follows a reference to Riemann's paper of 1 November 1850 in the pedagogical seminar "Über Umfang, Anordnung und Methode des naturwissenschaftlichen Unterrichts an Gymnasien." ("On the scope, order, and method of teaching the natural sciences in gymnasiums"). In this paper Riemann spoke of the "continuously filled space actually given to us" (von "dem uns wirklich gegebenen continuierlich erfüllten Raume"), and thereby adopted very early the position of natural philosophy on the notion of space. (We cannot agree with Dedekind, who says in this connection that we can recognize here Riemann's striving towards a unified conception of nature. It is true that Riemann mentioned in one breath gravity, electricity, magnetism, and the equilibrium of heat, and said that we have for this a "mathematical theory that is perfectly complete in itself" (eine "vollkommen in sich abgeschlossene mathematische Theorie"), but this merely referred to the fact that in all these cases the Laplace-Poisson differential equation yields the corresponding mathematical theory. At this point, we cannot as yet discern the idea of a unified field theory. This is confirmed by the reference to the equilibrium of heat, which has nothing to do with a field theory.)

Earlier guesses about Herbart's influence on Riemann could take into consideration only the few references in the collected works. But now that E. Scholz has looked through and evaluated Riemann's *Nachlass* in Göttingen in 1982 we have a clearer picture. Numerous excerpts from Herbart's works and some relevant notes have been preserved. Scholz was able to determine that some of these items date from the time of the habilitation lecture of 1854 in which Riemann first publicly mentioned Herbart's name. The following account is based on Scholz's paper.

Riemann's excerpts cover the period from 1796, the publication date of Herbart's contributions to eidolology (Herbart's term for the doctrine of the self), to the time of the posthumous publication of his observations on methodology. In addition to these two areas of Herbart's work there are detailed excerpts from his ontology and synechology, the doctrine of the continuum and of space. These four are Herbart's four disciplines of metaphysics. There are no excerpts from either esthetics or practical philosophy, two areas which, like metaphysics, are also included by Herbart in philosophy. This fits in with Riemann's note (W. 508), known for a long time, in which he says of himself:

> "The author is a Herbartian in psychology and in epistemology (methodology and eidolology) but does not, for the most part, concur with Herbart's philosophy of science and the metaphysical disciplines (ontology and synechology) related to it."
>
> ("Der Verfasser ist Herbartianer in Psychologie und Erkenntnisstheorie (Methodologie und Eidolologie), Herbart's Naturphilosophie

und den darauf bezüglichen metaphysischen Disciplinen (Ontologie und Synechologie) kann er meistens nicht sich anschliessen.")

We recall some of the core statements in Kant's *Critique of Pure Reason*: "Space is not an empirical concept derived from external experiences ... Space is a necessary *a priori* representation ... The apodeistic certainty of all geometric propositions is grounded in this necessity a priori ... Space is ... a pure intuition ... Space is imagined to be a given infinite magnitude ... Geometry is a science that determines the properties of space synthetically and yet *a priori*... geometric propositions are one and all apodeistic, that is, are bound up with the consciousness of their necessity; for instance, that space has only three dimensions; such propositions, however, cannot be empirical, or, in other words, judments of experience, nor can they be derived from such judgments." ("Der Raum ist kein empirischer Begriff, der von äußeren Erfahrungen abgezogen worden ... Der Raum ist eine notwendige Vorstellung a priori ... Auf diese Notwendigkeit a priori gründet sich die apodiktische Gewißheit aller geometrischen Grundsätze ... der Raum ist ... eine reine Anschauung ... Der Raum wird als eine unendliche Größe gegeben vorgestellt ... Geometrie ist eine Wissenschaft, welche die Eigenschaften des Raumes synthetisch und doch a priori bestimmt ... die geometrischen Sätze sind insgesamt apodiktisch, d.i. mit dem Bewußtsein ihrer Notwendigkeit verbunden, z.B. der Raum hat nur drei Abmessungen; dergleichen Sätze können nicht empirische oder Erfahrungsurteile sein, noch aus ihnen geschlossen werden.")

Riemann could not agree with any of these pronouncements, including the very possibility of speaking of *the* space. For him (*the*) *space* (*der Raum*) was a *physical* reality, given in experience together with the metric field of a metric determination. For the purposes of mathematics he created many spaces, of which the space in which the Euclidean constructions are valid is just one. Whether or not it fits physical reality is a question requiring an empirical determination. Every Kantian must have regarded such views as revolutionary.

Herbart criticized the Kantian conception as a "completely contentless, empty, and inappropriate hypothesis" (eine "völlig gehaltlose, nichtssagende, unpassende Hypothese") which reduced space and time to the role of empty containers to be filled by the senses with their perceptions. According to Herbart, all concepts that serve as forms of experience derive from experience, and this is also true of space. Man acquires his space notions from his mobility in his surroundings. As well, the ideas of "similar continua" (von "ähnlichen Continuen") are generated "under similar conditions" ("unter ähnlichen Umständen"), the ideas order themselves "next to and between one another so that it is not possible to integrate them except spatially." ("dergestalt neben und zwischen einander, daß man sie nicht anders als auf räumliche Weise zusammenfassen ... kann.") These quotations have also been culled from Riemann's memos looked through by Scholz.

For Herbart things were bundles (Complexionen) of properties and each property, such as color, could vary within its qualitative continuum. It seems that he could conceive of at most 3-dimensional continua, and that he saw continua everywhere. By contrast, Riemann could conceive of higher-dimensional continua, namely in mathematics:

> " ... on the other hand there are in common life only such infrequent occasions to form concepts whose modes of determination form a continuous manifold, that the positions of objects of sense, and the colors, are probably the only simple notions whose modes of determination form a multiply extended manifold."
>
> ("... dagegen sind die Veranlassungen zur Bildung von Begriffen, deren Bestimmungsweisen eine stetige Mannigfaltigkeit bilden, im gemeinen Leben so selten, das die Orte der Sinnengegenstände und die Farben wohl die einzigen einfachen Begriffe sind, deren Bestimmungsweisen eine mehrfach ausgedehnte Mannigfaltigkeit bilden.")

Thus here Riemann did not concur with Herbart's synechology.

In Chapter 4 we will return to Riemann's concept of a manifold, and in this connection we will also discuss the issue of ontology.

Two more points must be emphasized. The study of Herbart introduced Riemann to Western philosophy and made it unnecessary for him to take courses in the subject. Then there is Riemann's terminology. Here too he was influenced by Herbart. This can be seen from his notes on, and excerpts from, Herbart's works. *Speculation* (*Spekulation*), a term frequently employed by Riemann, denotes the striving to solve problems. Philosophy materializes out of speculation and its subjects are concepts. It is in this sense that we must interpret Riemann's use of the term *concept* (*Begriff*). His view of mathematics was to resemble Herbart's. Herbart regarded mathematics, apart from its concern with the manipulation of formulas, as part of philosophy, for, like philosophy, it made *concepts its subjects* (*Begriffe zu ihren Gegenständen* macht). This characterizes the core of Riemann's ontology.

In addition to Scholz's work of 1982 we wish to mention Gregory Nowak's essay "Riemann's Habilitationsvortrag and the synthetic a priori status of geometry" in Rowe and McCleary 1989 Vol. I, 17–46. This essay also discusses the older literature, by now superseded by Scholz's findings. However, we think that it is inappropriate for a discussion of the habilitation lecture to focus on its deviation from Kant's concept of space. Scholz has shown that Riemann's study of Herbart had a deeper and wider aim. Now we see that the absence of formulas in the lecture was not a gesture aiming to accommodate the nonmathematical members of the faculty. Rather, it was an attempt on Riemann's part to exhibit mathematics, in Herbart's sense, as part of philosophy, as thinking in concepts.

4. Turning Points in the Conception of Mathematics

4.1 The historians' search for revolutions in mathematics

In the last few decades, after the appearance of T. S. Kuhn's *The Structure of Scientific Revolutions* (1962, 1970), it has become fashionable to discuss revolutions in mathematics. The book by D. Gillies 1992 (ed.) *Revolutions in Mathematics*, Clarendon Press, Oxford, to be referred to henceforth as G followed by a page number, is a collection of relevant papers. The opinions range from Michael Crowe's thesis that there are no revolutions in mathematics to the opposite opinion, represented by J.W. Dauben besides others, that almost every situation in the history of mathematics that was sensed to have been critical, every situation followed by changes viewed as radical, should be regarded as indicating a revolution. Gillies goes so far as to speak of a Crowe-Dauben debate.

It is striking that, apart from general discussions of non-Euclidean geometries, Riemann's name has seldom played a significant role in these disputes. Exceptions are the work of L. Boi, and, above all, of J. Gray (G. 226 ff.), whose reflections go philosophically deeper than most other contributions. Moreover, Gray gives a reason why the debates on foundations or on the philosophy of mathematics bypass truly significant issues, such as Riemann's innovations (G. 242): "[It] may be said of all the debates in the foundations of mathematics: either they are technical and accessible only to logicians, or they are epistemological and draw their examples from concepts we meet in school. The result in each case is a debate that does not interest, and does not seek to affect, working mathematicians."

Now Riemann is very much of interest to all mathematicians working in analysis, geometry, number theory, or mathematical physics, provided that the historian does not, as is unfortunately the case with most publications, limit himself to examples accessible to beginners, such as the integral, pathological functions, conditionally convergent series, and possibly — something encountered even today — elliptic geometry confused with Riemann's non-Euclidean geometry; these are, at best, superficial indicators of turning points (Wendepunkte) — Riemann's term, which we prefer to the term "revolutions" — in mathematics. We hope that we have prepared the reader for more substantial fare.

But the search for "revolutions" must not be dismissed as a passing fad. Rather, it is a concomitant of an essential change in our conception of the history of science. Today we are no longer satisfied with chronological listings of discoveries and inventions, with biographies, with descriptions of contacts among contemporaries, to say nothing of discussions about priority questions. Nor are we satisfied with the so-called history of mathematical concepts, which was of some importance for a number of decades, for it took too narrow a view by assuming that concepts had been the essential element in all phases of mathematics. What Riemann's work illustrates is the notion of mathematics as a form of thinking in concepts and, as such, it had to be developed anew.

Before we can properly evaluate Riemann's work, it is useful to consider some popular examples of "revolutions" in mathematics. A catalogue of examples shows far more clearly what is meant by a revolution in mathematics than any attempted definition.

There is a consensus that the following have been revolutions: the Greek discovery of incommensurability and the subsequent development of the Eudoxian theory of magnitudes; the analytic geometry of Descartes (1637); the invention of the infinitesimal calculus (around 1680); the "new rigor" (epsilontics) in analysis after Cauchy (1821) and Dirichlet (1829); non-Euclidean geometry (around 1830); and Cantor's transfinite set theory (approximately from 1872 on). We stop here in order not to go beyond the domain of Riemann's direct influence.

If we limit ourselves to the three examples from the 19th century, then we must conclude that, while all of them were indicators of profound changes, they were not themselves essential innovations. For discussions of epsilontics and rigor we refer the reader to Sections 0.4 and 2.1 respectively. While Cantor's set theory stirred up emotions, it was also a symptom, of relatively little interest for analysis, of a deeper change that is conceivable without his ordinal and cardinal numbers. We will return to sets in Section 4.4.

Let us take a closer look at non-Euclidean geometry. Morris Kline's enthusiastic pronouncement that "The creation of non-Euclidean geometry was the most consequential and revolutionary step in mathematics since Greek times" is frequently quoted, and is even the motto of a lengthy paper in G. 169. Gillies himself writes that "After the infinitesimal calculus, the next major candidate for a revolution in mathematics is the discovery of non-Euclidean geometry."

As Gray put it, we are dealing here with an example that can be formulated in terms of high school concepts. For centuries attempts were made to deduce the parallel postulate as a theorem from the remaining axioms of Euclid, and all of them ended in failure. The discoverers of non-Euclidean geometry attained success in the following manner: they kept the axioms of Euclid but replaced the parallel postulate by the assumption that there are several straight lines through a point P that do not intersect a straight line g. Working with this system they obtained a great many theorems. Had these yielded a contradiction, the parallel postulate would have been proved indirectly. The many theorems they found, and the fact that no contradiction ensued, convinced them that they had discovered a consistent geometry. János Bólyai thought that he had created a new world out of nothing. Then models were given for the non-Euclidean plane. Specifically, Beltrami showed (in 1868) that it can be locally realized on surfaces of constant negative curvature, and Klein showed that the entire hyperbolic plane can be realized in the interior of an ellipse provided that one puts the distance between two points P and Q equal, apart from a multiplicative constant, to the logarithm of the cross ratio of the four points P, Q, R, S, where R and S are the points of intersection of the ellipse with the straight line PQ.

Where is the revolution in this? Consider an analogous state of affairs, the story of the complex numbers. For centuries people assumed that they could add to the real numbers a "number" i such that $i^2 = -1$, and they thus obtained many results without running into contradictions. In the first half of the 19th century people exhibited models of the complex numbers. This story is qualitatively just the same as the story of non-Euclidean geometry, yet no one speaks of a revolution!

In the second half of the 19th century, the new geometry attracted attention after the publication of the Gauss correspondence. Then it was not so much the mathematicians who thought of it as a revolutionary discovery. *They* saw it as the resolution of the question of parallels and as an occasional auxiliary tool in complex analysis. (See R. Bonola, H. Liebmann: *Die nichteuklidische Geometrie*, 2. Auflage Leipzig 1919.) Moreover, the group of motions of the hyperbolic plane provided a new example of a continuous group. There *was* excitement, but it was limited to people educated in the neo-humanistic gymnasiums in the spirit of the Kantian philosophical tradition. It was they who were bound to regard the new "geometry" as heresy. But such people regarded even 4-dimensional space as heresy!

Has non-Euclidean geometry introduced anything genuinely new? It is a fact that it remained entirely within the framework of Euclidean construction methods, whereas Riemann left this area and constructed his spaces in a completely different way. For 19th-century geometers other areas, such as Poncelet's projective geometry of 1822 and, admittedly with a measure of delayed response, Klein's Erlangen Program, were more interesting. Moreover, as we saw, algebraic geometry was also a field with many interesting aspects.

We can choose to talk about a *turning point in the approach* to the problem of parallels. The attempts to solve the Euclidean parallel problem landed mathematicians in a blind alley and a new turn led them out of it. Looked at in this light, the discovery of non-Euclidean geometry has for us an exemplary character. More generally, in some cases we will find Riemann's occasional use of the expression a "turning point" preferable to the use of the term "revolution."

We draw attention once more to the anthology of L. Boi et al. 1992, in which the *evolutionary* character of the history of geometry after 1830 comes to the fore and in which there are no explicit references to revolutions. This volume discusses the relations of geometry to epistemology and physics, and thus comes closer in its approach to that of Riemann, as befits an approach that prefers not to treat innovations as an internal mathematical issue. We will return to such a more comprehensive viewpoint in our concluding remarks, but must first elucidate the internal aspects of mathematics that were decisive in the decades after Riemann.

4.2 Turning point in the conception of the infinite in mathematics

The term "turning point" ("Wendepunkt") was used by Riemann himself in a note that M. Noether included already in 1902 in the supplementary volume to the *Collected Works*. Until that time little attention seems to have been paid to this note. Since Riemann was usually very careful about the way he put things, an emphatic formulation such as "turning point in the conception of the infinite" calls for special attention. It therefore seemed proper to investigate carefully the note and its context. E. Neuenschwander kindly found the original page, which is reproduced here in facsimile by permission of the Göttingen library. Its analysis produced a number of surprises.

First we quote Noether's wording, which will have to be corrected and put in context:

> "15 Nov. 1855: 'The discovery of the fact that infinite series and multiple integrals split into two classes [depending on whether or not the limit is independent of the order (in the case of series), and on the way we let the domain grow when it extends to infinity (in the case of multiple integrals)] constitutes a turning point in the conception of the infinite in mathematics.' "

> ("15 Nov. 1855: 'Die Erkenntniss des Umstandes, daß die unendlichen Reihen und die mehrfachen Integrale in zwei Klassen zerfallen [je nachdem der Grenzwert unabhängig von der Anordnung, bezw. von der Art, wie man das Gebiet, wenn es ins Unendliche geht, wachsen läßt; oder nicht], bildet einen Wendepunkt in der Auffassung des Unendlichen in der Mathematik.' ")

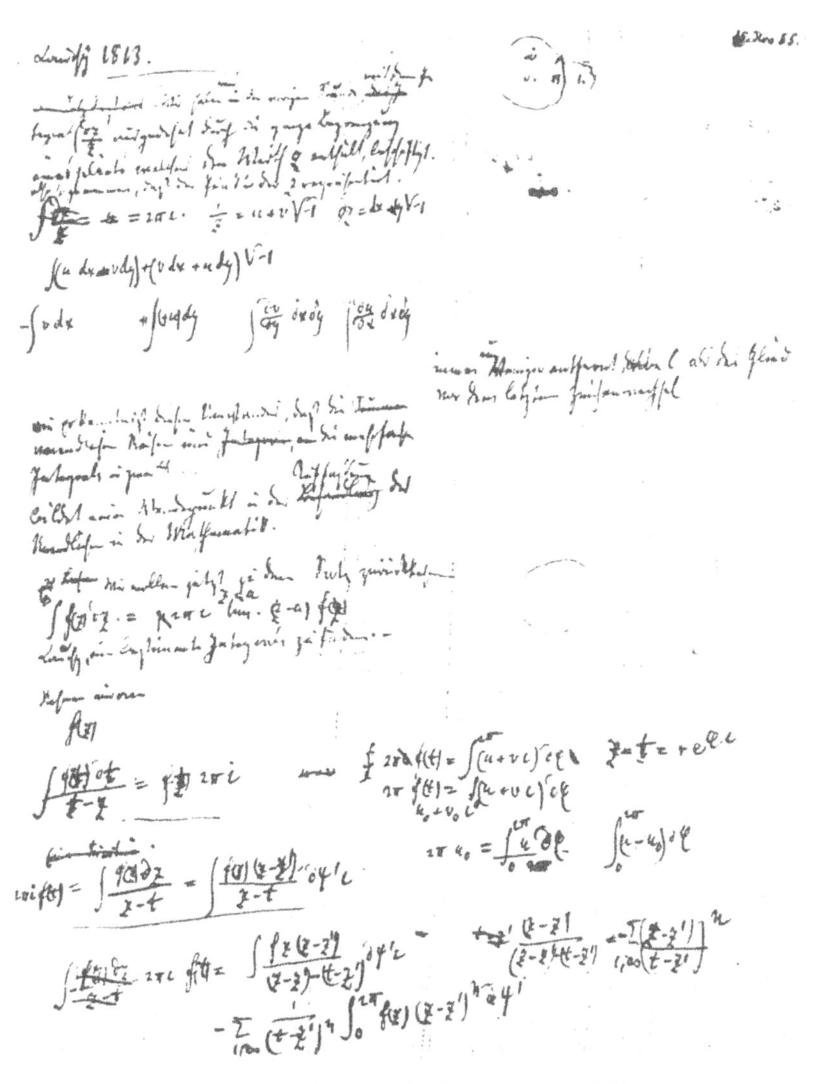

35: Riemann's note for the lecture of 15 November 1856

There is no doubt that the note pertains to the lectures on complex analysis. We know that at that time Riemann used to prepare every lecture, and especially its beginning, word for word. The relevant page is now in the Göttingen University Library, Cod. Ms. Riemann 23.4, fol. 22 verso. The text is as follows (the crossed-out words are in parentheses):

> "The discovery of this fact, that the (sums)
> infinite series and (integr....) the multiple
> integrals in two cl...

constitutes a turning point in the (treatment) conception
of the infinite in mathematics."

("Die Erkenntniss dieses Umstandes, dass die (Summen)
unendlichen Reihen und (Integr....) die mehrfachen
Integrale in zwei Kl...
bildet einen Wendepunkt in der (Behandlung) Auffassung
des Unendlichen in der Mathematik.")

Noether read Riemann's note somewhat inaccurately and added more than the long passage in brackets. The dates show that we are dealing with the same page; "of this fact" ("dieses Umstandes") emphasizes a direct reference to the context. Riemann wrote initially "treatment" ("Behandlung") and then replaced it by the stronger "conception" ("Auffassung"). The dots in cl... (Kl...) indicate that he had intended to insert here a phrase he was familiar with and had no need to jot down explicitly. The phrase in question can be found in the habilitation paper he submitted in 1854 (W. 235), and speaks of

the "insight that the infinite series split into two classes, depending on whether or not they remain convergent when one makes all the terms positive."

(die "Einsicht, dass die unendlichen Reihen in zwei Klassen zerfallen, je nachdem sie, wenn man sämtliche Glieder positiv macht, convergent bleiben oder nicht.")

Noether's addition in brackets also refers to this place.

The context is that of the relations, indicated in the upper half of the page, between line integrals and double integrals and their use in complex analysis. Obviously, the relation $\int \frac{dz}{z} = 2\pi i$, which holds if one circles the origin once, is introduced in order to discuss the Gauss-Green integral formula. Riemann puts $\frac{1}{z} = u + v\sqrt{-1},\ dz = dx + dy\sqrt{-1}$,

$$\int (udx - vdy) + (vdx + udy)\sqrt{-1}$$

$$-\int vdx \quad + \int (+u)dy \quad \int \frac{\partial v}{\partial y}\, dx\, dy \quad \int \frac{\partial u}{\partial x}\, dx\, dy,$$

followed by the passage just quoted verbatim. If we also note that Riemann wrote "Cauchy 1813" at the top of the page, then we get a clear picture that fits the remark on multiple integrals.

The hint can only refer to Cauchy's comprehensive paper "Sur les intégrales définies," submitted to the Academy on 22 August 1814 and published in 1827 with supplements, now in Cauchy *Œuvres* (1), 329–506. Pp. 394–396 contain what is probably the earliest discussion of the non-interchangeability of the order of integration in a double integral, one actually involving an integrand like Riemann's $\int \frac{\partial v}{\partial y}$ or $\int \frac{\partial u}{\partial x}$. At this point Noether's addition must be corrected. It is not that the domain of integration extends to infinity but that the integrand becomes infinite at a boundary point of the finite domain of integration. Cauchy wrote (I put y for his z): "Soit $K = \phi(x, y) = \frac{y}{x^2+y^2}$, et conçevons que l'intégrale

$$\iint \frac{\partial K}{\partial y}\, dx\, dy$$

doive être prise entre les limites $y = 0, y = 1, x = 0, x = 1$. Si l'on suppose les valeurs de y substituées avant celles de x, on aura

$$\int_0^1 \frac{\partial K}{\partial y}\, dy = \frac{1}{1+x^2}, \quad \int_0^1 \int_0^1 \frac{\partial K}{\partial y}\, dx\, dy = \int_0^1 \frac{dx}{1+x^2} = \frac{\pi}{4}."$$

Changing the order of integration yields the value $-\frac{\pi}{4}$. Cauchy remarked that this was due to the indeterminacy of the integrand for $x = y = 0$.

But what has this to do with a particular conception of the infinite? Luckily, we are helped here by statements from contemporaries. Poisson immediately reviewed the paper and his discussion is quoted in Cauchy, *Œuvres* (2) 2, 194–198. The report by the referees Lacroix and Legendre on 7 November 1814 (Cauchy, *Œuvres* (1) 1, 321–327), who devoted more than a page to the example, is even more interesting. What was presumably at issue was an easily explainable anomaly. If one evaluated the integral between the limits α and 1 on x and β and 1 on y, with α and β positive infinitesimal quantities, then, obviously, regardless of the order of integration, one obtained the result

$$\frac{\pi}{4} - \arctan\frac{\beta}{\alpha}.$$

Now everything depends on the quotient of the two infinitely small quantities! This conception of the infinite, in this case of the infinitely small, was common among the older French mathematicians; Cauchy adopted it from 1820 on. Gauss' discussion of French papers (*Werke* VI, 648) from 1828 is a relevant example) shows that this interpretation was not foreign to him.

Dirichlet and Riemann moved from infinitesimal mathematics to the notion of limits, which is probably what Riemann had in mind at this point. This conjecture is supported by his seemingly irrelevant statement bearing on conditionally and absolutely convergent series which we are about to take up.

The margin of the note contains the statement

> "removed from C by less than the term before the last change of sign."
>
> ("immer um weniger entfernt von C als das Glied vor dem letzten Zeichenwechsel.")

This becomes understandable in the light of the passage in the habilitation paper (W. 235) devoted to rearrangements of conditionally convergent series:

> "In fact, if in a series of the second class [i.e., a conditionally convergent series] we denote the positive terms successively by $a_1, a_2, a_3, \ldots$, and the negative ones by $b_1, b_2, b_3, \ldots$, then it is clear that both $\sum a$ and $\sum b$ must be infinite ... It is clear that upon a suitable arrangement of the terms the series will take on an arbitrarily prescribed value C. For if we take alternately so many positive terms of the series that their value is greater than C and so many negative terms that their value is less than C, then the deviation from C will never be more than the value of the term preceding the last change of sign."
>
> ("In der That, bezeichnet man in einer Reihe zweiter Klasse [einer bedingt convergenten Reihe nämlich] die positiven Glieder der Reihe nach durch $a_1, a_2, a_3, \ldots$, die negativen durch $b_1, b_2, b_3, \ldots$, so ist klar, dass sowohl $\sum a$, als $\sum b$ unendlich sein müssen ... Offenbar kann man nun die Reihe durch geeignete Anordnung der Glieder einen beliebig gegebenen Werth C erhalten. Denn nimmt man abwechselnd so lange positive Glieder der Reihe, bis ihr Werth grösser als C wird, und so lange negative, bis ihr Werth kleiner als C wird, so wird die Abweichung von C nie mehr betragen, als der Werth des dem letzten Zeichenwechsel voraufgehenden Gliedes.")

Part of the last sentence is almost the same as the marginal note.

In order to answer the as yet unanswered question of what this has to do with a new conception of the infinite, we must go back to two older notions of infinite series, both of which can be traced to Euler. I will explain both using the simplest example, namely the series for ln 2. The series

$$\ln 2 = 1 - \frac{1}{2} + \frac{1}{3} - \frac{1}{4} + - \cdots \qquad (i)$$

comes from the power series expansion of $\ln(1+x)$. The rearrangement

$$\frac{3}{2}\ln 2 = 1 + \frac{1}{3} - \frac{1}{2} + \frac{1}{5} + \frac{1}{7} - \frac{1}{4} + + - \cdots \qquad (ii)$$

has nothing to do with it. It is obtainable only by expanding other functions, such as, say, $\frac{1}{2}\ln(1+x^2)(1+x)^2$. In the Euler tradition (*ii*) is not a rearrangement of (*i*) but an altogether different series. In the case of Cauchy 1821, number series are still secondary. First one expands functions, and numerical series are obtained by substitution of particular values for the variable. As is well known, Euler says that the value of a series is the value of the expression from which it is obtained by expansion. (The idea of analytic continuation can be viewed as a revival of this conception.) In a series of functions, such as a power series or a Fourier series, each term has a definite position and there is no meaningful way of speaking of rearrangements. The commutative law was never applied to infinite sums.

For Riemann the functional expression was no longer the source of the series. This made the question of the role of the commutative law meaningful, and one had to realize that the "laws of finite sums" (W. 235) cannot be unrestrictedly applied to series. Implicit here is actually a new conception of the infinite, or a new way of dealing with it.

In addition to the algebraic interpretation of expressions and their expansions there is yet another tradition, which can be traced to Leibniz and to the youthful papers of Euler (Laugwitz 1986, 44–47), that actually involved operating with values of divergent numerical series. Euler used an infinitely large natural number, a *numerus infinitus* i, and obtained for the harmonic series the equality

$$1 + \frac{1}{2} + \frac{1}{3} + \frac{1}{4} + \cdots + \frac{1}{i} = \ln i + C_i \, ,$$

where, to within an infinitely small error, C_i is equal to the Euler constant $C = 0.57\ldots$. The rules of finite sums can be applied to such sums. This being so, we can obtain *(i)* from

$$\begin{aligned} &1 - \frac{1}{2} + \frac{1}{3} - \frac{1}{4} + - \cdots + \frac{1}{2i-1} - \frac{1}{2i} \\ &= \left(1 + \frac{1}{2} + \cdots + \frac{1}{2i}\right) - 2\left(1 + \frac{1}{2} + \frac{1}{4} + \cdots + \frac{1}{2i}\right) \\ &= \ln 2i + C_{2i} - (\ln i + C_i) \approx \ln 2i - \ln i = \ln 2. \end{aligned}$$

We obtain *(ii)* in a similar manner:

$$\begin{aligned} &1 + \frac{1}{3} - \frac{1}{2} + \frac{1}{5} + \frac{1}{7} - \frac{1}{4} + + - \cdots + \frac{1}{4i-3} + \frac{1}{4i-1} - \frac{1}{2i} \\ &= 1 + \frac{1}{2} + \frac{1}{3} + \cdots + \frac{1}{4i} - \left(\frac{1}{2} + \frac{1}{4} + \frac{1}{6} + \cdots + \frac{1}{4i}\right) - \left(\frac{1}{2} + \frac{1}{4} + \cdots + \frac{1}{2i}\right) \\ &= \ln 4i + C_{4i} - \frac{1}{2}(\ln 2i + C_{2i}) - \frac{1}{2}(\ln i + C_i) \approx \ln 4 - \frac{1}{2}\ln 2 = \frac{3}{2}\ln 2. \end{aligned}$$

This interpretation of the infinite reflects a Leibnizian (continuity) principle: the rules for the finite continue to hold for the infinite. The series *(ii)* contains summands other than those in the series *(i)* and so is not a rearrangement of it.

The rejection of this Leibniz-Euler conception of the infinitely large corresponds to the rejection of the infinitely small in the case of double integrals.

It is clear that Riemann was unaware of these methods for the computation of sums of series, for otherwise he could not have said that the inapplicability of the laws of finite sums to (conditionally convergent) series was

> "a circumstance overlooked by the mathematicians of the previous century...."
>
> ("ein Umstand, welcher von den Mathematikern des vorigen Jahrhunderts übersehen wurde...") (W. 235).

But I adopt the position that traditions are effective even if they are not substantiated by knowledge of the literature.

4.3 Turning point in the method: Thinking instead of computing

In a famous paper, in whose nearly three hundred pages Riemann's name was mentioned just twice, a great mathematician wrote: "I have tried to avoid the huge computational apparatus of X, so that here too Riemann's principle should be realized, according to which proofs should be impelled by thought alone and not by computation." ("Ich habe versucht, den grossen rechnerischen Apparat von X zu vermeiden, damit auch hier der Grundsatz von Riemann verwirklicht würde, demzufolge man die Beweise nicht durch Rechnung, sondern lediglich durch Gedanken zwingen soll.")

This sentence appears in Hilbert's so-called "Zahlbericht" of 1897 (*Ges. Abh.* I, 67), whose full title is *Die Theorie der algebraischen Zahlkörper* ("The theory of algebraic number fields"). X stands for Kummer. Neither the name nor the subject — to which Riemann made no direct contribution — matter very much here. The quoted passage brings out the clarity with which Hilbert recognized the change in the methodology of proof, so fundamental in Riemann's case, and the extent to which he regarded it as exemplary. Incidentally, Riemann's name is mentioned a second time two pages later in connection with his deep investigations of the frequency of primes, a topic mentioned in passing.

One should not misconstrue Hilbert's pronouncement as an endorsement of the pride of the thinker vis-à-vis the calculator. Like Riemann, Hilbert was a superb calculator, and he hastened to mention the deep insights into the theory disclosed by the works of Kummer and Kronecker. Moreover, we saw that while Riemann took pride in his ability to obtain earlier results "virtually without computing," he fully appreciated the contributions of "calculators" such as Kummer and Weierstrass (W. 67, 85, 101/102).

To us it seems obvious that proofs should be the result of thought, and we might well ask whether this has not been the guiding principle since the time of Greek mathematics. We are always aware, both at school and at the university, of this view of the essential character of mathematical proof, and, once a certain level of mathematical insight has been attained, we regard a conceptual proof as more appropriate and transparent than a computational proof, in which the final result tends to pop out suddenly and surprisingly

36: David Hilbert

after a great many transformations. Of course, many proofs combine thought inferences with transformations of expressions, and this is in particular the case with Riemann and Hilbert.

The reader will recall that in Part 4 of the introduction, as well as later, we saw on many occasions how different the situation was around 1850. The supreme mastery of computational transformations by Gauss, Jacobi, and Kummer was beyond doubt but had reached its practical limits. Kummer's treatment of the hypergeometric series used up close to 200 pages of *Crelle's Journal*, and assimilation of that much reading matter was virtually impossible. Continuation of this approach for series solutions of other differential equations was out of the question. A similar situation prevailed in the case of Abelian functions. People had ended up in a blind alley, and Jacobi had spelled this out with all necessary clarity. (See J. Gray in Rowe and McCleary I, 366.)

Riemann's dissertation marked a turning point in this development. A way out of the blind alley was consciously pointed out and taken. The program in §20 shows that from the very beginning Riemann was also aware of the practical significance of his approach (see Sections 1.2.1 and 1.2.3).

Klein (1926, 250) wrote persuasively about Riemann's period of study at Berlin: "It was probably Jacobi who provided him with the material stimulation; but Riemann did not adopt Jacobi's method. For him Jacobi was too much of an algorithmist. On the other hand, what attracted him to Dirichlet was a strong inner sympathy based on a similar mode of thought. Dirichlet liked to make theorems clear to himself by means of the intuitive substratum; besides, he dissected the foundations with logical precision and avoided long computations whenever possible." ("Wohl hat ihm Jacobi die stoffliche Hauptanregung gegeben; Jacobis Methode aber übernimmt Riemann nicht. Jacobi ist ihm zu sehr Algorithmiker. Mit Dirichlet verbindet ihn eine starke innere Sympathie ähnlicher Denkweise. Dirichlet liebte es, sich die Theoreme am anschaulichen Substrat klar zu machen; daneben zergliedert er logisch scharf die Grundlagen und vermeidet tunlichst lange Rechnungen.")

(Of course, Klein's addition of the "intuitive substratum" dilutes the statement. Similarly, the logical analysis does not stand "beside"; it is the essential core.)

In his memorial address for Dirichlet, Minkowski said (*Ges. Abh.* 460/461): "While the sharp sword, called by Riemann the Dirichlet Principle, may have first been swung by William Thomson's youthful arm, the modern period in the history of mathematics dates from the other Dirichlet Principle: to impel problems with a minimum of blind computation and a maximum of insightful thought." ("Mag auch das von Riemann Dirichletsches Prinzip benannte scharfe Schwert zuerst von William Thomsons jungem Arm geschwungen sein, von dem anderen Dirichletschen Prinzipe, mit einem minimum an blinder Rechnung, einem maximum an sehenden Gedanken die Probleme zu zwingen, datiert die Neuzeit in der Geschichte der Mathematik.")

4.4 Turning point in the ontology: Mathematics as thinking in concepts

A temporary characterization, short and superficial, of the change in the view of what were the *objects* of mathematics, a change that came about in the course of the 19th century, can be formulated as follows. At the beginning of the century arithmetic and algebra dealt with numbers and formulas, and analysis and geometry had as additional objects variables and figures respectively. At the end of the century, in Hilbert's case, mathematics dealt just with sets, and their structures were also given by sets. In the foundations of geometry and in algebra the structures were relations, i.e., subsets of product sets, and in analysis (with its convergence structures) they soon took the form of subsets of power sets. For the sake of brevity we have interpreted Hilbert's work in the Bourbaki sense, and have thus anticipated the tendency of the first half of the 20th century.

4.4.1 General concepts and the modes of their determination

Riemann occupied an intermediate position, and not just in terms of time. We claim that for him the objects of mathematics were no longer just formulas but not yet sets. They were concepts. We saw in the previous section that a turn to conceptual proofs as a method was imminent during his years of study. For this change to occur, concepts had to be viewed as the objects of theorems and proofs. Riemann reached this position between 1851, when he was awarded the doctorate, and 1854, when he passed his qualifying examination (Habilitation). This development was probably due to the influence of Herbart's philosophy. One indication of the appropriateness of our choice of these dates is the change in Riemann's view of functions of a real variable. In 1851 he thought that the concept of continuity coincides with that of a functional dependence given by a law (W. 3/4; see Section 1.2.3). He abandoned this view, at the latest, when he was writing his paper on trigonometric series.

The general formulations were carefully thought through in the habilitation lecture (W. 273/274). In the passages quoted below it is tempting to translate *concept* as *set* and *mode of determination* as *element*. The result is what the commentators tended to make of the lecture. The reader may do this at first, and then try, together with us, to understand Riemann's view, which is different from our own.

> "Notions of quantity are possible only where there exists already a general concept that allows various modes of determination. According as there is or is not found among these modes of determination a continuous transition from one to another, they form a continuous or a discrete manifold; the individual modes are called in the first case points, in the latter case elements of the manifold."
>
> ("Grössenbegriffe sind nur da möglich, wo sich ein allgemeiner Begriff vorfindet, der verschiedene Bestimmungsweisen zulässt. Je nachdem unter diesen Bestimmungsweisen von einer zur andern ein stetiger Uebergang stattfindet oder nicht, bilden sie eine stetige oder discrete Mannigfaltigkeit; die einzelnen Bestimmungsweisen heissen im ersten Falle Punkte, im letztern Elemente dieser Mannigfaltigkeit.")

Thus first there is the *concept*; by its different modes of determination it generates a manifold, a term we can replace without reservations by "set." Whether the manifold associated with a concept is continuous or discrete is implicit in the concept. Unlike Riemann, we now think of a set as given, and subsequently impose on it a (topological) structure in order to make it into a continuous manifold. This also explains the, for us initially strange, disjunction of concepts into those whose associated manifolds are continuous and those whose associated manifolds are discrete.

Later Cantor said that a set is the comprehension of given things as a single whole. Riemann admits this, at most, in the discrete case:

> "Concepts whose modes of determination form a discrete manifold are so common, that for things arbitrarily given there can always be found a concept, at least in the more highly developed languages, under which they are comprehended (and mathematicians have been able therefore in the doctrine of discrete quantities to set out without scruple from the postulate that given things are to be considered as all of one kind)..."
>
> ("Begriffe, deren Bestimmungsweisen eine discrete Mannigfaltigkeit bilden, sind so häufig, dass sich für beliebig gegebene Dinge wenigstens in den gebildeteren Sprachen immer ein Begriff auffinden lasst, unter welchem sie enthalten sind (und die Mathematiker konnten daher in der Lehre von den discreten Grössen unbedenklich von der Forderung ausgehen, gegebene Dinge als gleichartig zu betrachten)...")

We see that Riemann let himself be carried away to the point of making risky statements about linguistic means of expression for the sole purpose of emphasizing the primacy of the notion of concept, even in the case of heaps of objects lumped together without rhyme or reason. How much this issue must have meant to him is also indicated by the fact that, in this lecture as well as elsewhere, he viewed discrete sets as of little account. If we used the standard post-Cantor cardinality arguments, then we could easily reduce Riemann's position *ad absurdum*; indeed, it is enough to bear in mind that the number of linguistic expressions is finite, whereas the number of finite subsets of the natural numbers alone is infinite (and the number of "all" their subsets is uncountable). However, if we did this, then we would be abandoning the philosophy represented by Riemann in his lecture. Clearly, for him the only sets (manifolds) admissible in mathematics were those arising from concepts, and the interesting ones were the continuous manifolds, which he invariably put first:

> "They are so rare in ordinary life that the positions of objects of sense, and the colors, are probably the only simple notions whose modes of determination form a multiply extended manifold. More frequent

occasion for the birth and development of these notions is first found in higher mathematics."

(Sie sind "im gemeinen Leben so selten, dass die Orte der Sinnengegenstände und die Farben wohl die einzigen einfachen Begriffe sind, deren Bestimmungsweisen eine mehrfach ausgedehnte Mannigfaltigkeit bilden. Häufigere Veranlassung zur Erzeugung und Ausbildung dieser Begriffe findet sich erst in der höhern Mathematik.")

Using the language developed later in topology we could say that Riemann's continuous manifolds are arcwise connected Hausdorff spaces, and that, on the basis of plausibility considerations, he singles out those that are locally homeomorphic to $\mathbb{R}^n$.

This is not the place for the mathematical details dealt with in Section 3.1.3, but we do wish to add a remark on infinite-dimensional spaces. Riemann mentioned specifically "the possible determinations of a function for a given region" (W. 276). Consider the concept of a real function on the interval $-\pi \leq x \leq \pi$ that underlies his habilitation paper. There is associated with it a continuous manifold. This is so because, given real functions f and g, $(1-t)f + tg$ for $0 \leq t \leq 1$ is also such a function, and this makes for a continuous transition. A modern functional analyst would ask for the topology of this function space. For us, in the infinite-dimensional case the answer to this question is not unique, whereas for Riemann the topology had to be implicit in the concept. In the habilitation paper he used the topology of pointwise convergence: $f_n \to f$ precisely if $f_n(x) \to f(x)$ for every x in the interval. By now we know that this does not get us very far, and indeed Riemann's paper was not one of his great successes. Another natural topology would have been the topology of uniform convergence in the sense of Cauchy, in which the arguments x are all real numbers in the interval as well as those infinitesimally close to them. But this would not have been advantageous for Riemann's purposes even if he had been prepared to adopt Cauchy's concept of number, for in this sense discontinuous functions are never representable by trigonometric series.

4.4.2 The primacy of the continuous over the discrete in Riemann's mathematics

All of Riemann's work deals with the mathematics of continua in analysis, field physics, and differential geometry. Hermann Weyl wrote (N. 740) that Riemann's epistemological motive was "the principle, to understand the world from its behavior in the infinitely small" ("Das Prinzip, die Welt aus ihrem Verhalten im Unendlichkleinen zu verstehen"). The mathematics of the discrete entered his field of interest through continua. The distribution of primes was to be understood from the behavior of functions of a complex variable. In topology, Riemann started from continuous manifolds, with which there are associated characteristic numbers.

What bestowed "reality" on the complex numbers was their representation in the continuum of the plane (Section 1.1.2); their *being* was secured by their representation in a continuum, in a continuous manifold. In turn, a continuous manifold derived its existential quality from the realm of the conceptual. We can assume that, analogously, for Riemann the real numbers derived their being from the linear continuum. This being so, it would be pointless to look in his work for a particular justification of the concept of a real number.

Riemann saw himself as agreeing with Gauss, whose pronouncements on the metaphysics of numbers we collected in Section 1.1.2. The agreement with Herbart's synechology was only partial, for the latter's continua were limited to at most three dimensions, and his references to continuable forms of series, suggestive of a lineup, may have contained elements of a discretization.

J. Gray (G. 235) too emphasizes that Riemann's mathematical conception of space was not logically prior to the concept of a multiply extended magnitude. What is more, he thinks that the reverse was the case, i.e., that the greater generality of such concepts of magnitude guaranteed their existence, and he points out that Riemann took twofold-extended manifolds as the foundation of complex numbers and functions. According to Gray (G. 233), Riemann developed a new ontology by stretching the old concept of space and by reformulating geometry as the study of point sets furnished with a concept of distance. If we consider just the mathematical consequences, then in the final analysis this is indeed what we end up with in Riemann's case. But we think that the intermediate step, thoroughly discussed by Riemann before the introduction of a metric function, must not be ignored.

This was already underlined by E. Scholz (1980, 30–34), who remarked that Riemann had wrested a qualitative aspect from continuous manifolds and had tendentiously opened the possibility of transcending the purely quantitative character of mathematics. With arithmetization as the catchword, the immediate general tendency of mathematicians was to go in the opposite direction. By his detailed treatment of continuous manifolds prior to the introduction of coordinates Riemann made it very clear that numbers were secondary and played an auxiliary role. Had this not been the case, he could have made things simpler for himself and his audience by starting immediately from an n-dimensional number manifold as a generalization of the 3-dimensional coordinate space of analytic geometry and the 2-dimensional Gaussian parametrization in the theory of surfaces.

4.4.3 Riemann's concept of a manifold in the philosophical tradition

Riemann drew a clear distinction between concepts associated with continuous and discrete manifolds respectively. This puts him in a philosophical tradition that goes back to Aristotle. Aristotle's continuum can be endlessly subdivided into divisible parts, which implies that it does not consist of indivisible elements. It is potentially possible to find infinitely many points of division, but the continuum does not coincide with an actually infinite collection of points. Dedekind and Cantor abandoned this tradition.

Under the heading "antinomies," notes were found in Riemann's *Nachlass* that attest to his analysis of the problematic nature of the continuum. The specific juxtaposition of thesis with antithesis follows schematically, and to a large extent also thematically, that of Kant in his *Kritik der reinen Vernunft* (*Critique of Pure Reason*).

> "Thesis" is associated with the heading: "Finite, imaginable"; "Antithesis" with: "Infinite, systems of concepts that lie at the boundary of the imaginable."
>
> (Zur "Thesis" gehört die Überschrift: Endliches, Vorstellbares"; zur "Antithesis" gehört: "Unendliches, Begriffssysteme, die an der Grenze des Vorstellbaren liegen.")

The thesis "finite time and space elements" is juxtaposed with the antithesis "the continuous." This pair interests us here. Riemann has also the pairs freedom/determinism, God acting temporally (world rule)/timeless God (Providence), and immortality/temporality.

The only systems of concepts directly accessible to our thought are those of the thesis; in mathematics it is only the finite that we can deal with directly. The systems of concepts of the antithesis, in particular the continuable and therefore also the continuous manifolds,

> "are, it is true, concepts that are firmly determined by means of negative predicates [such as 'infinite'] but are not positively imaginable. But they can be thought of as lying at the boundary of the imaginable... "
>
> ("sind zwar durch negative Prädicate [wie unendlich] fest bestimmte Begriffe, aber nicht positiv vorstellbar. Sie können aber als an der Grenze des Vorstellbaren liegend betrachtet werden ... ")

Then Riemann tries to give a philosophical justification of the method of limits:

> "The method used by Newton for the justification of the infinitesimal calculus, and acknowledged from the beginning of this century by the best mathematicians to be the only one that yields certain results, is the method of limits. What this method consists in is that instead of a continuous transition from one value of a magnitude to another, from one place to another, or actually from one mode of determination of a concept to another, one considers first a transition through a finite number of intermediate steps and then lets the number of these intermediate steps grow in such a way that, without exception, the distance between any two successive steps decreases indefinitely" (W. 519). "Apart from ratios of magnitudes, passage to the limit leaves the system of concepts unchanged. However, at the limit itself, some of the correlative concepts of the system, namely those that bring about the interrelation among other concepts, lose their conceivability."
>
> ("Die Methode, welche Newton zur Begründung der Infinitesimalrechnung anwandte, und welche seit Anfang dieses Jahrhunderts von den besten Mathematikern als die einzige anerkannt worden ist, welche sichere Resultate liefert, ist die Grenzmethode. Die Methode besteht darin, dass man statt eines stetigen Üebergangs von einem Werth einer Grösse zu einem andern, von einem Orte zu einem andern, oder überhaupt von einer Bestimmungsweise eines Begriffs zu einer andern zunächst einen Uebergang durch eine endliche Anzahl von Zwischenstufen betrachtet und dann die Anzahl dieser Zwischenstufen so wachsen lässt, dass die Abstände zweier aufeinanderfolgender Zwischenstufen sämtlich ins Unendliche abnehmen" (W. 519). "Von den Grössenverhältnissen abgesehen, bleibt das Begriffssystem bei dem Uebergange zur Grenze ungeändert.. In dem Grenzfall selbst aber verlieren einige von den Correlativbegriffen des Systems ihre Vorstellbarkeit, und zwar solche, welche die Beziehung zwischen andern Begriffen vermitteln") (W. 520).

The last remark is a reminder that at the limit inequality can turn to equality and continuity of the functions of a sequence to discontinuity of the limit function. This shows that at this point Riemann's conception differed from that of Leibniz, who thought that the rules for the finite retain their validity for the infinite and that, say, rest should be regarded as an infinitely small motion. In ontological terms, Riemann thought that the continuous manifold derives its existence from the concept, and that it can be investigated only indirectly.

Only the finite is accessible to reflection in a manner that is mathematically precise and complete.

4.4.4 Thinking in mathematical concepts before Riemann

We recall Hilbert's view that Riemann should be regarded as a model proponent of the sentiment that, whenever possible, proofs should be the result of thought rather than of computation. This must not be taken to mean that there were no thought-proofs before Riemann. In order to evaluate his achievement in this area we must look at his predecessors, beginning with Euler. In analysis we must mention, first and foremost, Bolzano and Cauchy.

Another caveat. It is obvious that computational and geometric-constructive proofs also call for thought. As mentioned earlier, Hilbert was tireless in his praise of Kummer's achievements. But we are interested here in thinking in concepts.

It is often overlooked that Euler had clear ideas concerning such fundamental concepts of analysis as convergence and continuity (see Laugwitz 1985). But these concepts were neither objects of his mathematics nor tools in his proofs. Rather, numbers and functions were given by formulas, and functions also by means of their graphs. In Euler's mathematics the objects were formulas and figures but not concepts. This meant that conclusive proofs involved only certain admissible transformations of expressions and geometric constructions.

In order to follow the gradual change, we focus on the transparent example of the intermediate value theorem for continuous functions. Euler and his contemporaries formulated this theorem for just one important class of expressions, namely, polynomials. A computational proof called for a solution in the form of a (numerical) expression, and all attempts at such proofs from the time before Gauss must be regarded as failures. Gauss himself used the presumably obvious assertion, found already in Leibniz, that two "continuous" curves must have a point of intersection if one of them has points on both "sides" of the other. (Gauss used this assertion in his dissertation of 1799, where, like some of his predecessors, he wanted to prove the existence of a complex root of a complex polynomial.)

Proofs that we can regard as valid were first given by Bolzano (1817) and by Cauchy (1821). Both discarded formulas and resorted to concepts. They proved the intermediate value theorem not just for real polynomials but for continuous functions of a real variable defined in the modern sense.

Both noted that the proof requires the use of some completeness property, i.e., the existence of the limit, or of the least upper bound, for a monotonically increasing bounded sequence of real numbers. The difference between the two was that while Bolzano tried stubbornly to justify this property conceptually (in later work published only in the 20th century), Cauchy could almost read off this lemma, as well as others, from the decimal representation of the real numbers.

The key objects in Cauchy's textbooks of 1821 and 1823 were continuous functions and infinite series of such functions, in particular power series. In addition to other results, Cauchy proved the existence of the definite integral of a continuous function.

Until Riemann's time, Bolzano was ignored by mathematicians. As for Cauchy, hardly any attention was paid to his thinking in concepts, the more so since he seldom achieved in his later extensive works the same conceptual clarity that marked his early textbooks. Mathematicians exploited the algorithmically usable parts of his writings, such as his convergence tests, but his convergence criterion and his theorems on continuous functions were either ignored or misinterpreted, as was, for example, his theorem on the continuity of the sum of a convergent series of continuous functions. These were, to use Freudenthal's phrase, premature discoveries. Riemann avoided all reference to Cauchy's real analysis.

Dedekind claimed repeatedly that no one before him had proved that $\sqrt{2} \cdot \sqrt{3} = \sqrt{6}$. Cauchy could have included this as a simple exercise in the section of his *Cours d'analyse* devoted to fundamental concepts. We set down a proof in Cauchy's style in order to provide another example to be used for the clarification of the conceptual dislocations that occurred in the 19th century. The functions x^2, $\sqrt{x}$ $(x > 0)$, and $f(x, y) = xy$ are continuous. $\sqrt{2}$ and $\sqrt{3}$ are given by their respective decimal expansions, i.e., by rational sequences $a_n \to \sqrt{2}$, $b_n \to \sqrt{3}$. Because of the continuity of x^2 we have $a_n^2 \to 2$, $b_n^2 \to 3$, and because of the continuity of $f(x, y) = xy$ this implies that $(a_n b_n)^2 = a_n^2 b_n^2 \to 2 \cdot 3 = 6$. Because of the continuity of $\sqrt{x}$ we have $a_n b_n \to \sqrt{6}$ as well as $a_n b_n \to \sqrt{2} \cdot \sqrt{3}$. It follows that $\sqrt{2} \cdot \sqrt{3} = \sqrt{6}$.

From Dedekind's viewpoint this was not an acceptable proof because the argument used continuous functions, and, for him, such functions were alien to the arithmetic of real numbers. According to Dedekind, first one has to master the real numbers and only then can one introduce functions of a real variable. For Cauchy, however, the continuous functions, i.e., the variables, were the primary objects of analysis and the numbers were merely particular values that could be taken on by the variables. In addition to the real numbers, the *valeurs particulières* included $\pm\infty$ as well as the infinitely small magnitudes and the infinitely large numbers. According to Cauchy, the two latter kinds of values were actually deduced from the concept of a variable: an infinitely small magnitude was a variable with limit 0 and an infinitely large (natural) number was a (natural) number variable with limit ∞.

The change from Cauchy's approach to Dedekind's approach was already imminent in Riemann's view of functions as mappings, a view which gave primacy to the preimage and image (sets), notions implicit in the idea of a mapping. The mathematicians of the second half of the 19th century had no appreciation of Cauchy's approach. Cauchy belonged to an old tradition closely tied to physics, a tradition that went back to Leibniz. For Leibniz too magnitudes such as velocity were variables.

4.5 The ontology and methodology of mathematics after Riemann

Today the term "foundations of mathematics" is taken to mean, above all, set theory and mathematical logic, in the sense of Part 5, "Logics, Set Theory, and the Foundations of Mathematics, " in Grattan-Guinness 1994. If one adopts Freudenthal's suggestion and tries to understand Riemann in the role of a philosopher, then one must investigate the philosophical influences that originated with him and determine what discussions may have been triggered by his work. We will do this, but will limit ourselves to the second half of the 19th century and discuss Riemann's direct influence on Dedekind and Cantor and his indirect influence on Hilbert. For contrast, we will also discuss the "antimetaphysical" attitude at Berlin.

4.5.1 The primacy of number in the case of Dedekind

The 22-year-old Richard Dedekind presented his habilitation lecture at Göttingen on 30 June 1854, a few weeks after Riemann. Present were four professors of the Philosophical Faculty, including Gauss and Weber. Later Dedekind mentioned that Gauss had approved the aim of the lecture, and the manuscript had been carefully stored. Emmy Noether published it from the *Nachlass* in 1932 (Dedekind, *Werke* III, 428–438). Already here Dedekind sketched an approach that contrasted with Riemann's. The lecture consisted of the extensions of the number system, beginning with the natural numbers, and included the remark that the "permanence" of the laws of computation was decisive in this context: "The *law* teaches us here too how to interpret the *concept* for its greatest effectiveness" ("Das *Gesetz* lehrt auch hier, wie man den *Begriff* auffassen soll, auf daß er am wirksamsten werde") (p. 436, italics in the original). Dedekind expressed his dissatisfaction with the existing treatments of the irrational and complex numbers. He was to come back, persistently and repeatedly, to the problems dealt with in this lecture and he retained his fundamental position throughout his life.

For Riemann the primary concept was that of continuous magnitudes. For Dedekind the primary concept was that of number, which he regarded "as completely independent of the notions or intuitions of space and time" ("für gänzlich unabhängig von den Vorstellungen oder Anschauungen des Raumes und der Zeit"), "as a direct product of the pure laws of thought" ("für einen unmittelbaren Ausfluß der reinen Denkgesetze") (from the introduction to Dedekind 1888). For him it was "something obvious and not in the least new that every theorem of algebra and higher analysis, however remote, can be expressed as a theorem about the natural numbers, an assertion which I also heard time and again from the mouth of Dirichlet." ("etwas Selbstverständliches und durchaus nichts Neues, daß jeder auch noch so fern liegende Satz der Algebra und höheren Analysis sich als ein Satz über die natürlichen Zahlen aussprechen läßt, eine Behauptung, die ich auch wiederholt aus dem Munde von Dirichlet gehört habe.")

Like Riemann, Dedekind based his views, which were fundamentally different from those of his friend, on the authority of his two teachers, Gauss and Dirichlet!

While such a transcription of all of mathematics in terms of propositions about natural numbers was possible in principle, it was not meant to be actually undertaken in every case; it really pertained to the "creation and introduction of new concepts" (die "Schöpfung und Einführung neuer Begriffe"). For Dedekind the natural numbers too were not givens but "free creations of the human spirit" ("freie Schöpfungen des menschlichen Geistes") (Dedekind 1888, Introduction). On 24 January 1888 he wrote to H. Weber: "We are a divine species and undoubtedly possess creative power not only in material things (railroads, telegraphs) but especially in spiritual things" ("Wir sind göttlichen Geschlechtes und besitzen ohne jeden Zweifel schöpferische Kraft nicht bloß in materiellen Dingen (Eisenbahnen, Telegraphen), sondern ganz besonders in geistigen Dingen") (*Werke* III, 489). Dedekind jotted down for himself the following remark (memo in the *Nachlass*; see H. Mehrtens in Harborth 1982, 19): "Of all the aids created until now by the human spirit for the alleviation of its life, i.e., of the work that thinking consists in, none is as rich in consequences and so inextricably connected with man's inner nature as the concept of *number* ... even if he does not feel it, every thinking person [is] a number-person, an arithmetician." ("Von allen Hilfsmitteln, welche der menschliche Geist zur Erleichterung seines Lebens, d.h. der Arbeit, in welcher das Denken besteht, bis jetzt erschaffen hat, ist keines so folgenreich und so untrennbar mit seiner innersten Natur verbunden, wie der Begriff der *Zahl* ... jeder denkende Mensch [ist], auch wenn er dies nicht deutlich fühlt, ein Zahlen-Mensch, ein Arithmetiker.")

The difficult problem, after the relatively easy construction of the rational numbers out of the natural numbers, was that of the creation of the irrational numbers. Dedekind attached importance to the fact that the idea for it came to him on 24 November 1858, when he was about to teach for the first time an introductory course on the differential calculus. This construction was published only in 1872, shortly after the appearance of his theory of algebraic integers and of his justification of ideals in a supplement to the second edition of Dirichlet's lectures on number theory, edited by him and published in 1871. His creative move was always the same: he would begin with given well-determined mathematical objects and go over to their system, i.e., he would combine them into a set. Certain subsets of this system would provide him with new well-determined objects which he would combine into a new system. Thus the Dedekind cuts in the system of rational numbers yielded the real numbers, certain subsets in systems of algebraic numbers yielded the ideals, namely the "ideal numbers" introduced earlier formally by Kummer, and (later) the algebraic function fields yielded the Riemann surfaces (see Section 1.3.8).

The question of justification of the natural and the real numbers, first posed by Dedekind, seems to us today so self-suggestive, and his answers so obvious, that it is only with difficulty that we can imagine how bold these accomplishments were at the time. The booklet in which, in 1872, he provided his justification of the real numbers disregarded the philosophical traditions in which Riemann was still enmeshed. In Section 4.4.3 we learned about Riemann's notes on antinomies in which the thesis of the discrete was irreconcilably opposed by the antithesis of the continuous. At least in mathematics, Dedekind offered a synthesis and showed that the familiar real numbers could be created out of his attempted synthesis.

Assume the rationals. Just as in philosophy, and just as in Riemann's case, the linear continuum served as a limiting notion, and Dedekind cautiously refrained from its arithmetization. But it suggested an analogy. While it is agreed that there are points in the linear continuum L, it is not postulated that this linear continuum is *equal* to the set of these points. However, "If p is a definite point in L, then all points in L fall into two classes, P_1 and P_2, each of which contains infinitely many individuals; the first class P_1 comprises all points p_1 that lie to the left of p, and the second class P_2 comprises all points p_2 that lie to the right of p; the point p itself may be assigned at pleasure to the first or second class" (Dedekind 1872, 8) ("Ist p ein bestimmter Punkt in L, so zerfallen alle Punkte in L in zwei Klassen, P_1, P_2, deren jede unendlich viele Individuen enthält; die erste Klasse P_1 umfaßt alle Punkte p_1 welche links von p liegen, und die zweite Klasse P_2 umfaßt alle Punkte p_2 welche rechts von p liegen; der Punkt p selbst kann nach Belieben der ersten oder der zweiten

Klasse zugeteilt werden"). The rational numbers with their usual ordering have the same property, but they lack the property of the linear continuum L that every division into classes, every "cut," is generated by an element. Dedekind saw the essence of the continuity of the straight line precisely "in the following principle: if all points of the straight line fall into two classes such that every point of the first class lies to the left of every point of the second class, then there exists one and only one point that produces this division of all points into two classes, this severing of the straight line into two portions" ("in dem folgenden Prinzip: Zerfallen alle Punkte der Geraden in zwei Klassen von der Art, daß jeder Punkt der ersten Klasse links von jedem Punkte der zweiten Klasse liegt, so existiert ein und nur ein Punkt, welcher diese Einteilung aller Punkte in zwei Klassen, diese Zerschneidung der Geraden in zwei Stücke hervorbringt") (p. 10).

In this sense, the linear continuum, as a limiting notion, is thought of as complete. The rational numbers lack this property of completeness. However, by analogy with the notion of the linear continuum, Dedekind proceeded as follows: "Thus whenever we have a cut not due to any rational number, we *create* a new, an *irrational* number, which we regard as completely defined by this cut." ("Jedesmal nun, wenn ein Schnitt vorliegt, welcher durch keine rationale Zahl hervorgebracht wird, so *erschaffen* wir eine neue, eine *irrationale* Zahl, welche wir als durch diesen Schnitt vollständig definiert ansehen.")

We can easily verify that the real numbers given earlier by infinite decimals correspond exactly to these cuts. But Dedekind gave a definition that was independent of particular representations. He *created* the real numbers. When combined into a set (in Dedekind's parlance a "domain" ("Gebiet")), they exhibited the properties of the continuous linear continuum that are essential for mathematics. He created the real numbers as a *synthetic* continuum, a substitute for the arithmetically inaccessible linear continuum. Dedekind showed immediately that the newly created domain of real numbers has the required property of completeness: here to every cut there corresponds exactly one number.

If we reconsider Riemann's deliberations on continuous manifolds discussed in Section 4.4, then we must agree with Dedekind that the argument contains a gap: at first the approach is purely conceptual, but then, to make possible an analytic treatment, there is a sudden introduction of real parameters or coordinates. Dedekind remarked (1872, 4): "It is often said that the differential calculus deals with continuous magnitudes, yet nowhere is this continuity explained ... " ("Man sagt so häufig, die Differentialrechnung beschäftige sich mit den stetigen Größen, und doch wird nirgends eine Erklärung von dieser Stetigkeit gegeben ... ")

As for the latter issue, we would, of course, have to point out to Dedekind that for a long time there had existed good textbooks of the differential calculus that got along without using the undefined term of a continuous magnitude. Just as we did at the end of Section 4.4.4, we again mention Cauchy, who based his development on the decimal representation of the real numbers. Given this representation, the theorem on the convergence of a monotonically increasing bounded sequence is easy to see, so that Dedekind would not have needed to "resort to ... geometric evidence" (zu "geometrischen Evidenzen ... Zuflucht nehmen") in his lectures, an argument he used as the motivation for his deliberations. Of course, Dedekind had no use for proofs that relied on special representations. In the case of ideal theory he also "worked for a great many years," but he published the theory only after he was able to set it down in a manner independent of a particular representation (Introduction to Dedekind 1871).

Dedekind was invariably concerned with the numbers themselves and not with their particular representations. We will find it necessary to return to this point in Section 4.5.3.

Dedekind emphasized that a theory had to be developed so that it allowed one to recognize "from the beginning the character of invariance" ("von vornherein den Charakter der Invarianz"), which in fact is part and parcel of a concept. By this he also meant independence of special representations. Actually, proving that analysis is independent of the special representation of the real numbers as expressions to the base 10 would require some deliberation. It is possible that Dedekind's reluctance to follow up Riemann's geometry was due to the fact that one could not eliminate from it the representation that relied on special coordinate systems (see Section 3.1.9).

We must mention that Dedekind's synthetic continuum of the real numbers nevertheless exhibited certain properties that were foreign to the philosophical concept of a continuum. Given the antinomial relation of the discrete and the continuous, this could hardly have been otherwise. In the continuum of the philosophers the division points exist potentially and every subcontinuum is again divisible. On the other hand, the synthetic continuum $\mathbb{R}$ is identified with the set of its elements. That $\mathbb{R}$ was designated as a continuum, and that, in particular, the question of the cardinality of its subsets was called the "continuum hypothesis" is not the fault of either Dedekind or Cantor, both of whom knew the philosophical tradition very well.

4.5.2 *From arithmetization to axiomatization: Hilbert 1897/1899*

David Hilbert (1862–1943) learned about Dedekind's ideas from Heinrich Weber when he was still at the university of his home town Königsberg. In 1895 he became a full professor at Göttingen. The scope and depth of his creative work made Hilbert the dominant figure of his time. As pointed out in this book on a number of occasions, he also brought about a deeper understanding of Riemann's work.

At the yearly meeting of the Deutsche Mathematiker-Vereinigung (DMV) in Munich in 1893 Hilbert gave a talk titled "On the decomposition of the ideals of a number field into prime ideals" ("Über die Zerlegung der Ideale eines Zahlkörpers in Primideale"). This was published in *Math. Annalen* 44, (1894) 1–8, and is a masterpiece of "modern" algebra. Here Hilbert acknowledged the contributions of Kronecker but used Dedekind's conceptions as a matter of course. The same can be said in connection with his famous "Zahlbericht," commissioned by the DMV. It appeared in 1897 as "Die Theorie der algebraischen Zahlkörper" ("The theory of algebraic number fields") in *Jahresbericht der DMV* 4, 175–546. Our quotations are based on *Ges. Abh.* I, 63–363. We are primarily interested in Hilbert's views on ontology and methodology in the last years of the 19th century and will not attempt to discuss the rest of the Zahlbericht.

While Riemann made no contribution to the subject proper of the Zahlbericht his name is mentioned twice, once in an offhand way, in connection with his "profound investigations" of the distribution of primes (p. 65), and a second time in the following remark (mentioned earlier in Section 4.3), which is a terse but appreciative characterization of his approach:

"I have tried to avoid the huge computational apparatus of Kummer, so that here [i.e., in number theory] too Riemann's principle should be realized, according to which proofs should be impelled by thought alone and not by computing." ("Ich habe versucht, den grossen rechnerischen Apparat von Kummer zu vermeiden, damit auch hier der Grundsatz von Riemann verwirklicht würde, demzufolge man die Beweise nicht durch Rechnung, sondern lediglich durch Gedanken zwingen soll.")

This amounts to an acceptance of Riemann's methodology, but in ontology his model was Dedekind. At that time, Hilbert failed to appreciate Riemann's qualitative and conceptual approach in complex analysis and in geometry.

Hilbert largely accepted the arithmetization of mathematics and wrote "that, if I am not mistaken, the whole modern development of mathematics takes place, first and foremost, *under the banner of number*: Dedekind's and Weierstrass' definitions of the fundamental arithmetical concepts and Cantor's general number formations lead to an *arithmetization of the theory of functions* [i.e., of analysis] and help to realize the principle that in the theory of functions a fact is regarded as proved when, in the last instance, it has been reduced to relations involving rational integers. The *arithmetization of geometry* is the result of the modern investigations of non-Euclidean geometry, which focus on its rigorous logical construction and on the most direct and completely flawless introduction of number into geometry" ("daß wenn ich nicht irre, überhaupt die moderne Entwickelung der reinen Mathematik vornehmlich *unter dem Zeichen der Zahl* geschieht: Dedekinds und Weierstrass' Definitionen der arithmetischen Grundbegriffe und Cantors allgemeine Zahlgebilde führen zu einer *Arithmetisierung der Funktionentheorie* [d.h. Analysis] und dienen zur Durchführung des Prinzips, daß auch in der Funktionentheorie eine Tatsache erst dann als bewiesen gilt, wenn sie in letzter Instanz auf Beziehungen für ganze rationale Zahlen zurückgeführt worden ist. Die *Arithmetisierung der Geometrie* vollzieht sich durch die modernen Untersuchungen über Nicht-Euklidische Geometrie, in denen es sich um einen streng logischen Aufbau derselben und um die möglichst direkte und völlig einwandfreie Einführung der Zahl in die Geometrie handelt") (p. 66; italics in the original).

The term "non-Euclidean," as used by Hilbert, includes geometries other than hyperbolic geometry but excludes Riemannian geometry. We know what problem he was concerned with when he was writing the Zahlbericht. There is his letter to Felix Klein, dated 14 August 1894, which Klein published in *Mathematische Annalen* 46 and which has appeared as Supplement I in various editions of *Grundlagen der Geometrie*. Hilbert generalized the model of hyperbolic geometry by starting with an arbitrary convex solid K in space, or a convex region in the plane, and by defining the distance between two interior points A and B as the logarithm of the cross ratio of A, B, and the two points of intersection of the straight line AB with the boundary of K. He showed that the straight line is the shortest line joining any two points of K. Obviously, the parallel axiom does not hold in any of these geometries.

More important, Hilbert provided an axiomatic characterization of these geometries; in fact, his letter opened with this characterization. He took points, straight lines, and planes as elements and set down axioms of connection, axioms of betweenness for points on a straight line, and a continuity axiom. Then he stated the following theorem: "With every point we can associate three finite numbers x, y, z, and with every plane a linear relation between the three numbers x, y, z, such that all the points for which the three numbers x, y, z satisfy the linear relation lie in the plane in question ... " ("Jedem Punkte kann man drei endliche reelle Zahlen x, y, z und jeder Ebene eine lineare Relation zwischen diesen drei Zahlen x, y, z zuordnen, derart, daß alle Punkte, für welche die drei Zahlen x, y, z die lineare Relation erfüllen, in der betreffenden Ebene liegen ... ") Then he construed the x, y, z as coordinates in $\mathbb{R}^3$ and obtained the result: "Thus our original space has been mapped onto the interior of a nowhere-concave solid in Euclidean space." ("Unser ursprünglicher Raum ist mithin auf das Innere eines nirgends konkaven Körpers des Euklidischen Raumes abgebildet.")

Thus the program of arithmetization of geometry addressed in the Zahlbericht can be summarized as follows: one is to set down axioms for points, straight lines, and planes that yield a real coordinate representation for the geometry they determine. Our quotations come from the introduction to the Zahlbericht completed by Hilbert on 10 April 1897.

During the next two years Hilbert's thinking underwent a fundamental change, which is reflected in his memorial address on the foundations of geometry on the occasion of the unveiling of the Gauss-Weber monument at Göttingen. This marked the beginning, both ontologically and methodologically, of 20th-century mathematics. The *Grundlagen der Geometrie*, repeatedly reprinted and provided with new supplements, was to become a classic bestseller for a whole century. It established the modern axiomatic method and made sets the exclusive objects of mathematics. (Initially, Hilbert, like Dedekind, spoke of "systems" ("Systeme").) The "arithmetization" of geometry, the introduction of real numbers as coordinates, became a secondary theme. The *Jahresbericht der DMV* 8 for 1900 featured Hilbert's paper "Über den Zahlbegriff" ("On the concept of number") in which the (real) numbers were investigated axiomatically. For Hilbert, numbers had lost their special ontological role. The paper became Supplement VI in the *Grundlagen der Geometrie*.

The difference between the discrete and the continuous, emphasized by Riemann, seems blurred in Hilbert's axiomatic approach. In connection with his axiomatization of the real numbers in "Über den Zahlbegriff" Hilbert said explicitly that his two continuity axioms contained "no statement about the concept of convergence or about the existence of a limit" ("keine Aussage über den Begriff der Konvergenz oder über die Existenz der Grenze" enthalten). Hilbert characterized $\mathbb{R}$ (of course, up to isomorphism) as the largest Archimedean-ordered field. For the concept of a manifold, as required for the Riemann surfaces of complex analysis and the spaces of differential geometry, one also needs set-theoretic topological conceptions. The difference between discrete and topological structures is that in the latter one singles out sets of (say, open) subsets. In every case, however, the "manifold" underlying the structure is regarded as a given set of individuals. The set of distinguished subsets plays the role of a substitute for the old concept of a continuum.

Beginning with Hilbert, the transition from arithmetization to axiomatization was not limited to geometry. Under his influence this process began in analysis as well. In this connection we go back to the fundamental consideration in Chapter 2, especially Section 2.1.3, as well as to the failure of Riemann's program to find a conceptual characterization for the representability of a function by means of a trigonometric series.

In 2.4.2 we gave a perfunctory account of how functional analysis surmounted the arithmetization of the function concept. We can use Hilbert space as a prototype for this development. Consider, say, the space $L^2[0, 1]$, which we can describe as follows: start with a set of functions f, defined on $0 \leq x \leq 1$, for which $\int f$ and $\int f^2$ exist. We could take as such the set of continuous functions, or, more simply, the set of bounded step functions, or the polynomials, or the trigonometric polynomials with period 1. If we take as norm the square root of $\int f^2$ and complete the resulting metric space, then we obtain in each case $L^2[0, 1]$. Initially people were interested in interpreting the elements of this space as functions, and such an interpretation, in the form of Lebesgue-square-integrable functions, was provided in 1907 by the Riesz-Fischer theorem. However, this interpretation is hardly of consequence either for analysis or for physical applications. One deals with the axiomatically described Hilbert space and one is glad that it contains such useful functions as the piecewise continuous ones.

Today, with the advantage of hindsight, we can see that the reflections on the concept of the integral (Riemann, Lebesgue) are just as superfluous as Riemann's program of real-valued representations of functions by (trigonometric) series. The integrability of the functions in the initial set is obvious, and one can proceed according to one's taste: for continuous functions, preferred by the analyst, one can use Cauchy's concept of an integral; for step functions the integral is geometrically obvious; and for polynomials we can make do with algebra. The linear functional $\int f$ can be extended to the space L^2, and in each case the special integrable functions form a dense subset. This is how every mathematical grownup thinks today. Nevertheless, we continue to burden our students with the historically conditioned detour.

The spaces of functional analysis are examples of continuous manifolds in the sense of Riemann's habilitation lecture, in which he came close to functional manifolds. Twentieth-century analysis showed the fruitfulness of this conception and the relative fruitlessness of the point-by-point discussion of the functional representations in the habilitation paper, which Riemann himself failed to follow up.

4.5.3 The role of Georg Cantor

The reader may wonder why we have failed so far to mention the name of Georg Cantor (1845–1918), celebrated as the inventor of set theory. Actually, the two previous sections should have made it clear that, from the viewpoint of our subject, Cantor is far less important than are Dedekind and Hilbert. Besides, Dedekind developed the set-based *mode of thought* before Cantor. Specifically, he formed subsystems with definite properties (cuts, ideals) of concretely given systems (the rational numbers, the algebraic number and function fields) and used operations on sets (union, intersection). He looked at sets primarily from the viewpoint of their utility. In his paper of 1872 he emphasized that he had been in possession of his theory for a long time, but had refrained from its publication because the thing was "of so little use" ("so wenig fruchtbar"), and that what induced him to publish his reflections was the publication of theories of irrational numbers, first by Heine and then by Cantor. Hilbert started with unstructured sets without properties and imposed structures on them. In the relation between Dedekind and Cantor, which was friendly for a time, the latter certainly received more encouragement and stimulation than he gave. Hilbert accorded Cantor public recognition. But in his case too, thinking in terms of sets played a more important role than did the concrete contents of Cantor's *transfinite* set theory.

There is a twofold connection between Cantor and Riemann. Cantor came to his theory of point manifolds by considering open questions in Riemann's habilitation paper (see Section 2.5), and, like Riemann, he took a deep interest in philosophy, which strongly influenced his mathematical work.

After about 1960, Cantor's name became popular outside mathematics in connection with the inclusion in the school curriculum of so-called set theory. Strictly speaking, this development owed more to Dedekind, Hilbert, and Bourbaki than it did to Cantor. This historical misunderstanding did not prevent it from furthering interest in the history of mathematics. We can refer the reader to excellent books on Cantor by Meschkowski, Dauben 1979, and Purkert and Ilgauds 1987. A possible reason for the general interest in Cantor is that it is easy to give laymen a clear appreciation of a naive version of his philosophy of mathematics.

Cantor was anything but naive. He grappled with the philosophical traditions and was fully aware that his acceptance of the actual existence of the infinite broke radically with many of those traditions, in particular the Aristotelian one. Meschkowski regarded him as "perhaps the last great representative of Platonic thinking in mathematics" (den "wohl letzten großen Vertreter des platonischen Denkens in der Mathematik"). This is certainly too sweeping a judgment (Purkert-Ilgauds 1987, 105 ff). On the other hand, there is no doubt that, for Cantor, the objects of mathematics, the sets with all the properties ascribed to them, existed in a world of their own, a quasi-Platonic heaven of ideas. In a note that probably dates from 1913 Cantor put it thus: "As I conceive of it, metaphysics is the science of *the existent*, or, what amounts to the same thing, of what is *here*, i.e., exists, and thus of the world as it is in itself, not as it appears to us. All that we perceive with the senses and imagine with our abstract thinking is *the nonexistent*, and therefore at best a trace of the existent in itself" ("Metaphysik ist, wie ich sie auffasse, die Lehre vom *Seienden*, oder was dasselbe bedeutet von dem was *da* ist, d.h. existiert, also von der Welt wie sie an sich ist, nicht wie sie uns erscheint. Alles was wir mit den Sinnen wahrnehmen und mit unserem abstracten Denken uns vorstellen ist *Nichtseiendes* und damit höchstens eine Spur des an sich Seienden") (quoted after Purkert and Ilgauds 1987, 198). Set theory made statements about the world in itself, about the true being, and for him it belonged "completely to metaphysics," and thus, according to the quoted note, to ontology. This being so, he, unlike Dedekind, never spoke about the creation of numbers or of any mathematical things. For him there was a transient reality of things that corresponded to the immanent reality in the heaven of ideas and was determined by it. We can discover this transient reality but we need not create it.

It is very remarkable that Cantor did not grant to other mathematical constructs that which he ascribed to his set theory. A letter to Veronese, dated 17 November 1890, first published in Purkert-Ilgauds 1987, 202, is revealing: "The talk in my arithmetical investigations of the finite and transfinite is solely about probing the real in what exists in nature and never about *hypotheses*. On the other hand, you, like the metageometers Riemann, Helmholtz and company, think that you can set up hypotheses *in arithmetic as well*, which is completely impossible ... Just as it is impossible to set up fundamental laws in the arithmetic of finite quantities other than those known from time immemorial for the numbers 1,2,3,..., so too it is impossible to deviate from the fundamental arithmetical truths in the realm of the transcendental." ("Von *Hypothesen* ist in meinen arithmetischen Untersuchungen über das Endliche und Transfinite überall gar keine Rede, sondern nur von der Ergründung des Realen in der Natur Vorhandenen. - Sie hingegen glauben nach Art der Metageometer Riemann, Helmholtz und Genossen, Hypothesen *auch in der Arithmetik* aufstellen zu können, was ganz unmöglich ist ... So wenig sich in der Arithmetik der endlichen Anzahlen andere Grundgesetze aufstellen lassen, als die seit Alters her an den Zahlen 1,2,3, ... erkannten, ebensowenig ist eine Abweichung von den arithmetischen Grundwahrheiten im Gebiete des Transzendenten möglich.")

37: Georg Cantor

The allusion to Riemann as a metageometer is too vague to comment on. What matters here is that Cantor regarded his transfinite arithmetic as something real and existing in nature. What exactly had Veronese done? At the time, he was finishing his *Fondamenti di geometria*, which was to be published in 1891, a book in which non-Archimedean number systems appeared as coordinates in the geometry. His student Levi-Cività soon made this precise and classified non-Archimedean fields in papers which were extended by Hans Hahn in his dissertation *Über die nichtarchimedischen Größensysteme*, Vienna 1907, and thus provided the impulse for the theory of order structures in modern algebra. In the meantime Hilbert too dealt with non-Archimedean fields in his *Grundlagen der Geometrie*. Of course, Veronese did not talk about *quantities*. He dealt with *field elements* which were greater than finite natural numbers, something that fitted in with the turn to modern algebra initiated by Dedekind and continued by Hilbert. This Cantor closed his eyes to. It turns out that he did not think of set theory in the way we do, who follow Dedekind and Hilbert.

Moreover, Cantor thought that he could prove that the concept of infinitely small magnitudes is self-contradictory. The reference to rational function fields failed to undeceive him.

For Veronese and Levi-Cività see Laugwitz 1986, 221/222. For Cantor and Veronese see Dauben 1979, 233 ff.

Cantor disagreed with Herbart's philosophy of the infinite and poked fun at it, for he saw in it a restriction to the potentially infinite (Purkert-Ilgauds 1987, 115/116). What is more, he believed that he could deduce the existence of the actual infinite from that of the potential infinite: "For such a variable magnitude to be usable in a mathematical deliberation, the 'region' of its variability must, strictly speaking, be known already from a definition; but this 'region' cannot itself again be something variable ... which means that this 'region' is a definite actually infinite set of values" ("Damit eine solche veränderliche Größe in einer mathematischen Betrachtung verwertbar sei, muß streng genommen das 'Gebiet' ihrer Veränderlichkeit durch eine Definition vorher bekannt sein; dieses 'Gebiet' kann aber nicht selbst wieder etwas Veränderliches sein ... also ist dieses 'Gebiet' eine bestimmte aktual-unendliche Wertmenge") (Purkert and Ilgauds 1987, 115). Thus Cantor puts the *scope* of a concept ahead of its *content*, and this is precisely the crucial point. Herbart and Riemann belonged to a tradition that preferred to think in concepts and did not feel compelled to actually construct the "set" of all objects falling under a particular concept.

4.5.4 *The Berlin tradition*

On 1 July 1852, Dirichlet delivered before the Berlin Academy a memorial address for his colleague and friend Carl Gustav Jacobi, who had passed away on 18 February 1851. He said: "While the ever more pronounced tendency in the newer analysis is to replace computations by ideas, there are nevertheless certain areas in which computing retains its rights. Through his technical mastery, Jacobi, who so substantially promoted that tendency, made admirable contributions to this area as well" ("Wenn es die immer mehr hervortretende Tendenz in der neueren Analysis ist, Gedanken an die Stelle der Rechnung zu setzen, so gibt es doch gewisse Gebiete, in denen die Rechnung ihr Recht behält. Jacobi, der jene Tendenz so wesentlich gefördert hat, leistete vermöge seiner Meisterschaft in der Technik auch in diesem Gebiete Bewunderungswürdiges") (in Jacobi, *Ges. Werke* I, 21).

Dirichlet saw his own tendency as in the general direction of the development of analysis and acknowledged that Jacobi not only was a master of calculation but had also recognized the newer trend. During his Berlin years, the student

Riemann had been confirmed in his own tendency toward conceptual thinking by Dirichlet rather than by Jacobi.

The decades after Dirichlet's departure in 1855 brought a strengthening of the old algorithmic tradition in Berlin. First, Ernst-Eduard Kummer (1810–1893) was called from Breslau to the capital; then Leopold Kronecker (1823–1891), his gymnasium student and subsequently his lifelong friend, became a member of the Academy and, from 1861 on, lectured regularly at the university; finally, Karl Weierstrass (1815–1897) was nominated to the Academy in 1856 and immediately began to teach at the university, becoming a full professor only in 1864. This triumvirate of outstanding mathematicians instituted for the first time at a German university a program in advanced mathematics. Many students came, and Berlin mathematics gained tremendous influence in the country and abroad.

Notwithstanding individual differences, the three influential mathematicians had one thing in common: their view of mathematics was much closer to that of Jacobi and Eisenstein than to that of Dirichlet. While they acknowledged the results of Riemann and Dedekind, they viewed their methods with suspicion. To repeat Hilbert's dictum, they preferred to "impel proofs by means of computations." This stance was accompanied by an attitude towards changes in the ontology of mathematics that was at best temporizing and at worst dismissive.

The mathematical atmosphere at Berlin around 1860 was described by Leo Koenigsberger (1837–1921) in his book *Mein Leben* (*My life*) (Heidelberg 1919; esp. pp. 21–57). He began his studies at Berlin in 1857, advanced rapidly under Weierstrass' tutelage, and was duly offered an associate professorship at Greifswald. On p. 54 we read: "As well, all of us younger mathematicians had at the time the feeling that Riemann's views and methods no longer belonged to the rigorous mathematics represented by Euler, Lagrange, Gauss, Jacobi, Dirichlet and their like ... " ("Auch wir jüngeren Mathematiker hatten damals sämtlich das Gefühl, als ob die Riemannschen Anschaungen und Methoden nicht mehr der strengen Mathematik der Euler, Lagrange, Gauß, Jacobi, und Dirichlet u.a. angehörten ... "). A detailed and documented account is found in K. Biermann 1988.

The example of the hypergeometric series, presented in Section 1.3.1, gives us an idea of the relation of Kummer's mathematics to Riemann's. In all areas of his work, analysis, number theory, and finally also geometry, Kummer was the "algorithmist" par excellence. But he made no great fuss about his view of mathematics. It was different with Kronecker, whose dictum that God made the whole numbers and the rest is the work of man is more widely known than his mathematical work. The propaganda of Weierstrass' students made Kronecker appear to have been a contentious enemy of innovation. H. Edwards has

38: Leopold Kronecker

corrected this caricature in substantial publications (e.g., in Rowe-McCleary 1989, I. 67–68).

Kronecker's view is clearly expressed in his letter to Cantor, dated 21 August 1884 (Meschkowski and Nilson 1991, 196): "... that I, immersed very early under Kummer's guidance in philosophical studies, later, like him, realized the uncertainty of all those speculations and sought refuge in the safe haven of real mathematics. What was more natural than that I myself should try to identify its manifestations or its truths, as free as possible of all philosophical conceptions. I have therefore aimed to reduce everything in pure mathematics to the science of whole numbers, and I *believe* that this will succeed without exception. In the meantime this is just my *belief*. But where it has succeeded, I see in it true progress, although — or because — it is a regression to the simplest, even more, however, because it proves that the new conceptions are not *necessary* ...

"That I will state such objections only from time to time is due to the fact that I attach to them at most secondary value. I acknowledge true scientific value — in the field of mathematics — only in concrete mathematical truths, or, to state it more pointedly, 'only in mathematical formulas.' The history of mathematics teaches us that these alone are imperishable. The assorted theories of the foundations of mathematics (such as that of Lagrange) have been blown away by time, but the Lagrange resolvent has remained!"

(" ... dass ich, sehr früh unter Kummers Anleitung in philosophische Studien vertieft, nachher gleich ihm die Unsicherheit aller jener Spekulationen erkannt und mich in den sicheren Hafen der wirklichen Mathematik geflüchtet habe. Was war natürlicher, als dass ich in dieser Mathematik selbst nun mich bemüht habe, ihre Erscheinungen oder ihre Wahrheiten möglichst frei von jeden philosophischen Begriffsbildungen zu erkennen. Ich bin deshalb darauf ausgegangen, Alles in der reinen Mathematik auf die Lehre von den ganzen Zahlen zurückzuführen, und ich *glaube*, dass dies durchweg gelingen wird. Indessen ist dies eben nur mein *Glaube*. Aber wo es gelungen ist, sehe ich daran einen wahren Fortschritt, obwohl — oder weil — es ein Rückschritt zum Einfachsten ist, noch mehr aber deshalb, weil es denn beweist, dass die neuen Begriffsbildungen nicht *nothwendig* sind ...

"Dass ich jene Einwendungen nur gelegentlich machen will, beruht darauf, dass ich denselben nur einen höchst secundären Werth beilege. Einen wahren wissenschaftlichen Werth erkenne ich — auf dem Felde der *Mathematik* — nur in concreten mathematischen Wahrheiten, oder schärfer ausgedrückt, 'nur in mathematischen Formeln'. Diese allein sind, wie die Geschichte der Mathematik zeigt, das Unvergängliche. Die verschiedenen Theorien für die Grundlagen der Mathematik (so die von Lagrange) sind von der Zeit weggeweht, aber die Lagrangesche Resolvente ist geblieben!") (italics in original).

This reads very much like Dedekind's remark of 1888 that every mathematical theorem can be stated as a theorem about whole numbers, and very much like what Dirichlet is supposed to have said. But these two felt free to *form concepts* based on the whole numbers, and this Kronecker rejected as an unnecessary philosophical speculation. To this very day this difference remains the source of an irreconcilable antagonism; we are reminded of Siegel's letter to A. Weil.

Dedekind could afford to continue his work without being affected by Kronecker's ontological and methodological rules (for information about Dedekind's views see H. Edwards, O. Neumann, W. Purkert, " Dedekind's 'colorful remarks' on Kronecker's 'foundations'," *Archive for History of Exact Sciences* 27 (1982), 49–85). But in Cantor's case Kronecker touched a vital nerve.

There was little difference between the views of Kronecker and those of his Berlin colleague Weierstrass, who also felt that representation by means of formulas was the mark of what was of permanent value in mathematics. The Weierstrass lectures of 1886 date from the height of his personal controversy with Kronecker. They have the quality of an intellectual testament, which suggests that Weierstrass thought of them as his last ones. On p. 176 we find the following summary: "At first, the purpose of these lectures was to properly determine the concept of analytic dependence; to this there attached itself the problem of obtaining the analytic forms in which functions with definite properties can be represented ... for the representation of a function is most intimately linked with the investigation of its properties, even though it may be interesting and useful to find properties of the function without paying attention to its representation. The *ultimate* aim is always the representation of a function" ("Der Zweck der Vorlesungen war zunächst, den Begriff der analytischen Abhängigkeit gehörig festzustellen; daran knüpfte sich die Aufgabe, die analytischen Formen zu ermitteln, in denen Funktionen von bestimmten Eigenschaften dargestellt werden können ... denn die Darstellung einer Funktion ist mit der Erforschung ihrer Eigenschaften aufs innigste verknüpft, wenn es auch interessant und nützlich sein mag, Eigenschaften der Funktion aufzufinden, ohne auf ihre Darstellung Rücksicht zu nehmen. Das *letzte* Ziel bildet immer die Darstellung einer Funktion") (Weierstrass 1886, 156; in this connection see Laugwitz 1992).

The main content of the lectures was the famous approximation theorem, discovered a year earlier. Weierstrass viewed it as a representation theorem: every real continuous function on a closed interval is representable as a uniformly convergent series of polynomials with rational coefficients. He called this an *arithmetical expression*. Obviously, this arithmetization of analysis could not satisfy Kronecker. A full century of attempts by intuitionism, constructivism, and operationalism to obtain a development of analysis in Kronecker's sense that could be taught to beginners shows what a Sisyphean task Weierstrass had undertaken. Had he succeeded, the equivalence of the concepts of continuity and arithmetical expression, which Riemann still believed in at the beginning of his doctoral dissertation, would have been established but useless: to prove, say, the intermediate value theorem we need only the concept and not the expression. This was realized already by Bolzano and Cauchy. By relying on the concept of continuity they proved this theorem not just for polynomial expressions. The apparent ontological triumph of formal expressions could not avert their methodological defeat.

4.6 Concluding remarks

We have tried to portray the pace of the historical development in the different areas of Riemann's work and to present clearly his innovative contributions. We have also tried to trace the influences on him as well as the influence his works exercised over others. We will use the available materials to obtain a synopsis of Riemann's view of mathematics as well as of the whole of science. We can say that the concept of a continuous manifold shaped all of his scientific thinking and played for it the role of a supporting foundation.

As a guiding thread for a deeper understanding and for a comprehensive overview we will use the scheme of thesis *(T)* and antithesis *(A)* which Riemann himself set down (W. 518–520; see also Section 4.4.3).

As thesis Riemann has "the finite, the imaginable" ('Endliches, Vorstellbares") and for a corresponding antithesis "the infinite" ("Unendliches"), supplemented not by "the unimaginable" but by "systems of concepts at the boundary of the imaginable" ("Begriffssysteme, welche an der Grenze des Vorstellbaren liegen"). We will try to show that Riemann always tried to view science from the pole of the antithesis. In this basic scientific position he differed from the majority of mathematicians. Since he carried through this program, we may, like Freudenthal, regard him as a philosopher.

Among Riemann's contemporaries it was Kronecker who most clearly represented the opposite position. It was he who would let only the finite, embodied in the natural numbers, stand for "real mathematics" ("wirkliche Mathematik"), and who wanted to keep mathematics free from the uncertainty of philosophical conceptions and speculations. We recall his letter to Cantor, quoted in Section 4.5.4.

Obviously with physics in mind, Riemann first jotted down under *(T)* "Finite time and space elements" ("Endliche Zeit- und Raumelemente") and under *(A)* "The continuous" ("Stetiges"). In physics Riemann opted for *(A)*, continuous manifolds and fields as the physical magnitudes. This was the conceptual framework for his physics. It was not entirely inconceivable, but it required not just the finite, and could be approximated from the direction of the finite only as a limit. He rejected action at a distance, and took as a model of local action not the pressure and impact of particles but differential laws, which describe the interaction of fields at infinitesimally close positions in space. He regarded this view as empirically secure. In its favor was not only the mathematical physics of the Parisian scientists and of his teacher Gauss but also the fact that it could accommodate Newton's law of gravitation: r^{-1} is a fundamental solution of the differential equation for potential.

There is no reference in Riemann's note to mathematics, but we will try, by analogy, to extend his reflections to mathematics. This is a legitimate attempt because for him there was coherence in being as well as in thought, and it is plausible that mathematics, which has demonstrated its fruitfulness in a physics thought through from the ground up, is rich in content independently of physics.

In the setting of mathematics we will consider rather closely the following list of pairs (T_n, A_n) for $n = 1, \ldots, 10$:

(1) finite - infinite
(2) discrete - continuous
(3) formula (term, expression, figure) - concept
(4) calculus (rule-based term transformation) - conceptual proof
(5) number - variable
(6) epsilontics - infinitesimal argumentation
(7) construction - speculation
(8) quantity - quality
(9) rigidity - flexibility
(10) pure mathematics - mathematics in the natural sciences

In this schema the two sides must not be viewed as being of the same kind, and it would not make sense to try to effect a synthesis by introducing an in-between notion or compromise, something like a scale of greens between blue and yellow. Rather, the poles of the antithesis denote regulative principles that provide direction to thought (especially in Riemann's case) when it is not satisfied with the fundamentally accessible poles of the thesis (Kronecker's "real mathematics"). In the philosophical tradition one speaks of potentiality (*potentia*) and reality (*actus*), and this corresponds to the poles of antithesis and thesis. The styles of thought in mathematics differ in the level of influence exerted on them by the poles of the antithesis. Just as the setting of the development of physics is the tension between the concepts of particle and field, so too the history of mathematics shows a permanent state of motion between the poles of thesis and antithesis. *Riemann's work is characterized by the endeavor to come as close as possible to the poles representing the antitheses.*

On the other hand, to the extent to which it was influential and fruitful, mathematics remained close to the poles of the *thesis* from antiquity until the middle of the 17th century. This was also true of Euler as a mathematician, of Lagrange, of Gauss in number theory and non-Euclidean geometry, of Cauchy in algebra, of Jacobi, Kummer, Kronecker, Weierstrass, and Siegel. Those who tried to come closer to the opposite poles were Euler as a physicist, Fourier, Cauchy in analysis, Gauss in differential geometry and physics, and Hermann Weyl. Like Euler, Gauss, and Cauchy, other great mathematicians, such as Newton, Leibniz, Dedekind, Poincaré, and Hilbert, cannot be unambiguously located in this scheme without a more detailed analysis. Moreover, what we are saying must not be taken to mean that those on the thesis side made no use of, say, infinitesimal analysis in number theory; we need only think of Siegel.

The ten pairs in our schema are not sharply separated from one another. The pairs (1) to (4) are quite close, and so are (5) and (6), and (7) to (9). Other similarities, such as those between (2) and (5), are easily recognizable. While commenting, we will not be able to adhere precisely to the order of the pairs.

We begin with the pairs (1) to (4) and with the observation that mathematics was never satisfied with the finite alone. At least as a possibility, i.e., potentially, the infinite turned up in Greek mathematics: one can go on counting and one can extend constructions. When so doing, one need not think of an actually completed totality of natural numbers or of the constructible points on the plane.

The distinction (1) played no role in Riemann's mathematics. As for (2), he paid attention only to the continuous manifolds and not to the discrete ones. We will follow him to the extent of ignoring the finite but not the discrete. Otherwise, we would hardly be in a position to see him in relation to others.

The modern method of dealing mathematically with what Riemann called a continuous manifold has been known as analysis since 1696, when L'Hôpital published the first relevant textbook, the *Analyse des infiniment petits*. Initially, the term denoted problem-solving, but later it acquired its modern meaning. The works of Leibniz, one of the founders of this analysis, contain profound reflections which can serve to clarify our schema. The relevant writings are now available in a bilingual edition: G.W. Leibniz, *Schriften zur Logik und zur philosophischen Grundlegung von Mathematik und Naturwissenschaft* (*Writings on logic and on the philosophical foundation of mathematics and natural science*), Wissenschaftliche Buchgesellschaft, Darmstadt, 1992. Of course, it is advisable to critically check in particular cases the German translation of the publisher H. Herring against the synoptic version of the original. We refer to this edition as L.

The name of Leibniz turns up often in Riemann's writings in reflections on topology or *analysis situs*; Leibniz himself also used the term *geometria situs*, but in the 19th century the term "Geometrie der Lage" ("Geometry of position") denoted projective geometry. In the *Nachlass* volume of 1902, Noether noted (N. 710) that a few sheets of Riemann's historical-literary remarks on Leibniz had been preserved at Göttingen, and one can conclude that Noether found no technical notes. The existence of these remarks warrants the assumption that Riemann had access to Leibniz' writings. Gerhardt's edition of the mathematical writings of Leibniz began to appear in 1849, and important writings such as those on the principle of continuity, including the letter to Varignon of 2 February 1702, were already available in older editions. But E. Neuenschwander, who prepared a list of borrowings from the university library, was unable to give us any data pertaining to Leibniz.

In any case, there is no doubt that Riemann had an affinity with Leibniz. In accordance with a philosophical tradition going back to Aristotle, both thought that the continuum, the continuous manifold, must not be conceived as a collection of individuals or points, and that the continuous takes precedence over the discrete, and that conceptual thinking ranks above the calculus. Leibniz had certainly worked in the discrete, and had developed effective calculi (*Characteristica universalis, the differential calculus*); but justification of these comes from conceptual principles. In Riemann's time calculi flourished to such an extent that he wanted to try to contain them within appropriate bounds and to put conceptual thinking more in the foreground.

Riemann thought that mathematics and physics belong closely together; so did Leibniz. Leibniz viewed his continuity principle as fundamental, not only for mathematics and logic and for the ordering of thought and cognition but also as the fundamental law of the world order (*natura non facit saltus*) — cognitive principles are ontological principles. This was already expressed in the title of his essay of 1687 devoted to the continuity principle (L. 227 ff.): *Principium quoddam generale non in mathematicis tantum sed in physicis utile*, a general principle useful not only in mathematics but also in physics.

Consideration of the pair (6) will concretize our discussion of the pairs (1) to (4). We saw in Section 4.4.3 that Riemann referred to Newton's method of limits. He could have found a clear presentation in Leibniz' letter to Varignon of 2 February 1702 (L. 249 ff.). For Leibniz, the rules of the finite held for the infinite without restrictions. Riemann distanced himself from this position: in the limiting case some of the correlative concepts ceased to be conceivable (W. 520).

Riemann would probably not have accepted the Leibniz trick of regarding rest as a special case of its opposite, as infinitely slight motion, coincidence as an infinitely small distance, and equality as the extreme case of inequality (L. 254–257). True, he occasionally used infinitesimal arguments heuristically as well as for less formal communication, but he regarded them as nonrigorous. Leibniz sympathized fully with mathematicians who, proceeding like Riemann, wanted to show that it is not necessary to make mathematical analysis dependent on metaphysical controversies (L. 251). Epsilontics can be viewed as one of the possibilities, in the sense of Leibniz' arguments, for the elimination of infinitesimals. Following Leibniz, one can look at the rules and formulas of the differential calculus as follows: one can replace infinitely small magnitudes by arbitrarily small finite magnitudes and thus show an opponent who would contradict us that the error in our formulas is always smaller than the error set by him (L. 253). Today we could proceed differently and exploit mathematical possibilities unknown to Riemann. One computes in a calculus with infinitely small numbers in a manner made possible by so-called nonstandard analysis. In the final result one reduces everything modulo the infinitely

39: Gottfried Wilhelm Leibniz

small magnitudes. Behind all this is the concept of a homomorphism, which makes it possible to reduce an infinitely small distance to coincidence in a formally correct way. Of course, one should not think that it is possible to adequately describe the continuum by means of a number concept that makes available, in addition to the real numbers, infinitely small numbers, for here too, just as in the case of $\mathbb{R}$, we are dealing with a *set of numbers*. But what we do get is new models that come closer to (A_6).

Riemann aims to come as close as possible to the poles *(A)* via the "real" mathematics of the poles *(T)*. His statement that in mathematics one must reduce everything to equalities and inequalities contains the germ of the method, first used explicitly by Dedekind, of making do with algebraic and order structures. In fact, in Riemann's mathematics, order relations suffice for the inclusion of topological or convergence structures. Ordered fields that include $\mathbb{R}$ could have been readily accommodated in the schema had they been available. Dedekind's "modern" algebra could have also provided stimuli in this direction, but his interests centered on algebraic field extensions, and he was convinced that his cuts provided the only possibility for constructing a synthetic continuum. The prospect of non-Archimedean structures remained closed.

In Riemann's time, and with his collaboration, there came a long-lasting break with infinitesimals. But initially the continuous variable was retained. This Dedekind rightly objected to. In his habilitation lecture Riemann still spoke of a continuous transition from one mode of determination of a concept to another (W. 273) and then abruptly, and without comment, used real parameters for the continuous transitions. In this connection Weyl observed that "the continuum of the real numbers [was] only a single case, and not a very distinguished one" (dass "das Kontinuum der reellen Zahlen nur ein einzelner, nicht besonders ausgezeichneter Fall" sei), and that the desire to reduce the continuous manifolds to real coordinates was "objectively unjustified but expedient because of the ... calculational convenience of the number continuum" (Weyl 1988, 8–9) ("sachlich unberechtigt aber zweckmäßig wegen der ... kalkulatorischen Bequemlichkeit des Zahlenkontinuums"). But even Weyl made no attempt to deal with infinitesimal geometry by means of infinitesimal numbers.

In the setting of the space of physics Riemann spoke of the immeasurably small and the immeasurably large (W. 284); but he reverted in the next section (W. 285) to the terminology of the infinitely small. Of course, he did so in the sense of his discussion of thesis and antithesis: The

> "empirical concepts in which the spatial metric determinations are grounded, the concepts of a solid and a light ray, [seem] to lose their validity in the infinitely small,"
>
> ("empirische Begriffe, in welchen die räumlichen Massbestimungen gegründet sind, der Begriff des festen Körpers und des Lichtstrahls, [scheinen] im Unendlich kleinen ihre Gültigkeit zu verlieren,")

and it is conceivable that one would have to modify geometry if a simpler explanation of phenomena required it (such attempts are found, e.g., in the work of Wheeler).

Was the abandonment of the infinitely small as a mathematical method a historical accident, perhaps the result of a reluctance to justify these procedures rigorously, or was it a historical necessity? We assert that it was not an unavoidable development, but that precisely in Riemann's case the decision was not just emotional or accidental.

Indeed, his decision can be explained on the basis of his thinking in concepts and of his need to "drive back" the formulas, and is thus again related to the pair (3). In the mathematics of formulas and rule-based transformations it was not important to know what a letter a or x meant, or whether it denoted a real number or an infinitesimal, a constant or a variable. (Besides, the term "constant," which is still in use today, goes back to a time when, in analysis, numbers were regarded as special cases of functions: a variable *may* remain constant.) Cauchy had expressed his opposition to the "generality of algebra" (la "généralité d'algèbre") in which one viewed formulas as valid even in domains for which they were not initially justified (see Section 4.2 of the

Introduction). Nevertheless, he had computed with infinitely small magnitudes as if they were real numbers, had written $f(x+i)$ with real x and infinitely small i, and had therefore been accused of having thought solely of functions given by expressions. Of course, this is irrefutable, for when Cauchy introduced a particular function he necessarily gave an expression for it. His definition that i is a variable with limit 0 was no longer understood; nevertheless people could have continued to work with this definition within the conceptual framework of functions as mappings.

We asserted that the elimination of the infinitely small was not unavoidable. The resumption of infinitesimal mathematics in the form it took around 1960 was conceivable a century earlier. However, one of the preconditions for such a development was the granting of a fundamental role to formulas, and this Riemann was not prepared to do. We will give a brief clarification of this precondition.

The fundamental idea of modern infinitesimal mathematics can be described as follows (see Laugwitz 1986 for a more detailed account). Let $A(n)$ be a formula in which n is a free variable ranging over the natural numbers. Let Ω be an as yet unused symbol which will immediately turn out to be the symbol for an infinitely large number. In accordance with the Leibnizian principle we make the following stipulation: if $A(n)$ is a proposition of classical mathematics that is true for almost all $n \in \mathbb{N}$, then $A(\Omega)$ holds true. Since $A(n) : n > 1000$ is true for almost all $n \in \mathbb{N}$, it follows that $\Omega > 1000$, and it is easy to see that Ω is greater than every classical natural number, and that $\omega = \Omega^{-1}$ is positive but less than every classical positive real number.

Obviously, one must specify precisely what is to be meant by a formula, and Riemann's dislike of expressions could also be explained by the fact that such levels of precision were difficult to achieve in his time. Today we can also admit the symbols of mathematical logic as building blocks for formulas, and it was only in this way that, beginning in 1960, A. Robinson managed to develop a satisfactory infinitesimal mathematics. Such deliberations would have been readily accessible in the middle of the 19th century to, say, members of the Cambridge Analytical Society. A. De Morgan's (1806–1871) book on formal logic appeared in 1847, and G. Boole's (1815-1864) *The Mathematical Analysis of Logic* followed in 1848.

If we use the infinitely small number $\omega = \Omega^{-1}$ to construct expressions $B(\omega)$, then we can also obtain Cauchy's variables with limit 0, such as ω, ω^2, $\sqrt{\omega}$, and $\sin \omega$.

We have seen on a number of occasions that Riemann was not anxious to take up Cauchy's mathematics. He could get along without variables with limit 0. More generally, for him the concept of a variable lost its significance as a fundamental notion of analysis. He distanced himself from (A_5), the concept of a variable, and introduced into analysis a new fundamental notion, that of a mapping. Mappings were determined not by expressions but by conceptual properties such as continuity and differentiability. The elimination of variables implied the elimination of the special variables with limit 0, the infinitely small magnitudes.

The autonomy of the mapping concept was shortlived. It was preempted by set theory, which decreed that a mapping is none other than its graph, and thus a set. The term "mapping" was just condoned. As a result of the arithmetization of analysis, a function $y = f(x)$, which had to be written down completely, including its variables, became the set $\{(x, f(x)); x \in D \subseteq \mathbb{R}\}$. However, the reader will know from his own experience that the notion of a mapping is fruitful.

We said that the notion of a manifold was basic for Riemann. But what was the fate of his paper on trigonometric series? After all, it was the source of the subsequent investigations of extremely discontinuous functions which left precious little of Leibniz and of continuity. And to think that these series came originally from nature, which according to Leibniz makes no jumps!

The infinitesimal mathematics of the 20th century has once more justified the continuity principle in a more refined form. If we sum a trigonometric series not to ∞ but to an infinitely large upper limit, then the sum function g is again uniformly continuous in a more general sense: to every $\epsilon > 0$ there is a $\delta > 0$ such that $|x - \hat{x}| < \delta$ implies $|g(x) - g(\hat{x})| < \epsilon$. Of course, δ can be infinitely small for a finite ϵ. A function can change on an infinitely small interval by a finite amount without making a jump.

This is not the place for an exhaustive treatment of this topic. We will just note that Riemann completed this paper in order to carry out Dirichlet's wish. While writing it he was only too glad to take refuge in his reflections on natural philosophy, which were of incomparably greater importance to him. For all that, his efforts, always oriented towards conceptual thinking, yielded important stimuli even in real analysis. We discussed them in Chapter 2. It is conceivable that, given the prestige of its author's name, this paper may have made analysis follow avoidable detours. After all, people soon admitted sets very different from those which, as in Riemann's case, integrated the mode of determination of a concept into a new whole. The sets separated themselves from the concepts. But Riemann for one did not pursue the matter further.

The paper on primes is associated with the pairs (2) and (5). It demonstrated with particular clarity the preeminence of the continuous over the discrete. It may have been a source of gratification for Riemann to have derived by means of analysis a closed expression characterizing the sequence of primes. This expression was a result and not a starting point, and moreover, the resulting expression was a confirmation of the program in §20 of his dissertation, for it contained only the elementary operations mentioned there.

We now turn to the pairs (7), (8), and (9), which also have methodological implications.

At one time it was thought that scientific status has been achieved when the qualitative has been expressed in mathematical and quantitative terms. Already Kant had expressed himself to this effect. In his *Metaphysische Anfangsgründe der Naturwissenschaft* (*Metaphysical elements of natural science*) of 1786 he said that "in every particular natural science one can find only so much true science as one can find mathematics." (dass "in jeder besonderen Naturlehre nur so viel eigentliche Wissenschaft angetroffen werden könne, als darin Mathematik anzutreffen ist.") Riemann too thought it well known that scientific physics had come into existence only after the invention of the differential calculus (see Section 3.2.2). Of special importance in his time was the attempt to deal quantitatively with physiology and psychology — recall the Weber-Fechner law and Helmholtz' theory of tone sensitivity. Riemann's approach to the physiology of the senses shows his desire to come as close as possible to the conceptually formulated qualities in a mathematical and quantitative way. But in his case construction in mathematics was subordinated to speculative thinking that started from these qualities. This opened new possibilities that went far beyond what was accessible to the old constructions. This was the core of his speculative approach: the rigidity of the construction schemata was dissolved and the mathematics became flexible.

We can see this with particular clarity in geometry, and to this end we go back to our earlier detailed discussion of this issue. Gauss, Bólyai and Lobachevsky did not leave the rigid scheme of Euclidean constructions. Approaches such as Klein's Erlangen Program were also stuck in the rigid conception of geometry as the study of congruence, except that now congruence was defined in terms of a group other than the group of Euclidean motions. On the other hand, Riemann's manifolds were not tied to prescribed constructions or groups. Of course the resulting flexibility came to fruition only half a century later, in the mathematics of the theory of point sets (of topological spaces) and in Einstein's general theory of relativity.

The pair (10) can also be fitted in here. If we start with nature given not conceptually but in terms of phenomena, then we are dealing here with an attempt to comprehend it approximately from the direction of pure mathematics. This (Riemannian) procedure was anything but widespread. What *was* widespread was the tendency of mathematicians who investigated physics to impose on nature existing rigid mathematical systems like that of Euclidean geometry:

the geometry had to be identical with the system of Euclid, and the space of physics was thus rigid and absolutely fixed. Here too Leibniz was an exception among modern thinkers, for he conceived space as an ordering of things which entailed no need to introduce from the beginning a definite mathematical structure. He rejected the vacuum, the empty absolute space. He closed his paper *De ipsa natura sive de vi insita actionibusque creaturarum* with the remark that one needed new axioms which would give rise to a system that would mediate between formal and material philosophy (mathematics and natural science) so that the two would be correctly conjoined and preserved (L. 308/309). Of course, at the time the figures of ancient geometry enjoyed brilliant successes (ballistic parabolas, Kepler ellipses), and Newton's *Principia* of 1687 "lived off" Euclidean geometric constructions.

Here too we must bear in mind that for Riemann everything was determined by the antithesis. It inspired the scientific and mathematical approximation directed from thesis to antithesis. Numbers came to light post factum as eigenvalues (in a general sense) of a continuum determined by certain conditions. In mathematics, what belongs to the manifolds of analysis is eigenvalues such as the genus, and eigenfunctions such as the algebraic functions that live on them. An analogous situation prevails in physics to this day: fields manifest themselves through the eigenvalues of observables. This viewpoint, stated in different terms, had long been known in Riemann's time. As Riemann told Schering (see Section 3.3.3), he thought of particles as singularities in a continuous field. This idea turned up again in Einstein. In complex analysis Riemann also tended to characterize functions through their singularities. There was no point in trying to apply the same approach to real analysis, and Riemann did not pursue it further; in real analysis there is no connection between the local and the global, and what is left of continuity is virtually unusable. In real analysis continuity forms no regulative principle, and post-Riemannian mathematics was forced to try to create a substitute in a variety of ways.

We conceived of flexibility as the antithesis (A_9) of rigidity. But the avoidance of unrestricted flexibility need not lead to rigidity. Rather, it may initiate a *consolidation*. We must think of it in this sense when Riemann speaks of a turning point in the approach precisely in connection with epsilontics. We devoted a great deal of attention to the pair (6), epsilontics and infinitesimal argumentation, because here we seem to have come across an exception to his tendency towards the poles of the antithesis. Not that Riemann let the method of epsilontics become a rigid dogma. He saw in it a solid foundation for the mastery of the infinite in mathematics. The method seems to be tied to the real numbers, which Dedekind was to view as a consolidation of the continuous variables. And there is no flexibility left for the number concept. It is the fixed domain of real numbers that one must commit oneself to.

Riemann did not promote unrestricted flexible expansion. In fact, he subjected everything to regulative principles. Already in the early phase, in §20 of his dissertation, i.e., when he still talked of functional expressions, he restricted the arbitrariness, inherited from Euler's time, of the operational generation of expressions to the application of *certain* operations, because he thought that he could thus obtain all the functions required by experience. Then, soon after, he took as his guiding principle not operations at all but concepts. He refrained from pursuing the possibilities that led others to the consideration of ever stranger functions (and sets, at first of real numbers). It is possible that behind Siegel's opinion, quoted at the beginning of the Introduction, was his dislike of the kind of wild growth that flourished in mathematics following the work of Riemann and Dedekind.

When we look at the historical development we must keep two things in mind. One is that in mathematics before Riemann and Dedekind there had been wild growth of which we are hardly aware today. The other is that when it comes to the future development of mathematics, not everybody has paid attention to the regulative principles of its two cofounders: in Dedekind's case, insistence on fruitful ideas, in Riemann's case, the search for philosophical points of anchorage.

Bibliography

(W.) *Bernhard Riemann's gesammelte mathematische Werke und wissenschaftlicher Nachlass*. Herausgegeben unter Mitwirkung von R. DEDEKIND von H. WEBER. 2. Auflage: Teubner, Leipzig, 1892. Reprint: Dover, New York, 1953 and in (N.).

(N.) *Bernhard Riemann. Gesammelte mathematische Werke; wissenschaftlicher Nachlass und Nachträge. Collected Papers*. Nach der Ausgabe von H. WEBER und R. DEDEKIND neu herausgegeben von R. NARASIMHAN. Springer/Teubner, Berlin/Leipzig, 1990.

AHLFORS, L. V. 1953 "Development of the theory of conformal mapping and Riemann surfaces through a century." In: *Contributions to the theory of Riemann surfaces. Centennial celebration of Riemann's dissertation*. Annals of Mathematics Studies, No. 30, 3–13, Princeton.

ARENDT, G. 1904. See DIRICHLET 1854.

BELHOSTE, B. 1991 *Augustin-Louis Cauchy. A Biography*. Springer, New York.

BIERMANN, K.-R. 1988 *Die Mathematik und ihre Dozenten an der Berliner Universität 1810–1933*, Akademie-Verlag, Berlin.

BIERMANN, K.-R. 1990 *Carl Friedrich Gauß. Der "Fürst der Mathematiker" in Briefen und Gesprächen*, C. H. Beck, München.

BÖHM, J.; REICHARDT, H. 1984 *C. F. Gauß / B. Riemann / H. Minkowski. Gaußsche Flächentheorie, Riemannsche Räume und Minkowski-Welt*. Teubner-Archiv zur Mathematik, Bd. 1. Teubner, Leipzig.

BOI, L.; FLAMENT, D.; SALANSKIS, J.-M. 1992 (eds.) *1830–1930: A Century of Geometry*. Springer, Berlin.

BOTTAZZINI, U. 1977 "Riemann's Einfluß auf E. Betti und F. Casorati." *Archive for History of Exact Sciences* 18, 27–37.

BOTTAZZINI, U. 1986 "The higher calculus: A history of real and complex analysis from Euler to Weierstrass." Springer, New York.

BOTTAZZINI, U. 1991 "Riemann in Italia." In: RIEMANN. 1991, pp. 31–40.

BOTTAZZINI, U. 1992 (ed.) *A. L. Cauchy, Cours d'analyse etc*. Editor's introduction. XI-CLXVII. CLUEB, Bologna.

BRILL, A.; NOETHER, M. 1892 "Die Entwicklung der Theorie der algebraischen Functionen in älterer und neuerer Zeit." *Jahresbericht der Deutschen Mathematiker-Vereinigung* 3 (1892–93), 107–566.

BÜHLER, W. K. 1981 *Gauss, A biographical study*. Springer, Berlin.

BURKHARDT, H. 1908 "Entwicklungen nach oscillirenden Functionen und Integration der Differentialgleichungen der mathematischen Physik." *Jahresbericht der Deutschen Mathematiker-Vereinigung* 10.2. xii, 1804 S.

CAUCHY, A. L. *Œuvres complètes* in two series, Paris 1882–1974. Quoted as (2) 3 for ser. 2., vol. 3.

CAUCHY, A. L. 1821 *Cours d'analyse de l'Ecole Royale Polytechnique. 1e partie: Analyse algébrique*. Debure. Paris. Reprinted in BOTTAZINI 1992. [We refer to the page numbers of these editions. Those in Œuvres (2) 3 are different.]

CAUCHY, A. L. 1823 *Résumé des leçons données à l'Ecole Polytechnique sur le calcul infinitésimal*, Paris. New edition *Œuvres* (2) 4, 5–261.

CAYLEY, A. 1880 "Note on Riemann's paper 'Versuch einer allgemeinen Auffassung der Integration und Differentiation'." *Mathematische Annalen* 16, 81–82 = *Collected Mathematical Papers*, vol. 11, 235–236.

CLEBSCH, A.; GORDAN, P. 1866 *Theorie der Abel'schen Functionen*. Teubner, Leipzig.

COURANT, R. 1950 *Dirichlet's principle, conformal mapping and minimal surfaces*, with an appendix by M. SCHIFFER. Interscience, New York/ London.

DARBOUX, G. 1875 "Mémoire sur les fonctions discontinues." *Annales Scientifiques de l'Ecole Normale Supérieure* (2) 4, 57–112.

DAUBEN, J. W. 1979 *Georg Cantor. His mathematics and philosophy of the infinite*. Harvard Univ. Press, Cambridge/Mass. Reprint: Princeton Univ. Press, Princeton, N. J.

DEDEKIND, R. 1872 *Stetigkeit und irrationale Zahlen*. Vieweg, Braunschweig.

DEDEKIND, R.; WEBER, H. 1882 "Theorie der algebraischen Functionen einer Veränderlichen." *Journal für die reine und angewandte Mathematik* 92, 181–290 = R. DEDEKIND, *Gesammelte mathematische Werke*, Bd. 1, 238–350.

DEDEKIND, R. 1888 *Was sind und was sollen die Zahlen?* Vieweg, Braunschweig.

DEDEKIND, R. *Gesammelte mathematische Werke*, 3 Bde. Vieweg, Braunschweig, 1920–1932.

DIEUDONNÉ, J. 1978 *Abrégé d'histoire des mathématiques*. Hermann, Paris.

DIRICHLET, P. G. L. *Gesammelte Werke*, 2 Bände, Hrsg. L. FUCHS und L. KRONECKER, Berlin 1889–1897.

DIRICHLET, P. G. L. 1829 "Sur la convergence des séries trigonométriques qui servent à représenter une fonction arbitraire entre les limites données." *Journal für die reine und angewandte Mathematik* 4, 157–169; *Werke* 1, 283–306.

DIRICHLET, G.L. 1854 *Vorlesungen über die Lehre von den einfachen und mehrfachen bestimmten Integralen*, Hrsg. G. ARENDT. Vieweg, Braunschweig 1904.

DUGAC, P. 1976 *Richard Dedekind et les fondements des mathématiques*. Vrin, Paris.

EDWARDS, H. M. 1974 *Riemann's Zeta Function*. Academic Press, New York/ London.

EULER, L. 1748 "Introductio in analysin infinitorum," t. I. Lausanne. In: *Opera omnia* (1) 14. Deutsche Übersetzung von MASER, Neudruck mit einem Vorwort von W. WALTER: L. EULER, *Einleitung in die Analysis des Unendlichen*. Springer, Berlin, 1983.

EULER, L. *Opera omnia* in vier Serien, seit 1911. Quoted as (I) 8 for ser. I, vol. 8.

EULER, L. 1983 *Leonhard Euler 1707–1783. Beiträge zu Leben und Werk*. Gedenkband des Kanton's Basel-Stadt. Birkhäuser, Basel.

FREUDENTHAL, H. 1975 "Riemann, Georg Friedrich Bernhard." In: *Dictionary of Scientific Biography* vol. 11. New York, 447–456.

GILLIES, D. 1992 (ed.) *Revolutions in Mathematics*. Clarendon Press, Oxford.

GRATTAN-GUINNESS, I. 1970 *The development of mathematical analysis from Euler to Riemann*. M. I. T. Press, Cambridge/Mass.

GRATTAN-GUINNESS, I. 1990 *Convolutions in French mathematics 1800–1840*. 3 vols. Birkhäuser, Basel.

GRATTAN-GUINNESS, I. 1994 (ed.) *Companion encyclopedia of the history and philosophy of the mathematical sciences.* 2 vols. Routledge, London/New York.

GRAY, J. 1979 *Ideas of space. Euclidean, non-euclidean and relativistic.* Clarendon, Oxford.

GRAY, J. 1986 *Linear differential equations and group theory from Riemann to Poincaré.* Birkhäuser, Boston.

GRAY, J. 1994 "On the history of the Riemann mapping theorem." Studies in the history of mathematics, I. *Supplemento ai Rendiconti del Circolo Matematico di Palermo,* ser. II, no. 34, 47–94.

HANKEL, H. 1870 "Untersuchungen über die unendlich oft oscillirenden und unstetigen Functionen." Reprinted in *Math. Annalen* 20 (1882), 63–112.

HARBORTH, H. 1982 (Red.) *Festschrift zum 150. Geburtstag von R. Dedekind.* Abhandlungen der Braunschweigischen Wissenschaftlichen Gesellschaft Band 33, Göttingen.

HATTENDORFF, K. See RIEMANN 1869, 1876.

HAWKINS, T. 1970 *Lebesgue's theory of integration. Its origins and development.* University of Wisconsin Press, Madison/London.

HELMHOLTZ, H. 1868 "Über die thatsächlichen Grundlagen der Geometrie." *Nachr. der königl. Gesellschaft der Wissenschaften zu Göttingen* 15, 193–221.

HENSEL, K.; LANDSBERG, G. 1902 *Theorie der algebraischen Funktionen einer Variablen und ihre Anwendung auf algebraische Kurven und Abelsche Integrale.* Teubner, Leipzig.

HILBERT, D. 1899 *Grundlagen der Geometrie.* Teubner, Leipzig. Several editions with supplements.

HILBERT, D. *Gesammelte Abhandlungen* Bd. 1–3. Berlin 1932–1935.

HURWITZ, A.; COURANT, R. 1929 *Vorlesungen über allgemeine Funktionentheorie und elliptische Funktionen von Adolf Hurwitz.* Herausgegeben und ergänzt durch einen Abschnitt über geometrische Funktionentheorie von R. COURANT. Grundlehren der mathematischen Wissenschaften, Bd. 3. 3. vermehrte und verbesserte Auflage: Springer, Berlin.

KLEIN, F. 1921–23 *Gesammelte mathematische Abhandlungen,* 3 Bde. Springer, Berlin, 1921–23.

KLEIN, F. 1926 *Vorlesungen über die Entwicklung der Mathematik im 19. Jahrhundert*. Band 1. Springer, Berlin.

KLEIN, F. 1927 *Vorlesungen über die Entwicklung der Mathematik im 19. Jahrhundert*. Band 2. Springer, Berlin.

KLINE, M. 1972 *Mathematical thought from ancient to modern times*. Oxford University Press, New York.

KNOBLOCH, E. 1983 "Von Riemann zu Lebesgue–zur Entwicklung der Integrationstheorie." *Historia Mathematica* 10, 318–343.

KOCH, H. 1986 *Einführung in die klassische Mathematik* I. Springer, Berlin.

KOLMOGOROV, A. N.; YUSHKEVICH, A. P. 1992 (eds.) *Mathematics of the 19th century. Mathematical logic, algebra, number theory, probability theory*. Birkhäuser, Basel.

LANDAU, E. 1906 "Euler und die Funktionalgleichung der Riemannschen Zetafunktion." *Bibliotheca Mathematica* (3) 7, 69–79, *Collected Works*, Vol. 2, 335–345.

LANDAU, E. 1909 *Handbuch der Lehre von der Verteilung der Primzahlen*, 2 Bde. Teubner, Leipzig/Berlin.

LAUGWITZ, D. 1977 *Differentialgeometrie*, 3. Aufl. Teubner, Stuttgart. English translation of the first German edition: *Differential and Riemannian geometry*. Translated by Fritz Steinhardt. Academic Press, New York 1965.

LAUGWITZ, D. 1985 "Grundbegriffe der Infinitesimalmathematik bei Leonhard Euler." In: FOLKERTS/LINDGREN, Hrsg.: *Mathemata. Festschrift für Helmuth Gericke*. Reihe Boethius, Bd. 12. Wiesbaden-Stuttgart.

LAUGWITZ, D. 1986 *Zahlen und Kontinuum*. Bibliographisches Institut, Mannheim.

LAUGWITZ, D. 1989a "Definite values of infinite sums: Aspects of the foundations of infinitesimal analysis around 1820." *Archive for History of Exact Sciences* 39, 195–245.

LAUGWITZ, D 1989b "Grundlagen der Analysis bei C. F. Gauß : Trigonometrische Reihen." *Mathematische Semesterberichte* 36, 159–174.

LAUGWITZ, D. 1991 "Cauchy-Zahlen in der Infinitesimalmathematik." *Mathematische Semesterberichte* 38, 175–213.

LAUGWITZ, D. 1992 " 'Das letzte Ziel ist immer die Darstellung einer Funktion': Grundlagen der Analysis bei Weierstraß 1886, historische Wurzeln und Parallelen." *Historia Mathematica* 19, 341–355.

LAUGWITZ, D. 1993 "Die Formel von Cauchy-Hadamard in Riemanns Nachlaß." *Mathematische Semesterberichte* 40, 115–120.

LAUGWITZ, D.; NEUENSCHWANDER, E. 1994 "Riemann and the Cauchy-Hadamard formula for the convergence of power series." *Historia mathematica* 21, 64–70.

LÜTZEN, J. 1982 *The prehistory of the theory of distributions*. Springer, Berlin.

MESCHKOWSKI, H.; NILSON, W. 1991 *Georg Cantor, Briefe*. Springer, Berlin.

MONASTYRSKY, M. 1987 *Riemann, Topology, and Physics*. Birkhäuser, Boston/ Basel/Stuttgart. (Second Edition, 1999.)

MONNA, A. F. 1972 "The concept of function in the 19th and 20th centuries, in particular with regard to the discussions between Baire, Borel and Lebesgue." *Archive for History of Exact Sciences* 9, 57–84.

MONNA, A. F. 1975 *Dirichlet's principle, a mathematical comedy of errors and its influence on the development of analysis*. Oosthoek, Scheltema & Holkema, Utrecht.

NEUENSCHWANDER, E. 1980 "Riemann und das 'Weierstraßsche' Prinzip der analytischen Fortsetzung durch Potenzreihen." *Jahresbericht der Deutschen Mathematiker-Vereinigung* 82, 1–11.

NEUENSCHWANDER, E. 1981a "Über die Wechselwirkungen zwischen der französischen Schule, Riemann und Weierstraß. Eine Übersicht mit zwei Quellenstudien." *Archive for History of Exact Sciences* 24, 221–255.

NEUENSCHWANDER, E. 1981b "Studies in the history of complex function theory II: Interactions among the French school, Riemann, and Weierstrass." *Bulletin of the American Mathematical Society, New Ser.* 5, 87–105.

NEUENSCHWANDER, E. 1981c "Lettres de Bernhard Riemann à sa famille." *Cahiers du Séminaire d'Histoire des Mathématiques* 2, 85–131.

NEUENSCHWANDER, E. 1983 "Der Aufschwung der italienischen Mathematik zur Zeit der politischen Einigung Italiens und seine Auswirkungen auf Deutschland." In: *Symposia mathematica* vol. 27, Istituto Nazionale di Alta Matematica, Rom. S. 213–237.

NEUENSCHWANDER, E. 1987 "Riemanns Einführung in die Funktionentheorie." Abhandl. Akad. Wiss. Göttingen, Math.-Phys. Kl. 3. Folge, Nr. 44. Göttingen 1996.

NEUMANN, C. 1865 *Das Dirichlet'sche Princip in seiner Anwendung auf die Riemann'schen Flächen.* Teubner, Leipzig.

NEUMANN, C. 1884 *Vorlesungen über Riemann's Theorie der Abel'schen Integrale.* Teubner, Leipzig. Zweite vollstandig umgearbeitete und wesentlich vermehrte Auflage. Teubner, Leipzig.

PURKERT, W.; ILGAUDS, H. J. 1987 *Georg Cantor 1845–1918.* Birkhäuser, Basel.

REICH, K. 1973 "Die Geschichte der Differentialgeometrie von Gauß bis Riemann (1828-1868)." *Archive for History of Exact Sciences* 11, 273–382.

REMMERT, R. 1991 *Funktionentheorie* I, II. Springer, Berlin.

RIEMANN, B. 1869 *Partielle Differentialgleichungen und deren Anwendung auf physikalische Fragen.* Vorlesungen von BERNHARD RIEMANN. Für den Druck bearbeitet und herausgegeben von K. HATTENDORFF. Vieweg, Braunschweig. 3. Auflage: Braunschweig 1882; numerous later editions, completely rewritten and enlarged, were edited by H. WEBER and by P. FRANK and R. V. MISES.

RIEMANN, B. 1876 *Schwere, Elektricität und Magnetismus.* Nach den Vorlesungen von BERNHARD RIEMANN bearbeitet von KARL HATTENDORFF. Rümpler, Hannover.

RIEMANN, B. 1899 *Elliptische Functionen.* Vorlesungen von BERNHARD RIEMANN. Mit Zusätzen heraugegeben von HERMANN STAHL. Teubner, Leipzig.

RIEMANN, B. 1991 *Conferenza internazionale nel 125° anniversario della morte di G. F.B. Riemann.* Atti del convegno. Istituto S. Maria, Cittá di Verbania, Lago maggiore.

ROWE, D. E.; MCCLEARY, J. 1989 (eds.) *The history of modern mathematics* I, II. Academic Press, Boston.

SCHARLAU, W. 1981 (Hrsg.) *Richard Dedekind 1831–1981.* Vieweg, Braunschweig/ Wiesbaden.

SCHARLAU, W. 1990 (Hrsg.) *Mathematische Institute in Deutschland 1800–1945.* DMV/ Vieweg, Braunschweig/Wiesbaden.

SCHOLZ, E. 1980 *Geschichte des Mannigfaltigkeitsbegriffs von Riemann bis Poincaré*. Birkhäuser, Boston/Basel/Stuttgart.

SCHOLZ, E. 1982a "Herbart's influence on Bernhard Riemann." *Historia Mathematica* 9, 413–440.

SCHOLZ, E. 1982b "Riemanns frühe Notizen zum Mannigfaltigkeitsbegriff und zu den Grundlagen der Geometrie." *Archive for History of Exact Sciences* 27, 213–232.

SCHOLZ, E. 1992 "Riemann's vision of a new approach to geometry." In: L. Boi et al. 1992, 22–34.

SCHUBRING, G. 1990 "Zur strukturellen Entwicklung der Mathematik an den deutschen Hochschulen 1800–1945." In: SCHARLAU 1990, S. 264–276.

SIEGEL, C. L. 1932 "Über Riemanns Nachlaß zur analytischen Zahlentheorie." *Quellen und Studien zur Geschichte der Mathematik, Astronomie und Physik*. Abteilung B: Studien, Bd. 2, 45-80 = *Gesammelte Abhandlungen*, Bd. 1, 275–310.

SINACEUR, M.-A. 1990 "Dedekind et le programme de Riemann. Suivi de la traduction de 'Analytische Untersuchungen zu Bernhard Riemann's Abhandlungen über die Hypothesen, welche der Geometrie zu Grunde liegen' par R. Dedekind." *Revue d'Histoire des Sciences* 43, 221–296.

SPEISER, A. 1927 "Naturphilosophische Untersuchungen von Euler und Riemann." *Journal für die reine und angewandte Mathematik* 157, 105–114.

STAHL, H. See RIEMANN 1899.

STILLWELL, J. 1989 *Mathematics and its History*. Springer, New York.

TORRETTI, R. 1978 *Philosophy of geometry from Riemann to Poincaré*. Episteme, vol. 7. Reidel, Dordrecht/Boston/London.

WEIERSTRASS, K. 1886 *Ausgewählte Kapitel aus der Funktionenlehre. Vorlesung, gehalten in Berlin. Mit der akademischen Antrittsrede, Berlin 1857, und drei weiteren Originalarbeiten von K. Weierstrass aus den Jahren 1870 bis 1880/86*. Herausgegeben, kommentiert und mit einem Anhang versehen von R. SIEGMUND-SCHULTZE. Teubner-Archiv zur Mathematik, Bd. 9. Teubner, Leipzig, 1988.

WEIL, A. 1979 "Riemann, Betti and the Birth of Topology." *Archive for History of Exact Sciences* 20, 9–96.

WEIL, A. 1980 "A postscript to my article 'Riemann, Betti and the birth of topology'." *Archive for History of Exact Sciences* 21, 387.

WEIL, A. 1989 "Prehistory of the zeta-function." In: *Number theory, trace formulas and discrete groups*. Symposium in honor of ATLE SELBERG, Oslo, Norway, July 14–21, 1987. Academic Press, Boston, 1–9.

WEYL, H. 1913/1955 *Die Idee der Riemannschen Fläche. Mathematische Vorlesungen an der Universität Göttingen*, Bd.5. Teubner, Leipzig/Berlin. Dritte, vollständig umgearbeitete Auflage: Teubner, Stuttgart. An English translation, titled *The Concept of a Riemann Surface*, was published by Addison-Wesley in 1955.

WEYL, H. 1918/1923 *Raum. Zeit. Materie. Vorlesungen über allgemeine Relativitätstheorie*. Springer, Berlin. Fünfte, erweiterte Auflage: Springer, Berlin. An English translation (by H. L. Brose), titled *Space–Time–Matter*, was published by Dover in 1952.

WEYL, H. 1927 *Philosophie der Mathematik und Naturwissenschaft*. Handbuch der Philosophie, Abt. 2A. Oldenbourg, München/Berlin.

WEYL, H. 1932 "Topologie und abstrakte Algebra als zwei Wege mathematischen Verständnisses." *Unterrichtsblätter für Mathematik und Naturwissenschaften* 38, 177–188 = *Gesammelte Abhandlungen*, Bd. 3, 348–358. An English translation (by Abe Shenitzer) of this paper appeared in two parts in *The American Mathematical Monthly* in May 1995, pp. 453–460, and in August–September 1995, pp. 646–651.

WEYL, H. 1968 *Gesammelte Abhandlungen*, 4 Bände. Springer, Berlin.

WEYL, H. 1988 *Riemanns geometrische Ideen, ihre Auswirkung und ihre Verknüpfung mit der Gruppentheorie* (written in 1925). Herausg. v. K. CHANDRASEKHARAN. Springer, Berlin.

ZYGMUND, A. 1935 *Trigonometrical series*. Monografje matematyczne, vol. 5. Warszawa-Lwów. Reprint: Dover 1955.

Name Index

Name Index

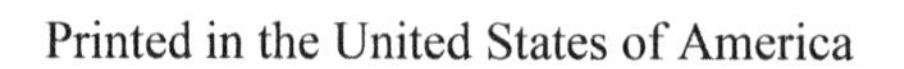
Printed in the United States of America

Lightning Source UK Ltd.
Milton Keynes UK
UKOW06f1201140715

255160UK00003B/13/P